Ingenious Principles Of Nature

E. W. Udo Küppers

Ingenious Principles Of Nature

Do We Reckon With Nature Or Nature Reckons With Us

 Springer

E. W. Udo Küppers
Küppers-Systemdenken
Bremen, Germany

ISBN 978-3-658-38098-4   ISBN 978-3-658-38099-1   (eBook)
https://doi.org/10.1007/978-3-658-38099-1

This Springer imprint is published by the registered company Springer Fachmedien Wiesbaden GmbH, part of Springer Nature.
The registered company address is: Abraham-Lincoln-Str. 46, 65189 Wiesbaden, Germany

I dedicate this book to
Lias, Zoe, Tomte, Mathilde
and all the grandchildren,
to which we – through our fault – leave a nature,
that makes their self-determined lives noticeably more difficult.

Preface

The content of this book – as obviously as the title points in one direction – is *not* intended to make you, dear readers, obsessed with nature. Even if the current threatening state of our basis of existence, the earth, and – closely connected to it – its richness of species or its interconnected biodiversity gives ample reason for many to long for a "return to nature." But nature and technology cannot be separated! Therefore, in addition to natural phenomena, political, economic, and social phenomena also permeate the book chapters.

On the other hand, what is wrong with remembering romantic verses like those in the "Abendlied" by Matthias Claudius, when we sacrifice forests that have grown over millions of years, as fundamental life-givers, for the sake of short-term filthy lucre and thus irrevocably destroy them?

> The moon has risen,
> The gold stars are shining
> In the sky bright and clear;
> The forest stands black and silent,
> And from the hills rises
> The white mist wonderful.
> *Matthias Claudius (1778)*

For today's dancing *lords of mankind* (see Chap. 8) around the golden economic "calf" of the earth and its increasing destruction, Matthias Claudius likewise holds ready in the fourth stanza a wisdom which, related to catastrophic forest conditions in the eighteenth century,[1] is not dissimilar to today's:

We proud human children
Are vain poor sinners
And don't know much at all;
We spin air webs,
And seek many arts,
And get further from the goal.

The insight that runs through this book is:

As ingenious as the principles of nature may seem to us, their forward-looking practical application in our living and working environment around the globe is a political issue.

Our gaze is directed towards the future, but without leaving out the past and the present, and this raises two central questions:

1. In view of the fact that everything on our planet has limits, but that these limits are transgressed in such a selfish way by the human species and have led and continue to lead to catastrophes of the greatest magnitude, the first question arises:

   How can this harmful activity be stopped? In other words: How can human reason be steered in the direction of sustainable development and the branches of life on which we sit be treated with considerably more respect, even though we are already in the process of sawing them off, in part with relish?

Examples of these local and global disasters visible to everyone in all walks of life are:

• Plastic-polluted oceans and resulting fish kills, as well as harmful micro-plastic particles incorporated along the food chain to humans
• Economically driven monocultures are partly responsible for the fact that former fertile land areas are now permanently barren, devastated and destroyed

---

[1] In the Outline of the Forest History of Central Europe, from 3000 BC to the present, the period of the eighteenth century is described as the last great clearing period of the forest, by thinned and plundered forests.   See   https://www.sdw-rems-murr.de/mein-wald/waldgeschichte/historisch/   (Accessed 12/10/2018).

- Increasingly poisoned air to breathe, due to emissions of carbon dioxide (CO_2), nitrogen oxide (NO_2), methane (CH_4), sulfur dioxide (SO_2), and particulate matter, especially in urban areas, and consequently increase in circulatory diseases
- Cheap but "tasteless" food produced by industrial mass production; clear trend towards a "throw-away society," not least due to industrialized "obsolescence" or planned installation of wear parts with a short service life
- Global climate change is clearly demonstrated by devastating, sometimes surprising effects in many areas of our social life and work

Politicians, as representatives of the people or state leaders, publicly present themselves as capable of action and concerned about the common good of the citizens. CEOs of national and multinational companies publicly emphasize their generous commitment through "declarations of self-commitment" to the protection of nature and the environment in the market, to the welfare of the workforce and customers. However, all of them are internally prisoners of their organizational structure, in two ways.

On the one hand, there is an extraordinary lack of necessary adaptability in an ever-dynamic, networked, and increasingly complex environment. For decades, this has been expressed primarily through entrenched and rigid hierarchical structures, in which senior executives live up to the *Peter Principle*[2] (Peter and Hull 2015).

On the other hand, these actors are characterized by a specific lack of self-reflection. This is not infrequently coupled with short-term, routine causal or monocausal success strategies, which ultimately fall victim to the dynamics of the environment. This is easily recognizable by the fact that socially burdensome consequences of various kinds, which occur due to a lack of change of perspective, mask the desired solutions for success. In sum, this juxtaposition leads to more burdensome effects than beneficial progress for our basis of life and work.

Asking for specific examples? Here are three dominant ones:

- The strengthening and realization of the socially so extraordinarily important education sector for all native and immigrant citizens of Germany has been on the drip of politics for decades, with minimal progress, whoever forms the government. But this does not prevent politicians of all stripes from proclaiming year after year with full fervor an "educational republic"

---

[2] The Peter Principle, original from 1970, named after Canadian teacher Laurence J. Peter states, "In a hierarchy, each employee tends to rise to his level of incompetence" (Peter and Hull 2015, p. 25).

of Germany. The opposite of an "uneducated republic" is shown with great regularity.

- The population has been drifting apart for years into a few rich or super-rich on the one hand and an unacceptable overwhelming majority of poor and existence-threatened people on the other, and so far the actors in the political arena have not come up with any goal-oriented sustainable solutions, except for occasional, *highly lazy* compromises. On the contrary! It seems that the social kit in society is becoming even more fragile at an accelerated pace. The far-reaching consequences of this policy would be disastrous.
- Politicians are fatally playing a risky game of aftercare for their population, often in close cooperation with industry, although risk prevention is the real order of the day. This policy is clearly recognizable in the handling of climate change (see below), which still a fanatical but powerful minority of climate deniers in politics and the *fossil* energy industry with their new tools *fake news do* not see or do not want to see.

Aftercare politicians recognize the seriousness of the situation too late due to their often static viewpoints and lack of change in perspective from emerging social problems. The child has fallen into the well, so we are trying to bring it up again. It would be better if it had never fallen in in the first place!

What kind of elected politicians are these, who without a well-founded *politician's apprenticeship*, passed by qualitative and quantitative examination criteria, feel qualified to make far-reaching decisions about millions of citizens, without well-founded, effective feedback control mechanisms? Anyone who, as an acting politician, has sworn to avert harm from the people and practices the opposite, often accompanied by enormous consequential problems and consequential costs, has missed any justification for exercising the profession. He should go or be gone!

What the Austrian alpinist and extreme climber Paul Preuß (1886–1913) formulated as the six principles – in particular the sixth – for safe climbing, also applies in a figurative sense to today's people in a position of social responsibility (italic additions refer to people acting in the present):

> Among the highest principles is the principle of safety *(sustainability)*. But not the spasmodic correction of one's own *insecurity (lack of self-reflection)* achieved by artificial aids, but that primary security which should be based in every climber *(every leader)* in the correct assessment of his ability to his will.[3]

---

[3] https://de.wikipedia.org/wiki/PaulPreuß/ (Accessed 11 Dec. 2018).

2. In view of the incontrovertible fact that, according to all scientific findings – actually, common sense is enough if we take a clear look at our
   immediate and wider environment – "business as usual" will not lead to
   any sustainable beneficial progress for the majority of people, the second
   question arises:

   *Which paths or detours do we have to take consistently in order to learn to under-
   stand the complex and highly complex interrelationships of our previous proposals for
   solutions better than we have been able to do so far? In addition to this, we can also
   ask: Do any role models exist that can guide us to find ways out of the self-created
   chaotic conditions of coexistence on our planet?*

Literally at this point, we encounter with our thoughts, our creative impulses,
and our intelligence, the long-established ingenious principles secured by the
highest quality control, the fundamental properties of nature. It is these that
have ensured the progress and survival of existences for billions of years up to
the present, and for about a quarter of a million years that of *Homo sapiens*,
the wise man.

The actual purpose of this book is to consult the ingenious principles of
nature for our man-made problems in order to create desirable strategies for
sustainable, resilient, and fault-tolerant products, processes, and organizations. This book is the last of a trilogy (Part 1: *Systemic Bionics*, 2015; Part 2:
The End of Indulgence, 2018, both Springer, Wiesbaden), through which runs
the common thread of a postulate that, with the title, and especially with the
subtitle of this book, expresses the core of our necessary thinking and acting
on our limited planet Earth.

Until far into the future, there is no adequate alternative to our evolution
on Earth. Therefore it is not only an imperative of reason but a fundamental
question of existence for all living beings on our planet to make use of the
ingenious principles of nature. Their perfect, evolutionary interaction – also
with inanimate nature – has led to the fact that our present "technical," "organizational," "economic," and "social" services have come into being and that
per se superior nature-compatible services are available free of charge.

We humans are readily capable of, and already on our way to, destroying
and annihilating this evolutionary treasure of immeasurable wealth with all
our might in the shortest possible time. In any case – with probability bordering on certainty – the progressive process of evolution, on its intricate and
interconnected developmental paths, will skillfully adapt and evolve in the
emerging Anthropocene and humanoid age – whether with or without human
intervention is of little importance!

The unchallenged fact that we, as part of evolution, have so far been able to preserve our native earth as *the* very basis of life, to fulfil our manifold goals and desires, and to realize progress of unimagined proportions, despite minor and major opposition, seems in many respects to be seriously endangered. It is, in the literal sense, a struggle for existence. A significant – if not the most significant – driver of this struggle is climate, or climate change.

Numerous results of scientific investigations impressively explained connections between politicians, industrialists of the fossil energy industry as "climate deniers" on the one hand and the proponents of climate change on the other hand. They show the concentrated power for the domination of opinion that is taking place. Worth mentioning are the reports of the Intergovernmental Panel on Climate Change[4] (IPCC), Hans Joachim Schellnhubers' work from 2015 on "self-immolation" in which he highlights the "fatal triangular relationship between climate, man and carbon," or that of climate researcher Michael E. Mann and cartoonist Tom Toles "The Madhouse Effect," German: Der Tollhaus Effekt (2018).

Are we already at a threshold with our Earth where perhaps the crossing of a multitude of complexly interconnected, so-called *tipping points*[5] leaves no door open for us to use the ingenious principles of nature for our continued existence – significantly more sustainably than before and perhaps for the last time?

Our evolutionary further development and climate change, which many people can recognize and sometimes painfully experience, are inextricably linked. Nature with its ingenious principles is a strong driver of progress on our Earth. Deniers of climate change in conjunction with people who cultivate a creationist[6] mindset, which is directed against evolution and thus also against the ingenious principles of nature, will not change anything about this.

With this book, I would also like to address all those who are not, or not yet, committed to sustainable progress on the only basis of life we have. There are many ways to do this:

- Through personal commitment to a way of life that is compatible with nature
- As part of an initiative against environmental and nature destruction

---

[4] http://www.ipcc.ch (Accessed 12 Dec. 2018).

[5] Tipping points are critical points or moments on often linear lines of development at which, due to feedback effects, further development takes a completely different, usually accelerated and destructive course. A return before the tipping point then seems impossible.

[6] Creationists believe in the literal interpretation of the biblical creation story and deny Darwin's theory of natural selection, which places our evolution by natural selection on a broad verifiable foundation.

- As a loud mouthpiece against political and economic ignorance with short-sighted misguided thinking and acting, such as the denial of indisputable facts
- And last but not least, by looking at the ingenious forms, structures, techniques, optimal strategies, and skillful fault-tolerant organizational processes that nature provides us free of charge for our adapted use of progress (Nachtigall and Wisser 2015 and 2013, respectively; Küppers 2015; Blüchel and Malik 2006; Malik 2007; Küppers and Tributsch 2002 and many others)

It depends on whether we are willing and able to decipher the *still* existing, unimaginably large treasure of helpful natural solutions and to use them for ourselves in a sustainable way.

Ingenious natural principles with sustainable progress, proven over millions of years, are juxtaposed with man-made activities which – in terms of time, geologically speaking, virtually in the blink of an eye – are in the process of shaping and destroying nature and the environment in a way that is unworthy of life. How will we decide with our conscious and subconscious minds: Do we reckon with nature or without it?

Bremen, Germany   E. W. Udo Küppers

References

Küppers, E. W. U. (2015) Systemische Bionik. Springer Vieweg, Wiesbaden

Küppers, E. W. U.; Tributsch, H. (2002) Verpacktes Leben – Verpackte Technik. Bionik der Verpackung. Wiley VCH, Weinheim

Malik, F. (Pub.) (2007) Bionics. Fascination of Nature. MCB, München (Original 2006: Blüchel, K. G.; und Malik, F. (Hrsg.) Faszination Bionik. MCB, München)

Mann, M. E.; Toles, T. (2018) The Madhouse Effect. Columbia University Press, New York, deutsch: Der Tollhaus Effekt. Solare Zukunft, Erlangen

Nachtigall, W.; Wisser, A. (2015) Bionics by Examples. 250 Scenarios from Classical to Modern Times. Springer Spectrum, Heidelberg N. Y. (Original 2013: Bionik in Beispielen. Springer Spektrum, Berlin, Heidelberg)

Peter, L. J.; Hull, R. (2015) Das Peter-Prinzip oder die Hierarchie der Unfähigen. 15. Aufl., Rowohlt, Reinbek b. Hamburg, Original 1969, Morrow, New York

Schellnhuber, H. J. (2015) Selbstverbrennung. Die fatale Dreiecksbeziehung zwischen Klima, Mensch und Kohlenstoff. 2. Edition., C. Bertelsmann, München

Contents

1

Introduction

Abstract Why does the evolutionary development in our nature lead to the fact that it has been developing steadily for about 4 billion years, despite the acceptance of five massive setbacks in the diversity of species?

The Wolf and the Goat.
Let's agree on an economic basis:
I will not eat your grass, and in return you give me your flesh in good.
Karel Čapek

Analogy: Man and Nature.
Let's agree on an economic basis:
I'm not destroying your progress, and in return you're giving me your biodiversity in good.
E. W. Udo Küppers

There may be times when we are powerless to prevent injustice.
But there must never be a time when we don't protest.
Elie Wiesel

I am not a linear existence.
Gerhard Polt

Karel Čapek, Czech writer (1890–1938). Citation source: Peace Library, Anti-War Museum, 10405 Berlin

Elie Wiesel Romanian-American writer, university lecturer and publicist. Citation source: Peace Library, Anti-War Museum, 10405 Berlin

E. W. U. Küppers, *Ingenious Principles Of Nature*,
https://doi.org/10.1007/978-3-658-38099-1_1

Gerhard Polt studied Scandinavian studies in Gothenburg and Munich. He is known as a cabaret artist, actor, poet and philosopher to a wide audience in and beyond Germany.

Why does the evolutionary development in our nature lead to the fact that it has been developing steadily for about 4 billion years, despite the acceptance of five massive setbacks in the diversity of species (Kolbert 2015, p. 24)?

Modern humans, who according to the latest findings have existed for around 250,000–300,000 years (Hublin et al. 2017), have played a very special role in this dynamically stable development path, which is still based on an inconceivably large diversity of species, of which we know only fractions. In more recent developmental history, this has led to the realization that humans have given themselves their own time period – presumably from about 1950 onwards – namely that of the Anthropocene (Crutzen 2002). Many human developments, which since the beginning of the "Industrial Revolution through the steam engine", in the late eighteenth century, up to the present of the "digitalization of man and machine" (Küppers 2018) have achieved enormous technical-economic progress, produce at the same time an enormous "rucksack" full of destructive influences on nature, the environment and thus also on our ability to survive.

In a geologically insignificant period of time, perhaps comparable to the blink of an eye to the average age of a human being of about 70 years, we are in the process of pulling the rug out from under our feet, in the truest sense of the word, through the creeping disappearance of species, the evolutionary foundation, without abrupt extinction.

We have risen to the position of apparent master over an evolutionary development lasting billions of years and actively intervene in fundamental networks, functionalities and, not least, natural principles with destructive effects and consequences.

With a strategy of monocausal short-sighted thinking and acting – short term missent – we are thus successively destroying the essence, the basis of our life, without which our evolutionary further development is also endangered.

The strong and numerically large heterogeneous or incompatible procedures, which will guide our biological further development – despite achieved and continued social, technical and economic progress by human intelligence – into stable dynamic tracks in the future, seems to be in danger. Decisively involved in this are developments in the field of digitalization and its willing anticipatory companions of algorithmic "Big Data", "Deep Learning" and "Artificial Intelligence"[1] (cf. Küppers 2018, Chaps. 3 and 4).

It is still not at all foreseeable what effects the artificial – man-made – intelligence will have on the natural intelligence of the evolutionary development in our nature.

In view of the ever more clearly recognizable natural and environmental destructions (cf. Schellnhuber 2015; Renn and Scherer 2015, among others), which are carrying out their catastrophic work through human ill-considered interventions in the basis of our lives, it is doubtful whether digitized strategies can at all stop, or at least slow down, anthropocene development in all habitats of our planet.

At this point we look back to the question at the beginning of the chapter. Would – despite all visible natural and environmental problems – an orientation towards natural principles help us to strengthen our ability to survive, thereby counteracting anthropocene effects? Could this be achieved by investing our creative human intelligence in developments that consistently follow the laws of a sustainable and prudent strategy – long term farseeing?

Based on the current inglorious state of our planet, which in retrospect has been created by an accumulation of human misguided monocausalities – paths of straightforward progress similar to the path of a horse with blinkers – it seems imperative and forward-looking to turn our dominant, all-swallowing, economic market strategies on their heads.

The philosopher Philipp Blohm speaks of a *market as a capricious God*, of a *market as a short-sighted God* and of a *market as a jealous God* (Blohm 2017,

---

[1] It means:

Algorithms: They are a unique well-defined sequence of computational rules or operations that lead to the solution of a problem. Algorithms can be implemented in computer programs. Thus, given a defined input, a defined output or solution is obtained. The value of a given algorithm determines its performance, the accuracy of its results, its scope, its compactness, and the speed at which it operates (Source: Penrose, R. (2002) Computational thinking. Spektrum Akademischer Verlag, Heidelberg, Berlin, p. 16).

Big Data: large amounts of data that can only be captured in powers of ten, e.g. 10^{13}.

Deep learning: English for "deep learning". Used as an optimization method in connection with artificial neural networks – KNN.

Artificial Intelligence: German for "Artificial Intelligence KI". AI is a branch of computer science. It is attempted to develop programs through algorithms that are modeled on the human neuron network and its processes. The aim is to simulate intelligent behaviour in machines or robots (source for Big Data, Deep Learning and Artificial Intelligence: Küppers, E. W. U. (2018) Die humanoide Herausforderung, Glossar, Springer, Wiesbaden).

p. 76–77). There is no hesitation in agreeing with this. In the midst of the analogue-digital transformation process "[…] it is hardly possible to buy into this market anew unless you are a mafioso, oligarch or app developer"[2] (Blohm 2017, p. 68).

It is far more beneficial for our future to take new paths and detours in order to ask: How can we anticipatorily avoid problems and risks for our further development in all areas of society, instead of always panting after new innovations with consequential problems on the well-trodden search paths with short-sighted causalities and entrenched routines.

It is the exact opposite of previous social solution strategies, which in an increasingly complex environment are themselves more likely to be recognized as a problem than as sustainable real solution alternatives.

Let us better reckon with nature and see in the following chapters what fundamental and workable principles nature has created. What extraordinary achievements the organizing of nature with its networked, innumerable individuals and species has driven their competitive and cooperative interaction to the highest qualities through principles that have stood the test of time. Chapters 2 and 3 give impressive examples of this.

If evolutionary development is capable of successively increasing its quality potential over billions of years, what should we humans – who are part of this development – wait for to use its ingenious development strategies and principles for our own survival in times of crisis? From the point of view of climate change and the Anthropocene, it would be better to speak of catastrophic effects and destructions of complete, formerly populated areas, which need to be reduced or avoided in a precautionary manner. The contents of Chaps. 4, 5 and 6 provide numerous answers for a nature-inspired, networked and sustainable practice in human design spaces.

---

[2] App developer is one of many new professions brought by the digitalization of our society.

They are people who create small "useful" programs for users of electronic media such as mobile phones or the Internet in order to offer them Internet search processes, medical self-protocols of their health and other facilitations in dealing with digitalization, regardless of any real benefit! (Source: Küppers, E. W. U. (2018) The Humanoid Challenge, Glossary, Springer, Wiesbaden).

References

Blohm, P. (2017) Gefangen im Panoptikum. Reisenotizen zwischen Aufklärung und Gegenwart. Residenz, Wien, Salzburg

Crutzen, P. J. (2002) Geology of mankind – The Anthropocene. Nature, 415, 23, Jan. 2

Hublin, J.-J. et al. (2017) New fossils from Jebel Irhoud, Morocco and the pan-African origin of Homo sapiens. Nature 546, 289–292

Kolbert, E. (2015) Das sechste Sterben. Wie die Menschheit Naturgeschichte schreibt. Suhrkamp, Berlin

Küppers, E. W. U. (2018) Die humanoide Herausforderung – Leben und Existenz in einer anthropozänen Zukunft. Springer Vieweg, Wiesbaden

Renn, J.; Scherer, B. (2015) Das Anthropozän. Zum Stand der Dinge. Matthes & Seitz, Berlin

Schellnhuber, H. J. (2015) Selbstverbrennung. Die fatale Dreiecksbeziehung zwischen Klima, Mensch und Kohlenstoff. C. Bertelsmann, München

Part I

The Inexhaustible Wealth of Evolutionary Adaptive Solutions

2

How Do We Get to Know Nature Better?

Abstract How do we get to know nature better? By presenting it in its wholeness or interconnectedness, at least as well as we understand it so far. We understand it only in its functional divergence and biodiversity similar to a blink of an eye in relation to our own lifespan.

One of the greatest, if not the greatest polymath, to whom even Charles Darwin paid homage, is remembered: Alexander von Humboldt. At the very least, however, Humboldt's travel descriptions inspired Darwin in his own work, Origin of Species (Werner 2009, pp. 68–95). To Alexander von Humboldt go back the insights of "interactions" in nature, which we still largely lack today in solving our problems.

Cycles and networks of effects, visualized using the example of an organism tree or a biocoenosis forest, show us life-sustaining interrelationships, which we humans destroy excessively and unreflectively with mindless growth compulsions and thus put our own survival at risk. Far-sighted problem-preventing thinking and action as opposed to short-sighted misguided thinking and action is therefore the central thread running through this book.

2.1 Alexander von Humboldt, a Naturalist and Polymath

This year, 2019, as this book is being written, marks the 250th anniversary of *Alexander von Humboldt's* birth on September 14. To commemorate his overwhelming achievements, especially on his nature explorations through Latin America, is not only due to his universal understanding of the processes of nature and further scientific knowledge from many disciplines that we use today as a matter of course. Manfred Osten, the long-time Secretary General of the Alexander von Humboldt Foundation in Bonn, summarises these as follows in his contribution in honour of the 250th anniversary of Alexander von Humboldt's birth (Osten 2019, p. 29):

> There are texts on some 30 scientific disciplines, including those disciplines – such as climatology, geography, ecology, oceanography, cartography, plant geography, ancient American studies, alpine studies and regional geography – that count Humboldt as one of their intellectual fathers.

It goes on to say (ibid):

> Humboldt proves to be an exceedingly "healthy one" (in allusion to the illness he endured, d. A.) for the future of our planet. For in his work he relentlessly predicts the symptoms of that "disease" which today reveals itself in the form of massive collateral damage as a result of anthropogenic interventions in *myriads of "interactions"* (italics emphasis mine) of nature. And Humboldt did not hesitate to name the real cause of this planetary disease: "He who does not feel nature will forever remain a stranger to it." Yes, he even ventured, with regard to the "interactions" of the "phenomena of nature," the hint of a future new anthropology: namely, that the, "phenomena of nature," "are at the same time moral for the heart that gratefully feels them." Humboldt is the yet to be discovered bearer of hope for a completely new "view" of nature. For what he, with the title of his bestseller of 1808, "Ansichten der Natur" (see currently Humboldt 2019; Wulf 2015; d. A.), urgently suggests to those born after him, is the securing of nature's prestige through its (sensual) reputation. It is the intertwining of its prestige with sensation, which is essential for nature's survival, and which manifests itself in the unity of admiring calculation and calculating admiration. It is the "view of nature" extended by aesthetic reason, in which the realm of objects unites with the realm of sensations in the "enjoyment of nature" – where data unite with poetry and a representational thinking to that sensual science that Humboldt exemplarily unfolds as a possible other science of nature, especially in the second "Kosmos" volume. With this, a quotation attributed to

Humboldt could be read anew – namely in the light of a new "view" of nature and the world, which Humboldt recommends to the 21st century: "*The most dangerous world view is the view of people who have never looked at the world*". (italics added by the author)

Two observations can be made about Alexander von Humboldt for today's crisis-ridden Anthropocene – with increasing destruction of nature – whereby it remains to be noted that these insights were already gained over 200 years ago, but humanity has obviously learned little or nothing from them:

> Nature lives and develops only through its *myriads of "interactions"* in animate and between animate and non-animate nature. Through our anthropogenic destruction we endanger ourselves to a great extent.
> Alexander von Humboldt: "The most dangerous world view is the view of people who have never looked at the world." (Source: East 2019, p. 29).

2.2 Thinking in Cycles and Networks of Effects

Finding an answer to the question that introduces in this chapter is as simple as it is incomprehensibly complex. The findings of Alexander von Humboldt on his Latin American travels show this very clearly.

It is simple because things happen before our eyes that are familiar to us over a long period of time. These include the ever-present seasons that make leaves on trees appear in varied shades of green in the summer. In autumn, leaves change color in fanciful yellow and brown patterns, falling off trees and becoming huge piles of biomass available to other living things for various purposes. For nature does not waste a thousandth of an ounce of material. Nature, as far as her intelligent economy of material processing is concerned, has been unbeaten for millions of years. As winter approaches, there is a pause in growth for the trees, which, with new shoots for leaves, prepare again the following spring to produce a luxuriant canopy of leaves under which many more plants and animals exist. Last but not least, we also benefit from the atmosphere of a forest, which offers us rest from everyday stress, relaxed hiking, food offerings such as mushrooms and much more.

The cycle of the forest is a typical, ever-recurring, and yet from year to year changing in many individual characteristics. The regularity of irregularity is what makes nature so fantastic when you get involved with it. For it is the declared aim of the author to accompany you, dear readers, on the way to a better understanding of nature with its unimaginably extensive and ingenious achievements.

In between, take a look at the following four Figs. 2.1, 2.2, 2.3 and 2.4, which are taken from Frederic Vester's window book "Ein Baum ist mehr als ein Baum" (A tree is more than a tree) (Kösel, Munich, 2nd edition, 1986). Figure 2.1 shows a tree in its living environment, while Fig. 2.2 directs the view to the overall interconnectedness of the tree.

Fig. 2.1   A tree in its living environment according to F. Vester (1986, p. 13), texts highlighted by the author

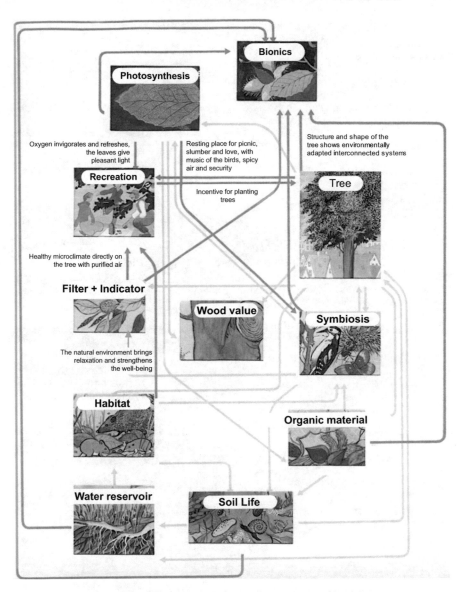

Fig. 2.2 Total cross-linking of a tree according to F. Vester (1986, p. 11 and 13), texts partly highlighted by the author

The same view can be seen in Figs. 2.3 and 2.4. Figure 2.3 shows a forest in its living environment, whereas Fig. 2.4 focuses on the overall connectivity of a forest.

What is a tree after all? This question is often heard from people who see the economic value of the tree as a piece of round wood from which boards

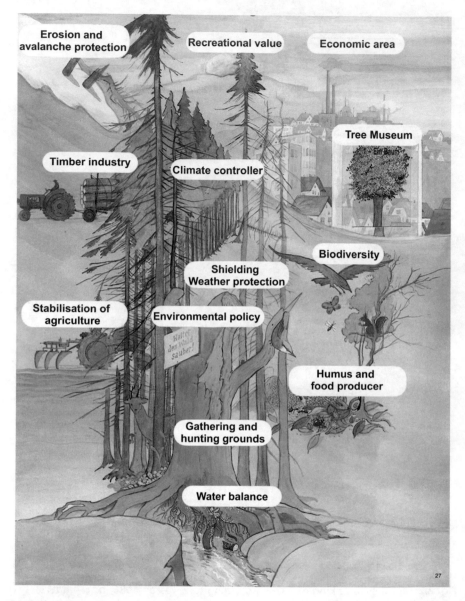

Fig. 2.3 A forest in its living environment according to F. Vester (1986, p. 27), texts highlighted by the author.

are made for all kinds of purposes. Thinner branches, with many leaves on them, only interfere with the economic exploitation, which is why they are reduced in size for transport or chopped up immediately after the tree trunk has been felled or sawn off. They are, so to speak, the worthless waste of the economic raw product log. The number of *solid* cubic *meters,* which is the

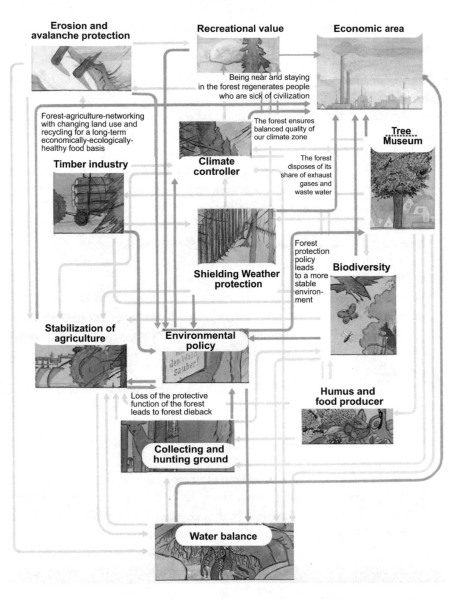

Fig. 2.4 Total connectivity of a forest according to F. Vester (1986, p. 25 and 27), texts partly highlighted by the author

measure of one cubic meter of compact wood mass, is the actual economic parameter for further processing and profit expectations. If the tree were able to charge us for its own performance – in view of the fact that it is connected with other living beings – then mankind would *certainly* have considered a more careful treatment of nature.

A much more realistic answer to the question: What is a tree? would be to look at Fig. 2.1:

- The tree produces life-giving oxygen (O_2) for all organisms in the biosphere through photosynthesis.
- At the same time, the tree stores the so-called climate gas carbon dioxide (CO_2).
- The tree is a multifunctional model for bionic solutions.
- The tree – and even more so the forest – provides people with a recreational value compared to noisy places to live and work in urban conurbations.
- As a filter and indicator for pollutants of the environment, the tree is well prepared.
- The economic value of wood has already been mentioned above.
- As a mutually fertilizing community, the tree is an ideal partner for animals and other plants.
- Many plants and animals use the tree as their own habitat and home.
- The organic material of the tree renews itself from year to year, whereby the fallen material is decomposed by an army of micro-organisms and micro-organisms in the soil without waste and is available as raw material for new life.
- Cavity-forming tree roots in the soil are an integral water reservoir at the base of the tree. From the roots to the tips of the leaves in the canopy, the tree is supplied with all the necessary nutrients.
- and many technical, physical and chemical performances more!

If one were to take the trouble and actually record qualitatively and quantitatively all the services provided by a tree – and even more so by a forest – according to holistic networked standards, people's respect for this evolutionary organism would have to be overwhelming.

Not nearly do products or processes exist in today's human technosphere that could match the output of a tree, let alone a forest. And yet, in spite of our intelligence and our technical progress, we are swinging ourselves up to be the shapers and controllers of nature. What indifference to life, to which we ourselves owe our existence.

Sarcastically, Samuel Longhorn Clemens (Mark Twain) (1835–1910)[1] could be stated in slightly abbreviated form:

---

[1] https://www.zitate.de/autor/Twain%2C+Mark?page=7 (Accessed 12/14/2018).

The right to be stupid is part of the guarantee of the free development of personality.

An even more impressive view of the performance of a tree is provided by the often invisible – because due to a lack of knowledge about nature it is almost impossible to grasp – links between energy strands, food webs and communication paths. Figure 2.2 can only superficially hint at the high complexity and density of networking. The achievements mentioned in detail in Fig. 2.1 would take on a far greater value if we were to record and analyse in detail the network of relationships between the tree and the living organisms around the tree roots, trunk and canopy. Hordes of scientists and people interested in nature have been engaged in unravelling the secrets of nature for a long time. And it will be a long time before even fractions of nature's secrets from unimaginably complex dynamic processes are recognized and applied beneficially to humans. The multidisciplinary field of knowledge of *bionics* (Küppers 2015; Nachtigall and Wisser 2015/2013; Müller and Müller 2003; Küppers and Tributsch 2002; Nachtigall 2002; v. Frisch 1974; Paturi 1974; Burkhardt et al. 1972; Gérardin 1968; Greguss 1985, to name just a few of hundreds of bionics authors) works precisely in this border area between nature and technology. Successes in the discovery of natural geo-inventions and technical applications have impressively demonstrated the efficiency and diversity of natural products, processes and organisational procedures for about 60 years.

Of course, the polymath Leonardo da Vinci (1452–1519) and his ingenious ideas on the technical imitation of natural models should not be missing among the authors, not least because he is regarded by many bionicists as the "father" of bionics (Mathé 1980).

60 years of systematic bionics research and development means that, despite all the successes – and failures – along the way, we have only taken a first small step towards exploring the complex secrets of nature. Therefore, as already indicated, a rich field of activity still awaits for generations to elicit the technical secrets from the unimaginable biodiversity of nature and to use them advantageously for future human life on earth.

That a forest is more than the sum of its trees, corresponds just as indisputably to the reality, as the sentence: A tree is more than the number of its solid cubic meters!

A look at the superficially indicated natural services of a forest, with an urban industrialized region in the background of Fig. 2.3, says something about its value for us humans and many forest dwellers. This value is even more evident in Fig. 2.4, where the *engineered* natural services are placed in an interconnected context. Literally nothing is isolated or even self-sufficient in nature! The forest network is the strategy and guarantor for strengthening the survivability of all its inhabitants. The same applies, of course, to natural networks in other habitats (biotopes).

In advance, there was talk of the *"value of the forest"*, which we can only guess at. The fact that forests, especially in times of climate change (WMO 2020; IPCC, Special Report 2018)[2] and the historically proclaimed Anthropocene (Crutzen 2000; Renn and Scherer 2015), which attributes environmental problems to human causes, have become fundamental helpers and guarantors of survival in a climate and environmental change that is already taking place, cannot be denied!

And yet the naivety, ignorance and ruthlessness or insensitivity of many decision-makers in our society, primarily in politics and business, in the face of a nature that has grown over billions of years, to which the decision-makers discussed above themselves owe their progress in life, never ceases to impress!

Where can we still find on our earth today increasingly grown, large-scale primeval forests, where indigenous people live their lives close to nature – undisturbed by any overexploitative civilization?

Where do companies – unselfishly and in *honest* cooperation – use the immense knowledge of primitive peoples about active plant substances for medical precautions or for effective and efficient healing processes in our civilized environment? Many of these nature-oriented examples of cooperative and beneficial action between groups of people who could not be more different show the effective power of nature, but unfortunately also the vicious cycles of exploitation, destruction of nature and extermination of indigenous peoples on a considerable scale.

In an article by the Executive Director of the UN Economic Commission for Europe, Christian Friis Bach[3] on "The Value of Forests", he writes:

The forest ecosystem provides many services from which everyone benefits. [...] It is difficult to attribute a monetary amount to them. Nevertheless, monetary attribution is one way to understand what is at stake when an ecosystem falls

---

[2] http://report.ipcc.ch/sr15/pdf/sr15_spm_final.pdf (Accessed 15 Dec. 2018).
[3] Bach, C. F. (2016) The value of the forest. All mistakes in dealing with ecosystems are still repeated today – for lack of knowledge. Süddeutsche Zeitung, 26. 8. 2016.

victim to, for example, construction (or the fossil fuel industry,[4] d. A.). Recent research (as of 2016) puts the value of services provided by forests at $16.2 trillion a year, and that of wetlands at $26.4 trillion. Together, this is more than half of the world's gross national product.

If one considers that the gross national product – today also called gross domestic product GDP[5] – also includes monetary benefits of e.g. costs for repairing traffic accident damage, repair costs of any kind, costs for repairing damage caused by climate disasters, etc., which do not produce any *new* products or services, then the value-cost comparison between natural services and human services is all the more impressive in favour of natural services (cf. also Grunewald et al. 2013).

There is no sign of a decisive trend reversal in recent years, which massively strengthens the forest and many other natural biotopes against the threatening effects of climate change – on the contrary. The euphoria that is spreading with new technologies such as humanless machine factories (Industry 4.0, 5.0, 8.8 etc.), artificial intelligence, humanoid robotics and the like, and which is leading more to a coming *energy trap* than to future sustainability, will demand even more effort from us to strengthen our survival in the evolutionary spirit of sustainable development.

The fact that in this context some *"digital prophets"*, in anticipatory obedience to the increasing power of algorithms and artificial intelligence, are already talking in parallel about "Life 2.0, 3.0, 6.5" etc., shows the entire short-sightedness and misguided overheated imagination of this group of people.

It is bordering on concentrated insanity when, on the one hand, people successively destroy and annihilate an environment worth living in, together with its fundamentally important biodiversity, as the basis for billions of years of progress, in order to finally create, by technical means, a new, fanciful "living world 2.0, 3.0 or 6.5", in which humanoid robots, "artificial intelligence"

---

[4] The fight of a few nature conservationists in Hambacher Forst, near the city of Cologne, against energy giant RWE, which wanted to expand its lignite mining area and thereby cut down the Hambacher Forst forest, has been ruled against by the Higher Administrative Court of North Rhine-Westphalia (OVW NRW – case number: 11 B 1129/18), see also: https://www.judid.de/ovg-nrw-stoppt-kahlschlag-im-hambacher-forst/ (Accessed 16.12.2018).

was decided by the courts in favour of the nature conservationists. This is a rare occurrence in Germany, given the close links between state-subsidised business and politics. See also: Pinzler u. Wolf: Holz gegen Kohle. Die Zeit, No. 39, p. 20, v. 20.9.2018; Schmitz: I feel a bit powerless. The RWE boss makes it clear: the energy company will clear-cut in the Hanbach forest. Interview in Handelsblatt No. 188, pp. 6–9, 28/29/30, 9. 2018.

[5] Gross Domestic Product GDP is the value of all goods and services produced within a country per year.

and gigantic artificial data networks generate new goals of progress beyond ecology.

The "silent" resistance of nature on our earth against the omnipotence fantasies of technical-economic progress believers is in full swing.

Biodiversity versus *algorithms* could be the *"new struggle for existence"*.

We will come back to this in various places in the book.

After the interconnected tree-forest interlude explained earlier, let us now turn to other obvious examples of nature that we can easily understand, but often – in the whole breadth of invisible connections – do not think about further. The pollination of plants by bees is one such. Here, too, a typical natural cycle between animal and plant becomes apparent:

1. The plants to be pollinated attract bees by their colourful flowers. (half cycle relationship from plant to animal).
2. The bees drink nectar from the flower and take plant pollen with them, which they deposit on other plant flowers, bringing male and female pollen together (half feedback loop relationship from animal to plant).

As obvious as this simple natural spectacle of plant pollination and food intake by bees may be, in reality it is connected to a multiple feedback loop network, with the participation of a number of other plants, animals, influences such as climatic conditions, etc., not mentioned here. It must also be clearly stated that the influence of man on the cycle of the pollination process, if we think of the application of toxic pesticides (including glyphosate) on the fields, has disastrous effects on a natural solution that has been proven and ingenious since time immemorial. We only know a tiny fraction of the visible and invisible interconnections of nature in order to be able to predict the real impact, not to say the catastrophic impact, that a human technical intervention in nature will actually cause. Also with this example I would like to underline how important and necessary it is to deal with nature more intensively than has been done so far; not as scientists who have been doing it for a long time in their special fields of biology, ecology, plant physiology etc., but as a person interested in nature who would like to experience the power and beauty of nature in all its facets tomorrow and the day after tomorrow and curiously asks: Why is that so?

A third simple and comprehensible example of ingenious natural cycles is that of the breeding business of birds. Whether it is outside the window at home, where local ravens or magpies begin to build their nests and then breed in the spring, or in the case of long-distance specialists such as wild geese, cranes and hummingbirds, which travel thousands of kilometres back and forth from Europe to Africa or other tropical countries every year, always breeding in the same places.

The young birds grow up to produce new generations of baby birds the following year. Nest-building, brooding, young birds, nest-building, brooding, young birds, etc., the perpetual cycle of new life, as it is with all other animals, plants and humans.

We can observe and immediately understand an unmanageable variety of simple, obvious cycles, such as those just described, without the need for specialist knowledge. But this only scratches the surface of what constitutes nature as a whole.

Understanding this whole, even approximately, leads us to the second part of the answer we are looking for to the question posed above (How do we get to know nature better?): Nature is incomprehensibly complex.

2.3 The Circular Principle as a Tool for Action and the Tragedy of the Commons

Before I accompany you into the breadths and depths of evolutionary development paths of nature, I would like to briefly familiarize you with a general feature of nature – thus also of ourselves – which should strengthen your understanding of natural phenomena: it is the *circulation principle* already mentioned several times before. Comparable terms are the *control loop principle* and the *feedback principle*.

The evolutionary progress of nature and the techno-economic progress of human inventive gifts have become more and more distant from each other in the past decades. Sustainability on the level of nature and excessiveness on the level of man have severely damaged the stability and progress of natural sustainable growth periods. The two – driven by human activities – work against each other instead of with each other. This is happening in a way that raises doubts as to whether humans are even anticipating the true problems of their environment and instead, through the lens of short-sighted misguided action, are only making their existence on Earth worse!

In the following two Sects. 2.3.1 and 2.3.2 we try to look at the evolutionary and the human-driven circular processes of survival, in order to bring the two together in Sect. 2.3.3 (Fig. 2.8), on the *tragedy of the commons*[6] See also (Hardin 1968).

2.3.1 The Cycle-Based Power of Evolutionary Nature

Nothing in nature happens in a straightforward isolated way or in straightforward communication between two or more living beings. Even the direct command of a superior to his subordinate may seem dominant for the moment. In reality, this unidirectional, straightforward communication – also referred to as the cause-effect principle – is integrated into an interconnected environment of suppressed or ignored influences, e.g. by the behaviour of other subordinates, by the influence of the specific environment in which the command is issued and affects the subordinate, by the current physical and mental states of all those involved in the communication, by the subsequent behaviour of the superior's superiors, by private influences, and so on.

Nature does not have a governing body, however designed, that issues rigorous orders or messages for others to carry out.

All three essential transport mechanisms of communication between living beings in nature, whether it concerns energetic, material or communicative processing, are part of a balanced, continuously adapting, i.e. adapting to the given environmental condition, network of interwoven circulation processes or feedbacks. Whether plants, animals or humans, they are all close or distant relatives and thus united by an impenetrable complex network that runs through all three spheres of life of water, soil and air.

The example of plant pollination by honey bees mentioned briefly above – in a far greater variety of species also by wild insects – showed for everyone immediately recognizable, the cycle or feedback principle. Plant pollination and food intake of bees and insects form a multiple feedback loop system, with the participation of a number of other networked cycles of nature and the environment not mentioned here.

We humans hardly seem to be interested in the extraordinary achievements of nature's circulation and feedback processes. We still rarely direct our thoughts in the direction of drawing our own conclusions for our survival from nature's inexhaustible genius, with fatal consequences.

---

[6] As the "tragedy of the commons" or tragedy of the commons we use – appropriate to the theme of the book according to Hardin an inevitable fate of humanity, would one look only for technological solution.

Excursus: Destroyed Natural Cycles Through Bee and Insect Mortality

Do we humans only care about insects when they attack us by the billions as pests, stinging unnoticed with sharp stingers – one of nature's many works of art – to weaken our health? Do we humans only care about insects when another deadly virus has been transmitted to humans in some region of the world? Or do we humans only care about insects when introduced insect species take over the reins of native insects, killing them and gradually tearing apart the locally grown natural networks – not least to the detriment of us humans? In addition, there are many technical means and industrial ways to increase economic yields (e.g. through monocultures, the use of pesticides and herbicides), from which those who stand at the beginning of the food chains suffer the most: *our* insects. The pronoun *our* is deliberately chosen because we humans are dependent, for better or worse, on the continued existence of the most species-rich class on earth, the insects!

Without insects no plant pollination and without plant pollination no food for us.

Important

Whatever we have been doing to insects for decades has currently led to the extent that three quarters of the world's insect population has declined (Philipp Stoddard, behavioural scientist, USA*). The situation is also dramatic for the reason that insects are a bioindicator of environmental change that directly affects us, as part of that environment, which we all too often forget! (Matthias Nuß, butterfly researcher, Germany*).

Not only Gael, J. Kergoat, biodiversity researcher, France* or Kolbert (2015) speak of the sixth great mass extinction, in which mankind has a decisive share! What must happen so that we do not saw off the branch on which we are sitting and preserve the biodiversity from which all living beings – not only we humans – owe their further development?

Horst Asböck, Austrian parasitologist*, also puts it in a nutshell by stating that, on a small scale, every human being can do something to become aware of species extinction, e.g. avoid food from huge monocultures and instead prefer regional food offerings. On a large scale, however, it is politicians and economic leaders who have the levers to stop species extinction permanently.[7]

---

[7] All scientists mentioned in the chapter paragraph contributed to the film by Torsten Mehltretter (2018): Das große Insektensterben, Mehltretter Media, commissioned by ZDF, in collaboration with Arte.

The sustainable preservation of vital cycles between plants, animals and humans concerns us all. In this context, physical size, which makes us humans seem more exalted than bees, butterflies and many other insects, does not play the slightest role. The opposite seems to be true, as we have seen before in the little excursus.

As a first visual introduction to the not always easy to understand work with circulation processes, nature as the mother of all circulation systems can be used (Fig. 2.5).

Behind the cycle of nature outlined in Fig. 2.5 are myriads of organisms that evolve over billions of years in as many network connections, in cooperative and competitive competition, thereby strengthening their ability to survive. The overarching principle of natural progress for *species survival*[8] is the *sustainability principle* (for classification of organisms, starting from the species, see Campbell et al. 2006, p. 12).

The principle of sustainability is a principle of action. Evolution probably hatched it from the beginning, billions of years ago, and strengthened it through its incomprehensible multiplicity of the interdependencies of organisms right up to the present day.

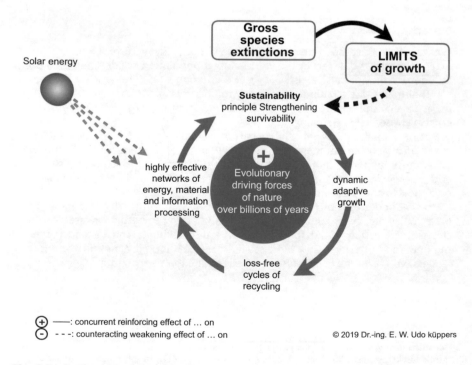

Fig. 2.5 Basic cycle process of nature

---

[8] In biology, species denotes the basic unit within a system. For human evolutionary taxonomy follows superordinate, after the group *genus* with the species *human, family, order, class, tribe, kingdom.*

Sustainability therefore means nothing less than the preservation of life's progress and its essential properties such as dynamic stability and the regenerative capacity of biological systems.

The only source of energy for survival, our sun, will continue to supply life on earth with sufficient energy for tens of millions of years, which will be used by plants and animals in ever more efficient networks for their own progress. A not insignificant factor in this successful interplay of species is the network system of energy. As in the food networks even the smallest fraction of substances is utilized, so no watt of energy is wasted uselessly (per year a solar irradiation power of 137 W/m^2 on average hits the earth's surface). Each organism converts the amount of energy that is optimal for its survival in association with other species.

Neither energetic nor material waste exists – in human parlance – for nature. Every living being optimizes its own *exergetic* (work-performing) system of order, but at the cost of limited lifetime and a globally increasing *entropy* (disorder, heat). And how do humans deal with energy and material from a global perspective? We will go into this more concretely in Fig. 2.6.

For the time being, let us look back at Fig. 2.5 and consider the various stations of the outlined natural cycle. On the one hand, there is the universal energetic driving source and power of the sun. Without it there would be no life on earth. The solar power source tirelessly drives the cycle of life, however it may change. After all, from tiny single-celled organisms that are said to have originated about 3.5 billion years ago, a multicellular wealth of species has arisen on Earth to this day, colonizing every space, no matter how inhospitable, from which armies of survival specialists have emerged and continue to emerge. From this highly effective network of energy-, substance- and information-processing processes (let us call it Station 1) in the cycle of nature (Fig. 2.5, left), progress takes place according to the principle of sustainability. This means that the ability of one's own species to survive – always within the framework of existing local networks of effects of other species – and environmental changes is strengthened (station 2) in the cycle of nature (Fig. 2.5, above). The complexity and dynamics of nature are based on growth that is highly adapted to environmental changes (station 3) in the cycle of nature (Fig. 2.5, right). Nothing is wasted or disposed of – according to human understanding. The last cycle station 4 of nature (Fig. 2.5, bottom) ensures loss-free because interconnected cycles of material recycling.

This globally sketched natural cycle, which in reality has myriads of cycles, is permeated by an equally highly effective communication network, which acts far beyond the species boundaries and cooperates with the material natural regularities in perfect harmony. The dynamic stability of these evolutionary processes is such that even major species extinctions may well cause evolutionary progress to experience prolonged periods of time disruption through species extinctions and

Connections between environmental changes caused by humans and their retroactive damage to themselves
– Excerpt –

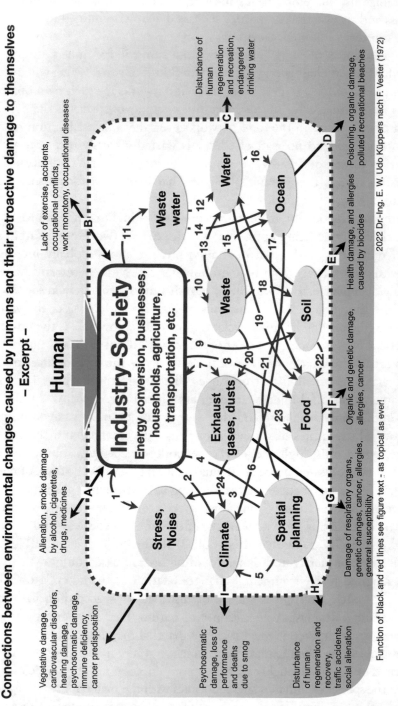

Fig. 2.6 Interrelationships between environmental changes caused by humans and the damage they cause. (Source: F. Vester The Survival Program (1972, p. 14). The links and their descriptions in Fig. 2.7 speak for themselves)

thus a slowing of evolutionary progress over millions of years (Kolbert 2015, p. 24). However, they cannot stop evolutionary progress completely.

As the central control organ in a complex environment, man is both the decisive driver for the damaging effects networked in Fig. 2.6 and the decisive precursor for avoiding the damage. This puts him in a predicament from which there is no escape!

2.3.2 The Circuit-Based Progress of Human Activities

Work has always been the basic drive of human survival. This principle of survival continues into the present and will almost certainly spread far into the future.

What is worked, how is it worked and what effects do work processes have on us humans and our environment were and are decisive questions that accompany our entire work process and its dynamics of progressive development throughout all periods of time.

Today, 20 years after the beginning of the twenty-first century, that can also be called – from a human perspective – the century of the "human-humanoid transformation", the "century of the Anthropocene,"[9] the "century of climate change", all the work and all the negative, partly catastrophic destructive effects associated with it are focused – more than ever – in one central statement:

> How do I ensure the continuity of nature and humanity through a *"survival program"?*.

The extraordinary complexity inherent in the above statement has been ignored by decision-makers on our planet for far too long – sometimes deliberately – thanks to their short-sighted misguided thinking and actions. The consequences of these deficiency activities are currently flooding the whole and only habitat we have, our Earth. Therefore, the defect removal – *repair principle* – has become a rapidly growing attractive industry today, feeding off the vicious circle:[10]

---

[9] The term Anthropocene refers to a new geochronological era in which humans intervene significantly in atmospheric, biological and geological processes and thus become an important shaper of changes in our environment. The term Anthropocene goes back to the climate researchers Stoermer and Crutzen (see Crutzen 2000; Crutzen and Stoermer 2002).

[10] The term "vicious circle" is used to describe those circular relationships between at least two influencing variables (or parameters or system elements) in which the effects of one system element on the other are continuously reinforcing or weakening. A continuous strengthening leads to a breakdown of the system under consideration, a continuous weakening results in a standstill of the system's functions. Systems can be any objects (houses, companies, computers, etc.) or any subjects (organisms such as humans, plants, animals) or any combination of the two. The opposite of a vicious circle is called an angelic circle (see explanation).

1. Vicious circle linkage
   The consequences of short-sighted thinking and acting lead to a higher economic yield and at the same time to an accumulation of earth-wide catastrophic destruction of the basis of life of nature and people.
2. Vicious circle linkage
   The accumulation of economic successes, despite accompanying earth-wide catastrophic destruction of the livelihoods of nature and people, reinforces the activities of short-sighted thought and action.

It seems to be in the nature of human activities that we often only think about the consequences of an action when it is too late to prevent the consequences that are actually considered avoidable from arising in the first place. This is especially true in a negative sense, but also in a positive sense. In a negative or destructive sense, the statement is understandable. In a positive or progressive sense, the statement is justified, but more difficult to comprehend, if we think, for example, of the early development of energetic nuclear power conversion, of humanoid robotics, or of artificially genetically modified, so-called "designer babies".

It is indisputable that complexity spreads to the smallest corner of our existence and that we have no other choice but to deal with it, which unfortunately we do not do with the necessary consistency of our thinking and acting. Dirk Baecker said it aptly in his little compendium on post-heroic management (Baecker 1994, p. 114–117):

"Complexity is not the problem but the solution!"

Let us return to the so-called survival program, in view of the state of our earth, which could not be clearer, namely increasingly unrealistic and life-destroying.

Since the publication of "The Limits To Growth" (Meadows et al. 1972), also looking back even further centuries (e.g. timber theft sixteenth century, H. K. v. Carlowitz, see Chap. 2.), we know through "The Limits To Growth" what the state of our planet is. We also know about the interconnectedness of habitats, which Meadows et al. determined through five basic, interrelated factors that essentially determine the limits of our existence on earth (ibid., p. 11–12):

1. Population or population growth
2. Agricultural production
3. Natural resources, by which is also meant biodiversityBiodiversity

4. Industrial production, which shows itself economically as progress with giant problems
5. pollution, with partial destruction of the life-sustaining ozone layer

Since then, many new initiatives and approaches have been developed, committed to understanding and acting for a future worth living; impossible to mention them all in detail here, but each one certainly a commendable action in its own right.

However, I would like to single out one approach, not least because "The Survival Program" was the title of a book and vocation for the author Frederic Vester (1972).

In his usual vivid manner, Vester shows the reality of the interrelationships on our planet. He does not exclusively address scientists of all disciplines, but also students, teachers, housewives who are involved in citizens' initiatives, employees and officials of local and state authorities (ibid., p. 7). Through the breadth of its addressed fellow citizens it shows at the same time the importance of networked problems, which cannot be understood and solved sustainably by so-called elites alone, but is the task and obligation of each individual citizen.

Without going into the details in detail, his graphic in Fig. 2.6 is not dissimilar to the previously mentioned five factors according to Meadows et al. The special value of the graphic in Fig. 2.6 at the time of its publication (1972) – and still today – is shown by the interconnected relationships of individual "influencing factors" (green circles) and the relationshsips to consequential problems (yellow circles) (ibid., p. 14). The reminder of Baecker's previously mentioned statement on *"complexity"* becomes understandable here.

Background Information

Even though almost half a century has passed since Meadow's five-factor approach and Vester's survival programme, and therefore there has presumably been sufficient time to orient oneself consistently in the direction of sustainable, environmentally compatible progress, summa summarum little has happened in terms of real practical progress. How else can we explain the fact that the world's oceans are full of plastic waste, that in the industrialized countries and elsewhere exhaust gases and particulate matter from traffic and their health consequences continue to increase, that the climate is steadily but clearly deteriorating, that the "green lungs" of the earth (such as the large contiguous Amazon forest area) are being sacrificed day after day to disdainful economic interests, and much more.

The human-initiated global cycle in the technosphere, as seen in Fig. 2.7, shows the cause of all evil.

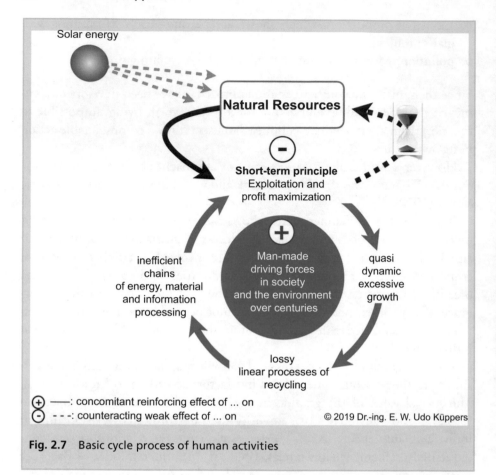

Fig. 2.7 Basic cycle process of human activities

As mentioned in Fig. 2.5, the cycle process of human activities as shown in Fig. 2.7 contains countless small interdependencies between work units that are only rarely formed into sustainable cycles – especially when they are networked. As a rule, these are causal or monocausal process chains that manufacture and design products of value, but at the same time, as the other side of the coin, are accompanied by waste streams that, in the worst case, are carelessly discarded into nature or the environment.

The human-designed driving forces of technical, economic processes primarily pursue the short-term principle of exploitation and profit maximization (in Fig. 2.7, central cycle, top). This results in quasi-dynamic growth of excessive proportions (Fig. 2.7, right). This in turn results in loss-making processes accompanying the excessive growth strategies, which cause enormous destruction in nature and the environment (Fig. 2.7, bottom). The continuation of these growth strategies leads increasingly to inefficient processing

(Fig. 2.7, left), which in turn – having reached the starting point of the cycle – again reinforces the principle of short-termism. A new round of vicious circle processes is ushered in.

An attached *stabilization cycle,* which is also called the angel circle or angel cycle[11] (upper part in Fig. 2.7), which on the one hand reduces the immeasurable natural wealth of resources through the reinforced short-term principle with a time delay (hourglass symbol), but on the other hand increasingly uses them, drifts more and more towards a one-sidedness at the expense of the natural wealth. The reduction of biodiversity or species diversity, a dominant and sensitive life-sustaining criterion, is thus insidiously destroyed.

In other words, the more the short-termism principle attacks natural resources, the more unstable nature continues to develop. From this increasing destruction of nature, the short-termism principle of human growth strategy continues to draw its natural resources, and increasingly so. According to human logic, at some point the time will come when natural resources no longer form a sustainable foundation for human growth excesses. Then the limits of any growth would be reached anyway.

Hope for sustainable progress, ironically, lies in nature itself, with its complexity and sophistication that is not easily understood, let alone controlled, by humans.

It is long past time to become humble towards nature and to appreciate its achievements in such a way that they also benefit humans in a sustainable way. The title of this book says nothing more. The question remains: How do we achieve this now, through correct, forward-looking practical action and not after endless – often superfluous – discussions and arguments? We have long known that education for all is a central key to sustainability and security of survival. But after more than 30 years of endless verbiage about education without significant progress within Germany, one of the richest countries on earth, strong hope can sometimes be eclipsed by wavering uncertainty.

As the saying goes:

> The need for preventive conservation and strengthening our ability to survive in the progress is inevitable. But the future is uncertain. This is our dilemma!

---

[11] The "angelic circle" is the name given to those circular relationships between influencing variables (or parameters or system elements) whose effect of one system element on the other is once reinforcing and once weakening. Thus the effects in the circle balance each other out, so that the functions of the system and the working system as a whole reach a stable state and do not pass over into destruction or standstill. The opposite of an angelic circle is called a vicious circle (see explanation).

2.3.3   The Tragedy of the Commons

From the previously described and separate consideration of global cycle functions of both spheres of life, on the one hand those of nature (biosphere) and on the other hand those of humans (technosphere), specific advantages and disadvantages could be identified.

In reality, however, the two spheres are inextricably linked. They have no choice but to work together. Technospheric impacts resulting from complex functional processes, such as the use of fossil energy resources, influence the biosphere just as much as the reverse is the case.

The increasingly alarming state of our living and working environment clearly shows that we have not yet understood how to use technospheric circulation processes for the sustainable benefit of us all. The six prime examples of misguided human causal developments mentioned in the preface show this indisputably.

Biospheric, complexly interconnected cycle processes that are responsible for concrete developments in nature, e.g. those between the oviposition of a beech bark beetle in the bark of a beech as host plant, are often connected by numerous links with other cycles in which countless animals and plants participate. This results in complex natural networks of action, which are not easy to understand. The dynamics and stability of these biospheric services of nature also influence technospheric processes! Without them, for example, many technical-chemical processes would not take place at all, due to a lack of natural raw material sources. The entire pharmaceutical industry and with it the medical drugs we use to maintain our health would be significantly reduced. We humans would literally be incapacitated and suffocated without forests and clean oceans to store carbon dioxide (CO_2 storage) and forests to produce oxygen (O_2 producers). This list could be continued indefinitely, although we are still only aware of a fraction of all nature's services.

Even though effect networks such as the one in Fig. 2.8 may seem complicated to some readers at first glance, the discomfort dissipates very quickly after some initial practice.

It should also be mentioned that models – in our case circuit and effect network models – of natural phenomena can only ever reflect sections of the true complexity of nature and are also subjectively influenced.

The art of dealing with natural phenomena is to understand their properties so well that we can understand them with a certain degree of scientificity and, if necessary, create models or practical solutions of them without losing sight of the complex reality and at the same time without interpreting things too simply.

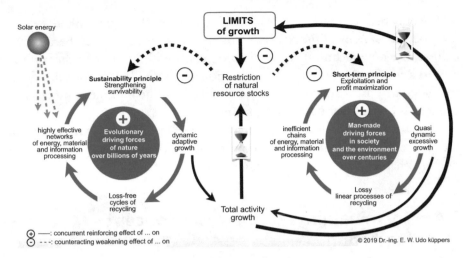

Fig. 2.8 Linked cycle representation between evolutionary and human development or progress strategy

It is therefore a balancing act on which we are moving, always with the right sense of proportion on our side. Let us therefore proceed to the action and resolution of the functions in the network of effects in Fig. 2.8. by starting with two individual circuits. Take a close look at the two circuits – green for nature and blue for human-initiated processes, which we have already learned about individually, now connected in Fig. 2.8. What do you recognize?

The cycle of nature described above (Fig. 2.5) and the cycle of human technology also described above (Fig. 2.7) are interconnected via the system parameter *total activity growth*. This has a direct effect on the *stock of natural resources* via a time lag, which becomes more and more limited with increasing growth from both spheres. While the natural cycle has learned through evolution to adaptively and skilfully adjust to deficiencies in its biosphere, the restriction of natural resource stocks through the system parameter total *activity growth* leads to a real harvest shortage for the processes of the technosphere in the long term, which weakens activities initiated by humans or brings them to a standstill. The always cultivated *short-termism* of humans through their individual and collective striving for power, technical economic supremacy and ability to progress, would have had its day. The influence of the *overall activity of growth,* with the less sustainable share from the technosphere, also massively and indirectly influences the limits of growth or the limits of biodiversity for its own benefit.

The only realistic consequence of the dilemma for the technosphere, which becomes the *tragedy of the Almende*[12] or *tragedy of the commons,* because we believe that only technical economic progress would bring us further on the way to a survivable future, is the realization: without close cooperation with the long-term proven technical achievements from the biosphere, sustainable progress is blocked for us in the long run.

Because of the universal significance of the term sustainability in all areas of life and work, which we will touch on in the course of this book, this introductory chapter deals with it in greater depth.

2.4 Sustainability: A Universal Concept for Action

A particularly obvious example of sustainability in nature is the regeneration of a forest, in connection with the technical economic use of wood. A certain Hanns Carl von Carlowitz plays a special role here. In 1713, his book with the "unwieldy" title: *Sylvicultura oeconomica oder Haußwirthschaftliche Nachricht und Naturmäßige Anweisung zur wilden Baumzucht* (Hamberger 2013, original 1713) was published.

In the same year, the book *Die Erfindung der Nachhaltigkeit (The Invention of Sustainability) was* published by the Sächsische Hans-Carl-von-Carlowitz-Gesellschaft e. V. in Chemnitz. In it Ulrich Grober writes (2013, p. 15–17, text transfer in "…"):

Hanns Carl von Carlowitz came from a noble family of the Electorate of Saxony, which for generations was involved in the "[…] management of the forests in the Saxon Erzgebirge […]" and earned their living with it. "[…] For the economy of the Electorate of Saxony, the secure supply of wood and charcoal to the mines and smelting works in the Erzgebirge was of strategic importance. […] For a long time, the supply of wood was primarily regarded as a transport problem. […] But in the decades following the Thirty Years' War, the resource crisis over wood came to a head. Certainly there was no general shortage of wood in Central Europe. […] What deeply worried this generation, however, was the prognosis of wood shortages, that is, a predictable general crisis that would occur in a generation or two if things continued as they were."

---

[12] Briefly reiterated at this point: the concept of the tragedy of the alpine pasture used is based on statements by ecologist Garrett Hardin (see literature source).

Who does not think when reading these lines from today's point of view, more than 300 years later, of comparable crisis scenarios, such as excessive consumption of resources with destruction of fertile soils, nature pollution through reduction of the diversity of species (biodiversity), environmental and health pollution through toxins from technical stationary and mobile plants or products, decades of educational misery, permanent energy waste of the highest magnitude, living in undignified and low-contact environments, social deficiencies such as old-age security, etc.? the socially burdening effects of which are not only effective in 1–2 generations, but already for several years (decades) and despite selective progress still show their "ugly" destructive side!

[…] Carlowitz formulated this later in his book very precisely. Already in the subtitle dem … *Grossen Holtz = Mangel … zu prospiciren*, i.e. *to foresee* the *crisis of* the "requisitum primum" (preliminary report) of *the central resource, and to prevent it* (italic emphasis d. d. A.), this can be seen well. (ibid. p. 17)

The invaluable achievement of Hanns Carl von Carlowitz was to have recognized that the destructive *routine* in the cycle of resource use of wood (Fig. 2.9) must be broken and instead a new kind of sustainable development through extensive resource use of wood is required, which also goes hand in hand with new saving techniques, as Fig. 2.10 suggests.

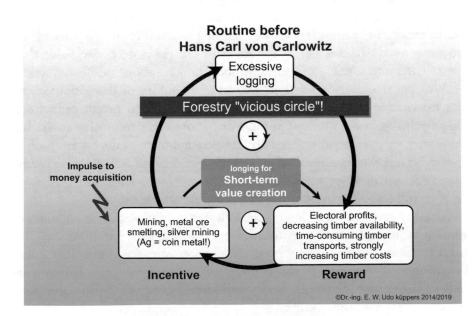

Fig. 2.9 Routine cycle for excessive forest management

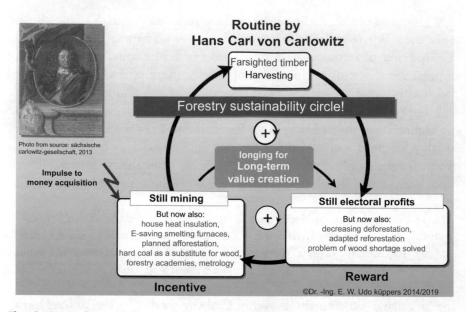

Fig. 2.10 Broken routine cycle according to Fig. 2.9 for sustainable forest management

At first, the acquisition of money was – and still is today – a universal trigger for the routine cycle that followed. At the time of Hanns Carl von Carlowitz, this trigger was aimed at mining with the extraction of the precious metal silver, from which coins were minted (see Fig. 2.9) The incentive was set. In the further course, woods were needed to stabilize the mining, which led to an excessive clearing of forests. This created a deceptive habit that degenerated into a matter of course without thinking about the consequences of this repetitive course of action. The only beneficiary of this vicious circle was the Elector himself (Reward), who derived electoral profits both from logging and from the minting of more and more silver coins. The desire for short-term value creation was the driving electoral force in the background.

However, it was foreseeable that both timber extraction in the forests and the extraction of silver ore would be limited. Hanns Carl von Carlowitz deserves full credit for recognizing this and seeking alternatives.

For every constructed vicious circle, whether it takes place as human intervention in processes of nature, technology or in the social environment, the habit or routine of an activity plays the decisive role. As a rule, only a few benefit from it (in Fig. 2.9, this is the elector), but the damage is borne by many (nature with destroyed forests and altered biodiversity, the obligated forest workers with their health).

Comparisons with economic, technical and social processes of the twenty-first century, which carry vicious circles, are the order of the day. Through routine work, profitable profits for entrepreneurs increase disproportionately, while workers and employees have to be satisfied with disproportionately low wage increases. Nature is still being destroyed – right down to the last corner of the earth – directly (deforestation of tropical forests (in South America and Asia) with decimation of endangered animal species and establishment of huge monocultures for excessive fodder cultivation for cattle, which we eat) and indirectly (exorbitant pollution of streams, rivers and seas by persistent – long-lasting – plastics and other pollutants).

The income gap between the poor and the rich population has been widening faster in a highly industrialised and rich country like Germany since 1990 than in other wealthy industrialised countries (OECD study 2016). In 2018, "Permanent poverty and entrenched wealth" prevail in rich Germany, as the WSI Distribution Monitor (2018) found.

> Accordingly, we need not worry about routine vicious circles in the twenty-first century; they will certainly accompany us for a long time to come. But vicious circles are only the consequence of human activity, which at worst – with a certain time delay – threatens our very existence and not the cause! Will autonomous humanoid robots one day in the distant future keep humans as service providers, as promised by supporters of "artificial intelligence" (Küppers 2018)? Would this make the vicious circles caused by humans and the immense efforts to overcome them through more sustainable adaptive habits or routines obsolete? Speculating about this would hardly do justice to the core concern of the book.

How did Hanns Carl von Carlowitz manage in the eighteenth century to break the damaging vicious circle (in Fig. 2.9) with his momentous habit or routine in order to still satisfy the Elector but at the same time to slow down the radical deforestation and forest destruction? Figure 2.10 shows the ingenious multifunctional solution!

Initially, the acquisition of money remains as the impetus for the circular process now functioning as the "sustainability circle". Mining is still the place of work from which raw materials for values or electoral profits are extracted. Through Carlowitz` reasoning, the timber needed for this was now obtained from extensive rather than excessive forest clearing. This was the decisive breaking of the previous routine or habit with disastrous consequences to a new routine or habit with sustainable forest growth. In other words, only as much timber was cut as would grow back elsewhere. It is true that the timber yield was lower because of the time it took for the trees to grow up before they could be harvested. At the same time, there was a shortage of firewood for heating homes. But the forest was preserved.

The disadvantage of a temporally reduced timber harvest for support and stabilization purposes in mining and the lack of firewood was more than compensated by new heat insulation of houses, energy-saving smelting furnaces, hard coal as a substitute for wood, a systematic planned reforestation and last but not least by the establishment of forest academies with a new measuring system.

Mining and forest developed into a process with long-term value creation. Carlowitz's goal of sustainable forestry was thus set in motion.

In the author's view, it was, among other things, two interlinked ingenious thoughts of Hans von Carolowitz that Dieter Füsslein (2013, p. 251) describes and which grant forestry sustainability, and thus fundamental *nature-sustainability*, a universal exemplary character:

1. nature (is) the foundation on which economy and society are built (today: strong sustainability), and for this he uses convincing ethical arguments.
2. Carlowitz does not add up economic, ecological and social resources, and certainly does not juxtapose three "sustainability pillars" for this purpose, but describes the reactions and correlations, their temporally, locally and functionally linked interconnections.

"Anything that is against nature will not endure in the long run" is a statement handed down by Charles Darwin. – It briefly and precisely describes the prerequisite of all life. The subtitle to this book speaks it in a different way, *"Do we reckon with nature – or nature [reckons] with us ...."*

Without a functioning natural network, maintained by countless species of organisms, whereby we humans allow ourselves to rise above the rules of nature, we are heading for the edge of our existence. Examples of huge plastic waste piles in the oceans already attacking natural food chains, permanent CO_2 pollution of the atmosphere that – in combination with other climate gases, such as methane CH_4, nitrous oxide N_2O and other substances – drive the so-called greenhouse effect (increase in temperature on Earth), clustered extreme weather events that flood entire coastlines all the way inland are the "saws" we use to actively sever the branches of life on which we sit.

Have we learned nothing after these self-inflicted and far from complete disaster scenarios, 300 years after Carlowitz ingeniously changed his mental routine model?

If we compare the controversial three-pillar model of sustainability from the 1990s, with its postulated and intended equal weighting of the economic, ecological and social pillars, with the pillar model of sustainability from today (2020), we really have not learned anything.

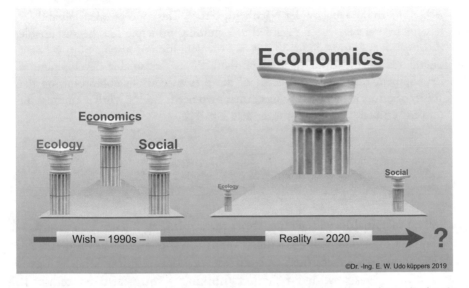

Fig. 2.11 Three-pillar model of sustainability: change in weightings from the beginning (1990s) to 2020

Figure 2.11 compares the weightings of the three-pillar model of sustainability from the 1990s – at the time of the 1992 UNCED conference in Rio de Janeiro – with those from 2020. No in-depth explanation of the changing pillar weightings over time is needed, because we feel their effects first-hand in our daily lives and work.

Hans Carl von Carlowitz's new far-sighted consideration to focus his forestry sustainable activities less on the organisms themselves and more on their reactions and correlations, on their temporally, locally and functionally linked interconnections, shows all his genius of recognizing natural principles.

It is therefore not surprising if we understand Hans Carl von Carlowitz's thoughts as *cybernetic, feedback or systemic,* at a time when these three terms were probably still unknown, and only in the 1940s by scientists such as Norbert Wiener, Jay W. Forrester *and* many others resulted in independent fields of science with the claim *"to think and act in a networked way"* (see E, W. U. Küppers 2018).

A few decades later, *sustainability* was again made accessible to a broad global public. The use of the term sustainability or *sustainable development,* which we take in this context from the Brundtland Report of 1987, became the basis for the United Nations Conference on Environment and Development (UNCED) in Rio de Janeiro in 1992.

According to the Lexikon der Nachhaltigkeit,[13] "The concept of sustainability" has for several years been regarded as a guiding principle for the sustainable development of humankind. To this day, the definition of sustainability is based on the text of the Brundtland Report of 1987: "Sustainable development is development that meets the needs of the present without compromising the ability of future generations to meet their own needs." (UN World Commission on Environment and Development 1987, p. 41)

> Sustainability in its original sense is too valuable to be left to individual interest groups with misconceived ambitions and one-sided goals.

When sustainability is lacking, short-term thinking – *short term missent* – is at play. Action is determined by permanent routines and relies on direct satisfaction of needs. In contrast, sustainable practice requires looking beyond one's own interests, with far-sighted thinking – long term farseeing. This requires a different kind of cognition. The complexity of reality, but also moral questions have to be appreciated.

A Brief Excursion into the Real World of Automotive Sustainability

Our political and economic decision-makers in Germany – more than 120 years later, since the invention of the automobile by the patent obtained by Carl Benz in 1886 – never tire of talking about the automotive industry as the so-called "key industry" of our country. The results of the rapid development of vehicles powered by fossil fuels are also encountered today in the form of resource-rich, high-priced, bulky, fast-driving but also exhaust-emitting off-road sedans or urban off-road vehicles, so-called "sport utility vehicles" (SUVs). The dichotomy between corporate goals and social or environmental responsibility is – despite publicly postulated social responsibility of corporate leaders – Corporate Social Responsibility (CSR) – rarely equal! The corporate striving for more growth and more profit still dominates, which seems to continue its short-sighted misguided economic trend unperturbed despite the real climate crisis and even more so due to the biodiverse destruction crisis of life (see Fig. 2.11).

A recent example of corporate responsibility in the automotive sector can be seen in an interview excerpt with a German manager of an automotive company (Interview with the CEO of BMW AG, Oliver Zipse, FAZ interview, 20 October 2019, p. 28, source: Braunberger et al. 2019):

In response to the FAZ staff's question on the subject of "SUVs" to company leader Zipse: "Won't you still have to bow to the green zeitgeist at some point and say: we're doing without SUVs?"

Oliver Zipse replied: "Why? We live in a social and, I would add, sustainable market economy. Customers decide which car they buy. If they choose an SUV, we'll deliver it."

---

[13] https://www.nachhaltigkeit.info/artikel/nachhaltigkeit_1398.htm (Accessed 20 Dec. 2018).

Zipse's answer – which is sure to meet with approval from his fellow board members at other companies – is interesting from several perspectives:

1. It refers to a so-called *sustainability DILEMMA*.

   The concept of sustainability is associated with completely different mental models and practical goals. It is extremely difficult for people to replace a mental routine cycle that has been acquired over years or even decades and is "calibrated" for short-term growth and short-term profit with a new mental routine that is adapted to reality, even if this is for reasons of survival. Because this is a *necessary prerequisite* for a practical adaptation of entrepreneurial products, processes – also organizations(!) – to a dynamic environment.

2. It points to a distinct lack of "backbone". As manufacturers and distributors of the SUV products referred to, the responsibility for the environmental pollution caused by SUV passenger cars is shifted onto the customers. Where is the much-discussed and "full-bodied" social responsibility of entrepreneurs?

   Behavioral economist Falk (2019, p. 31) has a clear opinion on this, in terms of cooperation for the benefit of the common good:

   > "Overconsumption of resources is morally wrong. Driving an SUV is like peeing in a stream that's drinking downstream. It's not just about costs and benefits, it's also about right or wrong. The morally relevant question, in my view, loosely based on Immanuel Kant, is how do I choose a consumption path that I *can want* 7.5 billion people to choose as well? What would it look like? How much room would there be for an SUV [...]?"

3. The problem of exhaust pollution from SUVs – but also from other vehicles – draws attention to another concept of sustainability, that of the *sustainability PARADOX*.

   The *sustainability paradox* is an expression of a product-oriented quality improvement for environmental relief – in this case: lower car emissions of pollutants that are harmful to the environment, but which is cancelled out by the quantity of the products and even causes the environmental impact to increase again.

   This is the well-known and infamous growth driver *"REBOUND effect"*.

   The automotive industry demonstrates its particular *sustainable environmental responsibility for* passenger cars by advertising less with particularly high individual CO_2-exhaust values due to fossil fuel propulsion energy and more with average, mostly estimated mean values of different passenger car types,[14] whose CO_2 value is significantly lower than the individual CO_2 values of the high priced passenger cars, including SUV models. The constraint imposed by the European Union's underlying new CO_2 limit[15] is palpable. The cocktail of other environmentally or climate damaging exhaust gases from vehicles of all kinds, whether PM (particulate matter), NO_x (nitrogen oxides), HC (hydrocarbons) from marine diesel, aircraft kerosene and many more still await the sustainable application of environmentally sound solutions.

Sustainability also means consideration. Consideration for what?

To future generations! They are to be left an unconditional space that they can use for their own ideas, their own concepts and their own practices. This consideration is a holistic one, encompassing both the preservation and continuation of nature, economic prosperity and social/cultural spaces. However, for several decades to the present, we have seen an increasing tendency to threaten these vital free spaces of creation and development.

What is the use of "top class" technical-economic progress, what is the use of future humanoids that relieve people of heavy physical work and offer services in (almost) every situation in life, if the basic prerequisite of an intact evolutionary nature is no longer assured?

Even if at present the emphasis on an economy that influences everything, with capitalist/turbo-capitalist development strategies, still covers up many actions, as can be seen from the numerous catastrophic disasters (including the US real estate bubble in 2007/2008, the Deep Water Horizon oil drilling platform in 2010 or, very recently, the Wirecard disaster in 2020) caused by economic and financially driven practices worldwide, a new wave of developments is nevertheless emerging that strengthens or helps to strengthen sustainability. Politicians have a special task here to set framework conditions for society and especially the economy. The usual short-term compromises only do harm and, in the long term, often turn into the opposite of what is intended by sustainability in terms of ecological equilibrium, social justice and economic security.

---

[14] https://www.vda.de/de/themen/umwelt-und-klima/co2-regulierung-bei-pkw-und-leichten-nfz/co2-regulierung-bei-pkw-und-leichten-nutzfahrzeugen.html (Accessed 6/28/2020).

Quoted from: Headline: CO_2 regulation for passenger cars and light commercial vehicles (regulation until 2020): "However, the CO_2 regulation does not set the overall European fleet value for each individual manufacturer at 95 grams for passenger cars or 147 grams for commercial vehicles. [...] Rather, a specific limit value is calculated for each manufacturer based on the average vehicle weight of the manufacturer fleets."

[15] https://ec.europa.eu/germany/news/20190415-co2-grenzwerte_de (Accessed Dec 16, 2019).

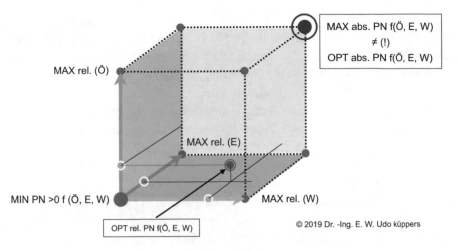

Fig. 2.12 Sustainability cube with the three dominant axes of action

The totality of Carlowitz's guiding principle, today called the sustainability triangle – i.e. social ethics (E), economy (W) and ecology (Ö), with mX as the unit of quantity – can be thought of abstractly and formally – i.e. initially not calculable – as the power product of sustainability P_N (also vectorially in three-dimensional space) (Fig. 2.12):

$$P_N \quad E^m E \quad W^m W \quad Ö^m Ö \qquad (2.1)$$

The multiplicative linking of factors protects thinking from subjective choice of emphasis and measures the influence of even non-measurable components on the correlating result. No factor may tend towards zero or towards minus – the estimation of the exponents (m = power) leads to a ranking of practical action. (Sächsische Carlowitz Gesellschaft [ed.] 2013, p. 251).

In Fig. 2.12, the three coloured impact areas of ecology Ö (green), social ethics E (blue) and economy W (yellow) are combined to form a sustainability space – N-cube.

The starting point of each activity on the three axes Ö, E and W is the lower left corner of the N-cube, whose axis values are by definition not equal to zero, because there is always such a small activity in the sense of ecological, social-ethical and economic action to be expected. The more intensively the specific, goal-oriented actions are located on the axes (symbolically marked by the points outlined in white), the more strongly the focused goals are pursued until a maximum is reached due to local system boundaries.

For the sake of simplicity, all three impact areas in the model in Fig. 2.12 were given identical lengths from "MIN PN" to "MAX rel." and equal weightings, whereas reality must be treated in a far more complex manner. This also includes comprehensible qualitative and quantitative, evaluable links between activities on the respective coordinate axes.

The absolute – theoretical – maximum in the model of the N-cube occurs when the actions in all three effect areas have simultaneously reached the most efficient value of their target approximation (in Fig. 2.12, upper right corner of the cube). However, as already mentioned, this is a purely theoretical value.

It is more realistic to assume that for every conceivable type of action, whether it takes products, processes or organisational procedures as the object of development, which are to be evaluated with ecological, social-ethical and economic criteria, a relative optimum will always result.

Such a relative optimum (red point *within* the N-cube) is shown as an example. A well-known example from the automotive industry can thus be easily understood:

An automobile manufacturer advertises a fossil fuel-powered SUV-vehicle[16] of the upper class, with a sporty appearance, optimal driving pleasure, high safety at maximum speed despite a mass of three tons and all kinds of accessories to financially strong customers in elegantly trimmed multi-colored advertising brochures and video films. Maximum sales revenue at minimum cost is the driving force of the manufacturer. The economic goal demands its price: maximum profits. Accordingly, the marking of the activity commitment on the W-axis in the N-cube (yellow point) is positioned near the relative maximum of possible W-performances.

Parallel to this, the activities of the car manufacturer for its product on the eco-axis appear to be of low efficiency, because as elegantly as the heavyweight bolide moves through road traffic, dangerous gases consistently pollute our air with every kilometre driven. Likewise, enormous amounts of energy are consumed along the manufacturing process, for a production of passenger car components from partly highly toxic raw materials. There is still no car model that offers consumers road safety with materials that are compatible with the environment as a whole, as well as appropriate revenues at affordable costs. The stumbling block is not so much the possibility of what is feasible, but the more than one hundred year old and outdated mobility system in which we find ourselves.

---

[16] SUV is an abbreviation from the English meaning "sport utility vehicle" or "sport and utility vehicle".

A low value can also be found on the activity axis of social-ethical performance or efficiency. Not least, the ubiquitous "diesel scandal" caused by German car companies shows the lack of empathy towards car buyers, some of whom have been deliberately deceived. The impending driving bans for diesel cars in numerous cities in 2019 may lead to considerable social injustice among the population, who will either be banned from driving in inner cities, forced to retrofit their cars at their own expense in a climate-friendly manner, or to forego individual transport altogether, which would also have a positive effect on air cleanliness.

What finally emerges in our model example in the sustainability cube is a relative optimum with dominant economic benefits, coupled with moderate benefits for nature and the environment and even smaller benefits for social-ethical concerns. With the exemplary SUV car, we are still a long way from a criterion of efficient sustainability or an efficient potential product of sustainability (PN, see Eq. 2.1 above).

Sustainability in the Carlowitzian sense is not always the outcome of development strategies and actions that one might expect. For example, the fulfilment of sustainable turnover and profit targets of companies, sustainable product quality, sustainable housing construction, sustainable monocultures, sustainable livestock farming and much more is anything but Carlowitzian holistic sustainability!

Mixing *sustainability* or *sustainable* with attached terms, strategies, actions or goals that are not linked to the holistic thought of Carlowitz lead nowhere. They can only be understood as expressions of diffuse linkages for the achievement of weak results of short-sighted perspectives. They are therefore not a progressive argument for the further development of our society, which brings us full circle to the beginning of this text. Countless publications exist on the subject of sustainability, of which the following can only provide a small glimpse of the variety of differentiated applications on our planet. See Moore et al. [eds] 2018; Osburg et al. [eds] 2017; Zimmermann [eds] 2016; Portney 2015; Heinrichs et al. [eds] 2014; Gruber 2010.

At the end of this chapter, let us not be misled or even seduced by those who believe that they can do without the fundamental services of nature in the anticipatory obedience of technical economic success. Just because nature has no *law firms* to protect it against overexploitation and blind destructiveness by short-sighted and misguided excessive globalization strategies, this does not mean that life – emerging from a long series of evolutionary, highly efficient cyclical processes – may be destroyed forever at whim.

With the same intensity as a minority which, by virtue of its material supe-riority, arbitrarily seizes the biodiverse wealth of life on our planet in order to market it for its individual interests – without regard for losses – we should feel obliged to the still existing natural wealth which secures our existence, more than all technical-economic successes put together.

> Redemption of the damage to nature, wherever evolutionary networks of life have been and are being massively destroyed by humans as polluters, is an essen-tial demand in the name of nature and the sustainable progress of life.

In conclusion, let us be inspired by Bob Marley's "Redemption Song" (album Uprising, 1980), which 40 years ago – as topical then as it is today – addressed the high-risk nature of atomic energy and in its lyrics asks us to *think for ourselves*. If we replace "atomic energy" with "global economy" in the song's lyrics, its message becomes all the more important when we consider not only the looming energy, nature, environmental and climate crises, but also the global communication networks controlled by a handful of corporations.

Redemption song (2nd verse)
Emancipate yourselves from mental slavery
None but ourselves can free our minds
Where! Have no fear for global economy (original: atomic energy)
'Cause none of them-a can-a stop-a the time'
How long shall they kill our prophets
While we stand aside and look?
Yes, some say it's just a part of it
We've got to fulfill the book
Won't you have to sing
These songs of freedom?
'Cause all I ever had'
Redemption songs
All I ever had
Redemption songs
These songs of freedom
Songs of freedom.

References

Baecker, D. (1994) Postheroisches Management. Merve, Berlin

Braunberger, G.; Meck, G.; Peitsmeier, H. (2019) BMW schaltet auf Angriff. Frankfurter Allgemeine Zeitung, Nr. 42, S. 28, 20. Oktober 2019,

Burkhardt, D.; Schleidt, W.; Altner, H. (Hrsg.) (1972) Signale in der Tierwelt. Vom Vorsprung der Natur. dtv 853, Moos, München

Campbell, N. A.;Reece, J. B.; Markl, J. (2006) Biologie, 6, Pearson, München

Crutzen, P. J. (2000) Geology of mankind. Nature, Vol. 415, S. 23

Crutzen, P. J.; Stoermer, E. F. (2002) *The "Anthropocene". IGBP Global Change Newsletter* 41, Mai 2000, S. 17–18

Falk, A. (2019) Ich und das Klima. Die Zeit, Nr. 48, S. 31, 21. November 2019

Gérardin, L. (1968) Natur als Vorbild. Die Entdeckung der Bionik. Kindler. München

Greguss, F. (1985) Patente der Natur. Technische Systeme in der Tierwelt. Biologische Systeme als Modelle für die Technik. Quelle & Meyer, Heideberg, Wiesbaden

Gruber, U. (2010) Die Entdeckung der Nachhaltigkeit. Kulturgeschichte eines BegriffsKunstmann. München

Grunewald, K.; Bastian, O. [Hrsg.] (2013) Ökosystem-Dienstleistungen. Springer Spektrum, Berlin, Heidelberg

Hamberger, J. (Hrsg.) (2013) Sylvicultura oeconomica oder Haußwirthliche Nachricht und naturmäßige Anweisung zur wilden Baum-Zucht. Oecom, München. Erstmals publiziert wurde das Buch auf der Leipziger Ostermesse des Jahres 1713

Hardin, G. (1968) The Tragedy of the Commons. Science 162 (3859), 1243–1248. https://doi.org/https://doi.org/10.1126/science.162.3859.1243

Heinrichs, H.; Michelsen, G. [Hrsg.] (2014) Nachhaltigkeitswissenschaften. Springer Spektrum, Berlin, Heidelberg

Humboldt, A. v. (2019) Ansichten der Natur. Nikol, Hamburg. Erstveröffentlichung 1808

Kolbert, E. (2015) Das sechste Massensterben. Wie der Mensch Naturgeschichte schreibt. Suhrkamp, Berlin

Küppers, E. W. U. (2018) Die humanoide Herausforderung. Springer Vieweg, Wiesbaden

Küppers, E. W. U. (2015) Systemische Bionik. Impulse für eine nachhaltige gesellschaftliche Weiterentwicklung. Springer Vieweg Wiesbaden

Küppers, U.; Tributsch, H. (2002) Verpacktes Leben – Verpackte Technik. Bionik der Verpackung. Wiley VCH Weinheim

Mathé, J. (1980) Leonardo da Vinci Erfindungen. Parkland, Liber SA and Edition Minerva SY, Fribourg, Genève

Meadows, D. H. et al. (1972) The Limits to Growth. Universe Books, New York

Mehltretter, T. (2018) Das große Insektensterben. Mehltretter Media. Im Auftrag des ZDF, in Zusammenarbeit mit ARTE

Moore, T.; Haan, F. J. de; Horne, R.; Gleeson, B.J. et al. [Eds.] (2018) Urban Sustainability Transitions. Springer Nature, Singapore

Müller G. K.;Müller C. (2003) Geheimnisse der Pflanzenwelt. Manuscriptum, Hoof, Waltrop, Leipzig

Nachtigall, W.; Wisser, A. (2015) Bionics by Examples. 250 Scenarios from the Classical to Modern Times. Springer, Cham, Heidelberg, New York. Siehe auch: (2013) Bionik in Beispielen, Springer Spektrum, Berlin, Heidelberg

Nachtigall, W. (2002) Bionik. Grundlagen und Beispiele für Ingenieure und Naturwissenschaftler. Springer. Berlin, Heidelberg

OECD (2016) Income inequality remains high in the face of weak recovery, Paris, France, OECD Publishing

Osburg, T.; Lohrmann, Chr. [Hrsg.] (2017) Sustainability in a Digital World. New Oportunities Through New Technologies. Springer, Nature. Switzerland

Osten, M. (2019) Die sinnliche Wissenschaft. Beitrag in: Die Welt, 2019, 29

Paturi, F. R. (1974) Geniale Ingenieure der Natur. wodurch uns Pflanzen technisch überlegen sind. Econ, Düsseldorf, Wien

Portney, K. E. (2015) Sustsainability. The MIT Press Essential Knowledge Series. Cambridge, Mass., London, England

Renn, J.; Scherer, B. (2015) Das Anthropozän. Zum Stand der Dinge. Matthes & Seitz, Berlin

Sächsische Carlowitz-Gesellschaft (Hrsg.) (2013). Die Erfindung der Nachhaltigkeit. Leben, Werk und Wirkung des Hans Carl von Carlowitz. München, oekom.

United Nations (1987) Report of the World Commission on Environment and Development. Our Common Future.

Vester, F. (1986) Ein Baum ist mehr als ein Baum. Kösel, München, 2. Auflage

Vester, F. (1972) Das Überlebensprogramm. Kindler, München

v. Frisch, K. (1974) Tiere als Baumeister. Ullstein, Frankfurt/M., Berlin, Wien

Werner, P. (2009) Zum Verhältnis Charles Darwins zu Alexander von Humboldt und Christian Gottfried Ehrenberg.HiN, Internationale Zeitschrift für Humboldt Studien, HiN x, 18, S. 68–95

WMO (2020) Statement on the state of the Global Climate. World Meteorological Organization, WMO-No. 1248. Geneva, Switzerland

WSI (2018) WSI-Verteilungsmonitor, Verteilungsbericht 2018, Hans-Böckler Stiftung, Düsseldorf

Wulf, A. (2015) Alexander von Humboldt und die Erfindung der Natur. C. Bertelsmann, München

Zimmermann, F. M. [Hrsg.] (2016) Nachhaltigkeit wofür? Von Chancen und Herausforderungen für eine nachhaltige Zukunft. Springer Spektrum, Berlin, Heidelberg

3

Basic Principles of Nature: Survival in an Open "Cooking Pot"

Abstract We have been living in close cohesion among organisms for billions of years. As organisms we are "open systems", which means: we use energy, substances and information from the environment and thus create order within ourselves. Finally, we give energy, substances and information back to the environment, only in a different form. The – by human standards – never-ending flow of solar energy is our only source of energy for life, the earth our only and finite source of raw materials. It should therefore require no discussion and be self-evident to treat our life-sustaining system Earth with care. The fact that this is happening less and less, and what consequences follow from this, we will learn by way of example in this chapter. It will also become clear that we can never win a "business as usual", a fight against nature. The evolutionary pressure that has been at work for billions of years cannot even be unhinged *en passant*. This grown strength of evolution is therefore discussed in detail, from which it should be clear once again that the overriding goal is and must be to respect nature in our further development and to cooperate with it, instead of destroying it senselessly – due to short-sighted desires.

3.1 Open Systems: Prerequisite for the Development of Life and How We Deal with It

Our earth is an extremely fragile structure. On the one hand, an ozone layer only a few kilometres thick in the stratosphere (15–20 km altitude) helps to preserve it. If it were missing, this would be harmful for all organisms on earth. Humans would be at risk of developing skin cancer or eye diseases, animals and plants would be exposed to harmful UV-B radiation (350–280 nm wavelength), which could even endanger the life-supporting photosynthesis of plants. Figure 3.1 shows schematically the principle of *open systems* on our planet.

Open systems are the prerequisite for the further development of living organisms, because they permanently absorb energy and matter from the environment to maintain their metabolism. The end products of the organismic metabolic processes are eliminated and returned to the environment for further cyclic processing.

The processing of energy and matter in the organism creates a stable dynamic equilibrium, a functional state of order, so to speak, which minimizes disorder or entropy that cannot be converted into work. The functional circulatory system in our body, with the participation of specific networked organ functions, is responsible for this. If the process of internal organismic work performance is disturbed, we become ill.

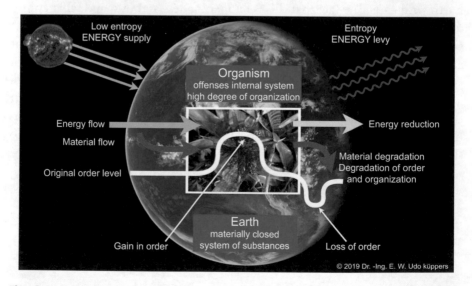

Fig. 3.1 Principle sketch of *open systems* based on Riedl (2000, p. 112), greatly expanded by the author. (Source "blue marble": https://earthobservatory.nasa.gov/blogs/elegantfigures/2013/04/22/earth-day-and-night/. Source "sun": https://www.nasa.gov/image-feature/active-regions-on-the-sun. Both sources accessed on Jan 5, 2019)

The fact that organisms have been able to develop at all, not completely independent of the inanimate environment, is due first and foremost to our only external source of energy, the sun, and to a progressive efficiency of the internal metabolism. In summary, one could say: the greater the energy metabolism and thus the ability to perform work in a differentiated manner, the more independent or free one's own creative possibilities are, while the complex dynamic environment becomes increasingly entropic, richer in disorder. Let us take a brief look at our own organism.

To what extraordinary power efficiency a human organism is capable of, consuming an average basic energy consumption of 1800 kcal daily, is shown by the power required by a human brain with about 86 billion nerve cells, each with about 10,000 nerve connections: only 20 W of power, no more than the power of an incandescent lamp (Küppers 2018a, p. 55). With a wattage value that is perhaps 3–4 times higher, our body's own energy metabolism consumes only a little more, – incidentally, energetic efficiency values that are unparalleled in the technosphere.

Every living being is, so to speak, a functioning *"working organism"* that minimizes entropy, i.e. energy that cannot be converted into work, due to its inner functionality. The term "working machine", which makes little sense in this context but was used in earlier times – occasionally even today – associates the human organism with a mechanically running machine in which all functions are rigidly connected with each other via a gear train, a metaphor that has long been outdated.

However, living organisms and inanimate machines have one thing in common: both are subject to a limited life span. The living organisms as limited dynamic stability and work performing islands of order in an environment increasingly characterized by disorder on the one hand, and machines on the other hand, by their limited working capacity due to inevitably increasing wear of their parts.

Open systems are the prerequisite for the development of life, but for the individuals themselves they have only a limited life span, maintaining order and doing work.

The fact that this limited life span over a tiny period of time – measured in geological and evolutionary terms – of about three human ages or about 200 years, since the beginning of the industrial revolution in the nineteenth century, has also led and continues to lead to enormous damage to our basis of existence on and the protective ozone layer around our Earth, is made particularly clear by the climate change that is taking place, our destructive attack on biodiversity and, not least, the new Earth age we have created, the Anthropocene, with its consequences.

This leads us to the central question: ***How do we treat our earth?***

3.1.1 How Do We Treat Our Earth?

Will we one day, if we continue as we are with short-sighted development strategies, burn ourselves out, as the title of a book (Schellnhuber 2015) admonishes? McKibben (2019) describes the ominous development of the Earth in his article about a shrinking planet due to various influences, such as weather extremes, years with new heat records, flood plains with "[…] humanity in retreat" (ibid., p. 53) and "Dying animals, declining harvests" (ibid., p. 55). The scenarios could not be worse.

What sustainable strategies must we implement to survive in the increasingly warm and open "cooking pot" of the biosphere? From the perspective of politics and nature, 2019 leaves little hope for fundamentally sustainable development progress of life. Two action items will illustrate this.

Strand 1: People in action against nature

Jair Bolsonaro, newly elected to the office of head of government of Brazil on January 1, 2019, intends to further destroy the largest "lung" on earth in terms of area, the Amazon rainforest, which is a fundamentally important CO_2 reservoir for our habitat on earth and threatens many indigenous peoples in their natural habitat (Blasberg et al. 2019; Habekuß 2019; Schweers 2018). Mining and cattle ranching on vast farmlands are to penetrate deep into the rainforest, systematically destroying lives irrevocably at the expense of mindless economic interests. The passion with which politicians and the relevant industries (Küppers 2018b) display ultra-short-sighted actionism, in a region that is vital for the survival of all humanity, with many so-called biological "hot spots",[1] is incomprehensible with common sense. Bolsonaro's Brazilian destructive actionism – which from today's perspective (mid-2020) has even drastically worsened – is thereby only the continuation of a crime against the livelihood of plants, animals and humans spread over the whole earth.

Of course, from a *systemic perspective,* we as critics of Brazil's current environmental policy are also partly to blame for the destruction of the rainforest. For we are the ones who demand cheap food in the supermarkets of industrialized nations – in this case, specifically meat from cattle farms in South America or Brazil. As a result, the expansion of cattle farms at the expense of rainforest destruction drives a diabolical regulatory cycle unabated. There are always two sides to a coin.

Another well-known example is the reduction of the habitat of orangutans in Borneo and Indonesia, which has been going on for years, with all its

---

[1] In the biological-ecological sense, "hot spots" are regions on our globe that have an extraordinarily high density of biological diversity – biodiversity.

consequences, due to the increasing cultivation of palm oil plantations, whose product, palm oil, is ecologically and economically senselessly misused for heating plants and as fuel for vehicles. In the report of the Bayerischer Rundfunk from 2018,[2] with the title: "*Palm oil plantations instead of jungle – orangutans on Borneo threatened*", which laments a massive decline of the orangutan by 150,000 individuals since 1999, Jamartin Sihite, executive director of the Borneo Orangutan Survival Foundation (BOSF) is quoted with the following words:

> Here you go to jail if you kill an orangutan.
> But if you cut down a whole forest and the orangutan dies from it, nothing at all happens to you.

This strategy of irretrievable destruction of nature and killing of the already decimated primate species of orangutans – practiced in the economic-political alliance – happens on the basis of a capitalist-driven *vicious circle strategy* according to the following pattern or compulsive sequence as follows:

1. The starting point is the economically driven sale of rainforest areas by the responsible government to global logging companies. ➔
2. The destruction of the fundamentally vital rainforest is accompanied by a loss of habitat for orangutans, with the cheap acceptance of the systematic decimation and killing of this primate species, which is at great risk of extinction. ➔
3. On the cleared open spaces, fast-growing economic plantations in the form of palm oil plantations are planted in monocultures, which – after a few years – are awarded a so-called palm oil certificate RSPO,[3,4,5] (Roundtable of Sustainable Palm Oil), so to speak, with a *first ecological "foolscap"*, without any sustainable value. ➔
4. In addition to the inexpensive palm oil admixture in countless food-stuffs (sweets, ice cream, margarine, etc.) and everyday products (soap, detergents, cosmetics, etc.), a second ecological "fool's cap" is put on the palm oil product by the state-ordered palm oil biodiesel admix-

(continued)

---

[2] https://www.br.de/rote-liste/orang-utan-menschenaffen-borneo-bedroht-vom-aussterben-durch-palmoel-plantagen-100.html. Accessed 12 Jan. 2019.

[3] https://www.rspo.org (Accessed Jan 12, 2019).

[4] https://www.beobachter.ch/umwelt/okolabel-der-palmol-maulkorb/ (Accessed Jan. 12, 2019).

[5] https://jakartaglobe.id/news/kalimantan-villagers-file-complaint-against-rspo-in-switzerland/ (Accessed Jan 12, 2019).

(continued)

ture in cars with (fossil diesel fuel.), a *second ecological "fool's cap"* is put on the palm oil product by state-ordered palm oil biodiesel admixture in passenger cars with fossil diesel fuel, which is supposed to suggest to passenger car diesel drivers that they are driving in an environmentally friendly way, although they are contributing to the destruction of the rainforest by the forced purchase of the palm oil admixture!

5. The vicious circle continues, with more and more destruction of rainforest and expansion of monoculture plantations with highly dubious ecological arguments but not at all dubious returns and profits for the commercial enterprises involved.

6. Here the global capitalist *vicious circle* closes to the benefit of a small minority but to the overwhelming detriment of the majority of people.

An additional well-known example of current value concerns Arctic ice masses melting due to climate warming. Christian Haas of the Alfred Wegener Institute in Bremerhaven, Germany, assesses the situation of cyclical polar ice melt in the conclusion of the 2018 article: Summer sea ice extent still on a low course – minimum in sight (meereisportal.de)[6] as "[...] increasing impacts of the warming trend [...]" of the Earth, in which humans are significantly complicit. Sea ice decline leads to increased consequences for the Arctic climate system. Not only animals such as polar bears and arctic foxes have to fear for their habitat, but also the indigenous Inuit are being pushed back from their accustomed habitat by technically prosperous actions of industrial nations in search of new oil and gas energy sources and other raw materials (Bartsch 2016; Remsmeier 2016).

After all, since the invention of plastics with long-lasting – persistent – properties in the mid-nineteenth century, our planet has become a global plastic waste dump, not only at the backyard ramps of supermarkets and on our doorsteps. Plastics spread across all oceans, through the air, across all soil surfaces and mountains, even into the Arctic and Antarctic regions, and ultimately through many food chains, becoming a deadly threat to the continued existence of life of all kinds. This realization has now matured into general knowledge.

---

[6] https://www.meereisportal.de/archiv/2018-kurzmeldungen-gesamttexte/meereisminimum-arktis-2018. Accessed on 12 Jan 2019.

We humans seem to be ingenious in using a strong focus on details to make one "vicious circle" after another work and to derive short-term economic success from it. At the same time, however, we are largely blind to recognizing the emerging burdensome consequences of these causal development strategies in the indisputably complex and dynamic environment, let alone avoiding them at an early stage – before damage occurs.

The list of life-destroying, dominant, economically led actions in favor of the few and at the expense of the many could be continued at will. The question is: How long can humanity afford such blinkered actionism? Or do we believe, in an emerging time, full of promises for a better future, that through digitalized work (buzzwords like "Industry 4.0" or soon "Industry 13.3", BMWE 2019) we can also take a new digitalized influence on nature (buzzwords "Life 2.0" or soon "Life 8.3", Assheuer 2010)?

Strand 2: Nature in action against people

Water is the most abundant chemical compound on our planet, where two hydrogen atoms combine with one oxygen atom to form a molecule H_2O. With its different states of aggregation – solid, liquid, gaseous – water is present in all regions of the earth. Quite incidentally, water is our life substance par excellence. Humans are only able to live without water for a few days, without solid food for several weeks.

Our knowledge about water is constantly growing. Our special way of dealing with water – despite a great affinity for it, in infancy we have a body water content of approx. 80%, in adulthood still approx. 60–65% (Schmidt et al. 2005, p. 706) – is of an impressively limited nature. Two recurring examples should support this claim.

Water as a liquid

Technical constructions for dealing with water exist in abundance. They help to store water supplies in times of shortage (dams) and to control and meter our daily water consumption through raw pipes or open channels. All water comes from nature and is returned to it in cycles. But woe betide those who interfere with these natural cycles and manipulate them for economic reasons. In times of ice melt or continuous rain, enormous and highly turbulent volumes of water briefly pour through engineered channels – which once had extensive natural floodplains – and have been straightened by humans for navigation. Despite all precautionary measures and closures against flooding, we remain powerless against these occurring forces of nature until further notice, as the regularly occurring hurricanes in the USA, which put whole cities and areas under water, impressively but with catastrophic consequences confirm again and again.[7]

In many cases, however, we have only ourselves to blame, because our interventions in the course of natural watercourses were and are focused on short-term economic goals, without having sufficiently calculated the delayed consequences that now seem to take us by surprise! Behind this there is also unambiguously *a logic of failure, a lack of strategic thinking in complex situations,* as the psychologist Dietrich Dörner (1989) has already impressively described for other areas of life. When intervening in complex situations in our environment, it is therefore essential to pursue not only one goal but several goals at the same time, not least to interlink these goals. Only on the basis of such a precautionary strategy could we succeed to some extent in mitigating the enormous damage that has occurred so far – in this case water damage – with catastrophic effects.

> "Fighting against the power of nature" is the goal we will *never* achieve – but harnessing the forces of nature where possible and appropriate remains the sustainable imperative.

However, our development of the science of technology has been built up over decades on the aftercare principle, often also called the risk principle. Potential damage is repaired when it occurs. The consequential cost principle is derived from this. In contrast to this is the precautionary principle. The declaration of the United Nations Conference on Environment and Development (UNCED) in Rio de Janeiro in 1992 concretizes the precautionary principle in chapter 35 paragraph 3 of Agenda 21[8] (BMU 1997):

Given the risk of irreversible environmental damage, a lack of full scientific certainty should not be used as an excuse for delaying measures that are intrinsically justified. In the case of measures relating to complex systems which are not yet fully understood and where the consequences of disturbances cannot yet be predicted, the precautionary approach could serve as a starting point.

---

[7] https://www.zeit.de/news/2018-09/15/wirbelstuerme-in-den-usa-180912-99-931139. Accessed 12 Jan. 2019.

[8] http://www.un.org/Depts/german/conf/agenda21/agenda_21.pdf. Accessed 13 Jan. 2019.

Water as frozen snow crystals

No matter what the state of aggregation of water is, it always carries surprises or dangers in special situations. What we call snow catastrophes are still predominantly single events, with periods of calm in between, usually lasting for years. Whether, in the course of climate change, we will have to adjust to permanent annual snow catastrophes, such as those that have subsequently occurred in German-speaking countries, seems more likely than unlikely.

Safety in everyday conversation in a village in the eastern Bavarian border region between a pensioner, a younger man and a commuter (Eisch-Angus 2019, p. 7):

> On the other side of the platform, a house is scaffolded; the roof is probably being repaired here after the snow disaster last February*. Two or three metres of snow that didn't melt away the whole winter; the older woman can't remember anything like that from her childhood.
>
> *In February 2006, the disaster situation was declared in Eastern Bavaria due to incessant snowfall. Experiencing and coping with this snow disaster together has firmly rooted it in the collective memory.
>
> The snow disaster in my eastern Bavarian research area in early 2006, in which buildings collapsed, endangered houses had to be evacuated, and localities were at times barely accessible, was very present in the collective memory of the region at the time of the interview and remains so to the present (2019, d. A.) (ibid., p. 287–288).

Snow disasters, such as the one in 1978/1979 in Schleswig Holstein[9] and the current one in the Austrian and German Alpine regions from 2019[10] will certainly continue to shape our collective memory about disasters in our close surroundings. However, whether we derive the right risk conclusions for ourselves from this (Renn 2014) remains an exciting question.

Our interaction with animate and inanimate nature has left much to be desired in many periods of time. The current geological period of the Anthropocene in particular is not spared from this – on the contrary!

Many scholars with foresight have identified and warned against potential hazards to life, interrelationships in nature, and the impact of human activities on nature. Among them were and are important polymaths of the twentieth century, such as Jay Wright Forrester (1971); Donella and Dennis

---

[9] https://www.ndr.de/nachrichten/schleswig-holstein/Fuer-Schneekatastrophe-besser-geruestet-als-vor-40-Jahren,schneekatastrophe148.html. Accessed 13 Jan. 2019.

[10] https://www.tagesschau.de/ausland/schnee-alpen-119.html. Accessed 13 Jan. 2019.

Meadows (2010); Meadows et al. (1972); Edward Osborne Wilson (1995, 1992); Gregory Bateson (1985) with only some of their extensive works and many others, who looked and look at our earth from a systemic perspective, whereby they recognized more clearly than others the indisputable connections in the biosphere between, plants, animals and humans.

Despite the obvious fact that we move in a thoroughly interconnected, species-rich world and remain inseparably connected to it as human beings, we are still blind to recognizing and appreciating nature's ingenious achievements for life and using them for our own benefit. This is particularly regrettable because such reflections also lead to humans recognizing themselves more clearly as part of a network of nature than has been the case to date. This leads us directly to a concept that is of central importance for all life on earth: species richness or biodiversity.

With our unreflective arrogance and arrogance – native to the Western hemisphere – we confront nature. Bateson pointed this out in 1972 in "Steps to an Ecology of Mind. Collected Essays in Anthropology, Psychiatry, Evolution and Epistomology" (see Bateson, German edition 1985, p. 627–633). He identifies this hubris (extreme form of arrogance or overestimation of oneself) as an important factor influencing the dynamics of ecological crises. Figure 3.2 shows hubris in a diagram of networked influences in societies.

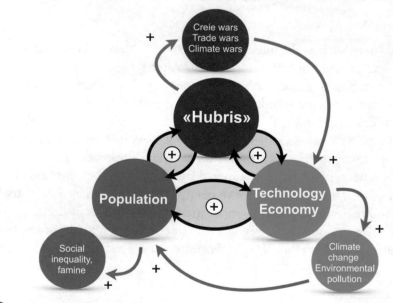

Fig. 3.2 Occidental hubris as the background of action. Graphic based on Bateson (1985, p. 630). The author's sketch supplements this with current social areas in common feedback networking

Bateson drew the "dynamics of ecological crises" as three interconnected autocatalytic phenomena, in the concurrent rotating "gears" of hubris, technology, and population, resulting in destructive effects on pollution, famine, and war. He states:

> The problem the world faces is simply to introduce any processes into this system that are counter-intuitive (uniform direction of rotation of the "gears" in his hubris graph, d. A.).

Starting from the most important problem in the 1970s, the population explosion (in addition to this, climate change, the Anthropocene, and possibly in the future digitalization will be added as general problems, d. A.), Bateson explains (ibid., p. 651):

> That the ideas which dominate our civilization at the present time (1970s, i.e.) derive in their virulent (dangerously affecting, i.e.) form from the Industrial Revolution. They can be summarized as follows:
>
> (a) It's us versus the environment.
> (b) It's about us versus other people.
> (c) It depends on the individual (or the individual society or the individual nation).
> (d) We can exercise unilateral control over the environment and we must strive for that control.
> (e) We live in an infinitely expanding "frontier".
> (f) Economic determinism is *common sence*.
> (g) Technology will do it for us.

Much of what Bateson listed under a-g has persisted, even intensified, into the present time. Much is based on confrontation with life, instead of switching to the more sustainable system change of cooperation, as we can learn from nature in a rich way, because there confrontation and cooperation are represented together.

It is indisputable that economic determinism, which is still doing its thing today, is conceivably ill-suited to unifying technology, the economy, society and nature. – *Nor* will technology alone "[…] do it for us already", as Bateson put it.

The fact remains: Without a "revolutionary", clear and sustainable turn to nature and its principles of life, (without falling into nature mania, but also without technical-economic advances for society), which offer us far more than just short-term risky solutions, as they are common in the technosphere, we will steer into a future problem field of unknown proportions. Preventing this is everyone's duty.

3.1.2 Species Richness: Conservation or Loss of Biodiversity?

Section 3.1.1 already clearly shows examples of how we treat our only basis of life, the Earth. A change, i.e. a reduction in species diversity or biodiversity, has an even more significant impact on our own survival because it is evolutionarily fundamental. Is ongoing additional land use by humans driving local terrestrial biodiversity into a corner and thus beyond the planetary "safety limit?" (Newbold et al. 2016). This is one of many questions of fundamental and central importance to our continued existence.

The vicious circle of agricultural monocultures with herbicide treatment (herbicides are chemical compounds for *weed control*), the consequence of the influence of herbicides on animal pollinators such as bees with their time-delayed species extinction has been having an ominous effect for several years. Monotonous agricultural land is thereby recognized as one of several causes of bee mortality (wiss. Paper among many others: Calatayud-Vernich et al. 2019; Kopit and Pitts-singer 2018; Forfert et al. 2017; Krupke et al. 2012; general information on bee mortality: BUND 2018). It is noteworthy that *weeds* refer to so-called accompanying vegetation, which enriches biodiversity but disturbs it for economic reasons of agricultural monoculture yield maximization and is thus deliberately destroyed by means of chemical environmental poisons. This reveals another vicious circle within a purely economically and technically driven agriculture, to the detriment of biodiversity and life. An example from South America shows the unfortunate chain of circumstances following the decimation of a single endemic bee species (taken from Küppers 2018b, p. VIII–VIX).

> Several years ago, man caused the death of a single bee species in South America, the large female orchid bee (Euglossa), by irresponsible and economically driven destruction of nature, by clearing virgin forest. It is a special creature in the global nature network, because only it can pollinate the Brazil nut tree. Human intervention led to the (almost) total loss of production of the endemic plant. Widespread clearing to this day also leads to the aguti being threatened with extinction because the South American rodent is almost the only animal that can break the hard Brazil nut capsule and thus distribute individual seeds in the soil, ensuring the growth of new Brazil nut plants over a wide area.[11] The result is that for years there have been no Brazil nut seeds in their shells and spherical

---

[11] https://www.regenwald.org/uploads/regenwaldreport/pdf/regenwald-report-022012.pdf (Accessed Jan 16, 2019).

husks in the earth-spanning trade, something we Europeans notice especially during the Christmas season. Wild bee populations are degraded by increased commercial economic pressure, in many ways to "service animals" of humans. For example, forest clearing and monocultures, which result in a drastic reduction of biodiversity, not only contribute to the reduction of bee species, but also promote it through the increasing occurrence of "[…] neonicotine(s), (which are) highly potent, synthetically produced substances used in agriculture to kill insects" (Guillén 2017, p. 22). Behind this are globalizing flows of goods combined with the well-known goal of economic turnover and profit maximization.

Edward Osborne Wilson, one of the world's best-known biologists and ant scientists, describes another example of how we ignore biodiversity at the expense of economically driven maximizing returns from agriculture, with all the consequences caused by short-sighted thinking and action. The example reveals a long conflict, reaching into the present, between the preservation of high biodiversity of crops and their monocultural excessive processing for human food.

After a compilation of numerous plant species whose active ingredients are used as medicines, Wilson (1995, p. 351) sees equally promising prospects in wild plants that can serve as food.

Few of the species that could potentially be commercially exploited are currently (1990s, n.d.) reaching world markets (presumably this hasn't changed much today, 2020, given the increasing GM intervention in a handful of crops that determine our basic diet, n.d.).

Probably 30,000 plant species have edible parts, and a total of 7000 species have been grown or collected for food throughout history. In contrast, ninety percent of the world's plant food production comes from only twenty species and over fifty percent from only three species – wheat, maize and rice. This thin cushion of species covers mainly the cooler climates, and in most areas of the world these few species are grown in monocultures that are susceptible to disease and infestation by insects and nematodes.

The disproportion is particularly evident in the example of fruits. A dozen temperate species – apples, pears, strawberries, peaches, and so on down the whole familiar list – dominate the markets of the northern hemisphere and are also strongly represented in the tropics. In contrast, 3000 other species are available in the tropics, but only two hundred of these are actually used. Some species such as cherimoyas, papayas and mango have recently gained the same status as an export product as the banana, while carambolas, tamarindo and coquito are showing promise in the market. But most consumers in the north have yet to enjoy such delicacies as lulos.(the "golden fruit of the Andes."), mamones, rambutanes and the almost legendary durian and mangostino, considered by aficionados as the best-tasting fruits on earth.

The radical reduction over the years to a few plants and fruit varieties is unmistakable in today's markets in the cities of the industrialized nations. Only idealists nowadays offer, for example, new apple varieties from long-forgotten apple cultivars, with an extraordinary taste that we can hardly perceive sensually thanks to the uniform taste of today's mass-produced fruit.

> Another aspect of biodiversity in our affluent economy is to put a cost on everything so that comparisons can be made, for example, between building an energy-efficient house and a food crop with hardy properties. This typical approach, dominated by economic thinking, is described by philosopher Bryan Norton (1992, p. 222–228) as follows, where the term *value of biodiversity* refers to potential costs, for example for the discovery of important properties of a plant currently considered useless with future marketing potential:

For most species, however, our current knowledge (1992, op. cit.) is not sufficient for a calculation (this is true – despite scientific progress – given the still overwhelming biodiversity up to the present time, op. cit.). Consequently, in addition to the known values they can give for a few species, economists calculate a *potential value* for those of unknown utility: this term refers to the value that may accrue to a species currently considered useless [...] through the possible discovery of a useful use in the future (Randall 1992, op. cit.).

If we eradicate a species today, we thereby preclude such discoveries in the future. According to the definition of Fischer and Hanemann (1985, d. A.), potential value is therefore the benefit that results from considering the possibility that a species currently threatened with extinction may prove useful at a later date. The question would thus be: how much are people willing to pay to preserve a species and thus keep open the option that new discoveries will later be made that prove the usefulness of that species?

[...] Expressing these potential values (of species, d. A.) in money seems a daunting task; in fact, the problem is even greater than it first appears. One cannot begin to calculate a potential value until one has identified a species, until there is conjecture about its utility, until one can somehow express that utility in monetary terms, and until one can estimate the probability of future discoveries related to that species (so that the evolution of value over time can be predicted). Only when all these conditions are fulfilled, one can try to convert the possible future values into dollars or marks (today Euro, d. A.).

I think it is safe to say that while there is a great deal of theoretical interest in ascribing utility and thus potential value to individual species-impressive progress has been made in formal elaboration-it will be a long time before it is possible to state the total value in present-day monetary units for even a single species. (Norton 1987)

In my estimation, it is virtually impossible that any calculation of the actual value of a species, especially within a complex dynamic environment, will ever lead to a relevant cost result with which economists or other interested parties can make monetary comparisons – not least profit calculations.

Here, a development process collides with a completely different thought pattern, on the one hand nature with its dynamic and complex networking between species, on the other hand the economy and technology, with their defined causal process flows and products, to which a clear cost factor should be able to be assigned. Even if an attempt is made to assign a cost factor to the value of natural species as *products* by means of probability calculations, there still remain the often unknown services that are available as a result of inter-connections between species and between species themselves. It is more than a Sisyphean task to try to calculate species of nature in monetary terms. To put it bluntly: it borders on human intellectual underachievement – on stupidity – to subjugate nature in such a primitive economic way.

Norton has also expressed his opinion on this by commenting on the "[…] difficult question of the value of biological diversity as follows" (Norton 1992, p. 226):

> […] perhaps it makes more sense to look closely at the question itself and consider why we are looking for an answer in the first place. It provides a lot of information about ourselves. It is a sign of our extraordinary arrogance, simply because we are the only species that holds conferences and writes books to address this question. And this impression is hardly mitigated when we look at the reasons why we usually ask about the value of biodiversity.
>
> Why do people insist on assigning a monetary value to every species? We are told that this is necessary because there are important decisions to be made that may lead to the extinction of other species. And those decisions should be based on certain analytical frameworks – that is, each species must be assigned a value according to our economic system. If a species has no monetary value, it is not considered at all. In other words:
>
> We are supposed to express the value of species in dollars or marks (nowadays euros, d. A.) so that it can be compared with that of land near reservoirs and kilowatt-hours from hydroelectric power stations.

We can justifiably assume that the statements on biodiversity made in the 1980s/1990s will continue to apply today and certainly well into the future. Time has a different meaning in nature than it has today in the context of a human lifetime.

A recent inventory of biodiversity on our planet "Living Planet Report 2018: Aiming higher" (WHO 2014, p. 6–7) cites the following figures:

The Living Planet Index also tracks the state of global biodiversity by measuring the population abundance of thousands of vertebrate species around the world. The latest index shows an overall decline of 60% in population sizes between 1970 and 2014. Species population declines are especially pronounced in the tropics, with South and Central America suffering the most dramatic decline, an 89% loss compared to 1970.

What is clear is that without a dramatic move beyond 'business as usual' the current severe decline of the natural systems that support modern societies will continue.

Furthermore, the report also brings up the economic aspect of species, which has already been dealt with in a differentiated manner (ibid., p. 6):

All economic activity ultimately depends on services provided by nature, estimated to be worth around US$125 trillion a year. As we better understand our reliance on natural systems it's clear that nature is not just a 'nice to have'. Business and the finance industry are starting to question how global environmental risks will affect the macroeconomic performance of countries, sectors and financial markets, and policy makers wonder how we will meet climate and sustainable development targets with declining nature and biodiversity.

In this context, and in order to conclude the chapter, we let the worldly-wise, soon to be 90-year-old biologist Edward O. Wilson have his say, and in an interview (Habekuß 2019, p. 33) he expresses a decided opinion on this matter. In response to Habekuß's question, "There are attempts to measure the value of nature. If we learn what its services are worth, will we stop destroying it?" To this Wilson:

I was excited about the concept of the bioeconomy for a while, I said to all these ambitious business bosses: save the environment and you'll get a lot back, an income you never expected.

By now I understand: Putting a monetary value on the diversity of life is a sure way to kill it.

Nothing more to add to Wilson's clear statement!

Ending the indulgence (Küppers 2018b) *of* our current risk-accumulating approach to nature is a categorical imperative!

At this point we leave the still topical question of our future dealings with the earth, in view of the obvious powerlessness, especially in the face of the forces of nature. However, we will come back to this in detail when discussing possible sustainable solutions that we want to derive from the ingenious principles of nature.

Let us now turn back to nature as our evolutionary guarantee of survival. We will now deal with exciting questions about the development of life beyond individuals, how species, which are considered to be the basic unit of today's biological systematics, come into being and form a biodiversity of unparalleled richness, which mechanisms play a role in this and much more, which enables us to survive in an evolution that has lasted for billions of years. Without question, evolution, the phylogenetic development of living organisms, is of paramount importance.

3.2 Evolution: Darwin's Origin of Species

Of course, we cannot treat the subject of evolution as intensively as scientists like Ernst Mayr (2001, 1998, 1954) or Theodosius Dobzhansky (1973, 1968, 1939), who are mentioned as representatives of many others.

As a guiding thought, however, we take with us into this chapter the statement of Theodosius Dobzhansky, who in 1973 headed a publication with the following sentence:

Tip

Nothing in Biology makes Sence except in the Light of Evolution.

The Origin of Species – original: On the Origin of species by Means of Natural Selection (Darwin 1859, current and complete new German translation for 100 years, Darwin 2018) – appeared at a time when today's genetic engineering with its tools for manipulation – genome editing, e.g.. CRISPR/Cas[12] – of DNA (deoxyribonucleic acid), as the carrier of genetic material, and epigenetic influences on genetic material (see Sect. 3.7) were still in the fog of the future.

---

[12] CRISPR/Cas is a very precise tool that makes pinpoint changes in the DNA sequence. See, e.g. https://www.transgen.de/lexikon/1845.crispr-cas.html and https://www.mpg.de/11032932/crispr-cas9-mechanismus. Accessed 13 Jan. 2019.

The apparently increasing turn to Darwin's thoughts on the origin of species and their development have received new impetus through the fields of genetic engineering or *epigenetics*. Epigenetics is a field of biology that, roughly speaking, analyses acquired characteristics of organisms that determine certain gene activities without – according to current theory – leading to genetic change through *mutation* and *recombination* (we will explain the individual technical terms in detail in the respective subchapters).

Using the text "Vom Sein Zum Werden" (From Being to Becoming), written by Joseph Reichholf (2018) in an afterword to the German edition of Darwin's Origin of Species (2018), essential excerpts are quoted below. They summarize in the necessary brevity what Darwin compiled in his book. The basis for this was his nearly 5-year voyage on the HMS Beagle, towards South America in 1932, which he embarked on at the age of 22. What he saw and recorded in writing, from which he discovered ingenious ideas for new strategies for life, analysed precise behaviour patterns and their causes through countless observations, not to mention the multitude of plants and animals he collected, which he described meticulously, from all this arose a universal theory of life that is still valid today. It is a theory of fundamental existence, but open to new developments, such as those currently taking place through genetics.

Table 3.1 gives an initial overview of the topics covered in Darwin's book on the origin of species, the detailed comprehensibility of which – as exciting as the individual chapter contents may be – can only be touched on superficially within the framework of this book. However, I would like to warmly recommend the exciting content to you as interested readers. After all, it is one of the most successful stock theories ever written.

Let us now follow, in excerpts and in a highly condensed form, Darwin's thoughts and experiences, which Reichholf has summarized so comprehensibly and succinctly in the epilogue to the 2018 German translation "Vom Sein zum Werden" ("From Being to Becoming") of Darwin's "The Origin of Species", as well as citing current developments beyond Darwin's horizon of knowledge. All quotations that follow in the context are from the source Reichholf (2018, p. 572–603), additional sources are cited op. cit.

Table 3.1 Structure of the chapters in Darwin's "The Origin of Species"

Darwin's "The Origin of Species" (German translation from 2018)		
Chapter	Headline	Outline – extract
Chapter I	Domestication variation	Cause of variability Characteristics of domesticated variability Unconscious selection
Chapter II	Variation in nature	Individual differences Doubtful species Species of the larger genus of a region vary more frequently than those of the smaller genera
Chapter III	Struggle for existence	Its importance for natural selection Geometric ratios of propagation General competition Complex relationships of all animals and plants within the whole of nature
Chapter IV	Natural selection or The survival of the best Adapted	Natural selection Sexual selection On the generality of crosses between individuals of a species Extinction determined by natural selection Progress of the organization
Chapter V	Laws of variation	Effect of changed conditions Acclimatisation Compensation and economics of growth Species of a genus vary analogously
Chapter VI	Difficulties of the theory	Difficulties of the theory of descent with modification Absence and rarity of transitional varieties Organs of ultimate perfection Transitional forms
Chapter VII	Various objections to the theory of natural Selection	Longevity Progressive development Causes, which can prevent the acquisition of useful Structures by natural selection Affect Reasons for doubt about large and immediate modifications
Chapter VIII	Instinct	Instincts comparable to living creatures, but with different origins Instincts graded Slave-holding ants Asexual and infertile insects
Chapter IX	Mixing	Infertility varies gradually, not universal, affected by close crossing, removed by domestication Cause of infertility of first crosses and of hybrids Comparison between hybrids and crossbreds irrespective of their fertility

(continued)

Table 3.1 (continued)

Darwin's "The Origin of Species" (German translation from 2018)		
Chapter	Headline	Outline – extract
Chapter X	On the incompleteness of the Geological evidence	On the present absence of intermediate varieties On the passage of time, concluded from the degree of erosion and deposition On the absence of intermediate varieties in all formations On the sudden appearance of groups of species Age of the habitable earth
Chapter XI	On the geological succession of living beings	On the slow and gradual appearance of new species About their different degree of change Groups of species follow the same general rules as individual species in their appearance and disappearance About simultaneous changes in life forms around the world
Chapter XII	Geographical processing	Today's spread cannot be explained by differences in external conditions Importance of barriers Centers of creation Means of widening by change of climate and land level and by occasional means Dispersal during the ice age
Chapter XIII	Geographical distribution – Continued	Distribution of freshwater organisms About the inhabitants of oceanic islands On the relationship between islanders and those on the nearest mainland
Chapter XIV	Mutual relationship of living beings, morphology, embryology, rudimentary organs	Classification, groups and subgroups Natural system Variety classification Extinction separates and demarcates groups Morphology between members of a class, between parts of an individual Embryology, its laws explained by variations and additions at young age and heredity at corresponding age
Chapter XV	Recapitulation and conclusion	Recapitulation of the objections against the theory of natural selection Causes of the general belief in the immutability of species How far the theory of natural selection can be extended Effects of their adoption in the study of natural history

3.2.1 Darwin and Lamarck

Reichholf begins by stating that recent research on evolution is changing our understanding of the course of evolution. This brings us closer to Darwin's thoughts again, after he had been "for more than a century […] regarded as a Lamarckist in disguise and also scorned […]".

> Lamarck's[13] notion that organisms strive for change on their own to achieve something (attractive), this "um-to explanation," Darwin clearly rejected. His natural selection had to proceed imperceptibly; as slowly as the earth-historical processes of weathering or the geological reshaping of landscapes and continents.
>
> Adaptation, adaptation, was for him not an active action of living beings, but the passively enforced consequence of selection brought about by the environment (due to dynamic and complex processes, d. A.).

However, it is this non-activity of organisms that is being questioned by recent research in genetics or epigenetics.

3.2.2 Struggle for Life: Favoured Races

"On the Origin of Species" was published by Charles Darwin in 1859 and titled in 1872:

> The Origin of Species by Means of Natural Selection, or the Preservation of Favoured Races in the Struggle of Life.

Already one year later, in 1860, the first German-language translation by the zoologist H. G. Bronn appeared with the title:

> On the Origin of Species in the Animal and Plant Kingdom by Natural Breeding, or the Preservation of the Perfected Races in the Struggle for Daseyn.

The obvious weakness of the phrase "struggle for existence" was recognized by Ernst Haeckel, probably Darwin's strongest supporter at the time and the mastermind of scientific ecology. In the 7th chapter of his *"General Morphology"* on the subject of *"Breeding and Selection"* (Vol. 1, pp. 238–239, in: Begon et al. 1998, p. XVII) he describes it in clear terms:

---

[13] Jean-Baptiste Pierre Antoine de Monet, Chevalier de Lamarck (1744–1829), French botanist, zoologist, and evolutionary biologist was Darwin's opponent. Lamarck based his theory of biodiversity on acquired characteristics of organisms.

If we wish to apply the concept of the struggle for existence in a sharply defined way, we must limit it to the mutual interrelations of organisms, to the necessary competition of organisms for the more or less indispensable necessities of life.

We emphasize this particularly because Darwin uses the term, however, preferably in this actual main meaning, but occasionally also in a further metaphorical extension, which harms its purity and easily leads to misunderstandings (as with Herbert Spencer* an English philosopher, sociologist, evolutionary theorist and contemporary of Darwin, the author). For he (Darwin, d. A.) also calls the dependence of organisms on organic and inorganic[14] conditions of existence a "struggle us Dasein," it says, for example, that plants and animals in states of scarcity wrestle with the necessary conditions of existence; and calls this "wrestling for existence," whereas only that wrestling should be called which takes place between several organisms for those necessary conditions of life.

*Herbert Spencer[15] very misleadingly reinterpreted Darwin's "survival of the fittest" as "survival of the best-adapted individuals", which gained dubious fame in connection with Social Darwinism, a theory in which aspects of Darwinism are mapped onto social developments.

In letters[16] to the Japanese politician Kaneko Kentano[17] in 1892, Spencer described his neo-Darwinian thoughts as follows:

> [...] To your remaining question, respecting the inter-marriage of foreigners and Japanese, which you say is "now very much agitated among our scholars and politicians," and which you say is "one of the most difficult problems," my reply is that, as rationally answered, there is no difficulty at all. It should be positively forbidden. It is not at root a question of social philosophy. It is at root a question of biology. There is abundant proof, alike furnished by the inter-marriages of human races and by the inter-breeding of animals, that when the varieties mingled diverge beyond a certain slight degree the result is invariably a bad one in the long run.

These neo-Darwinian thoughts of Spencer formulated in the letter to Kentano already show his attitude to transfer Darwin's elaborated evolutionary concepts completely unreflectively to human societies. Those of Herbert Spencer have little to do with the clarity of Darwinian concepts emphasized by Ernst Haeckel. The correspondence goes on to say:

---

[14] An example around the struggle for inorganic, mineral resources in barren regions is shown by ibexes climbing up the steep dam wall of the Cingino reservoir in northern Italy in search of salt, which they lick from the rocks. https://de.wikipedia.org/wiki/Cingino-Staumauer#cite_note-2. Accessed 14 Jan. 2019

[15] https://de.wikipedia.org/wiki/Herbert_Spencer. Accessed 18 Jan. 2019.

[16] http://praxeology.net/HS-LKK.htm. Accessed 18 Jan. 2019.

[17] https://de.wikipedia.org/wiki/Kaneko_Kentarō. Accessed 18 Jan. 2019.

[…] I have for the reasons indicated entirely approved of the regulations which have been established in America for restraining the Chinese immigration, and had I the power would restrict them to the smallest possible amount, my reasons for this decision being that one of two things must happen. If the Chinese are allowed to settle extensively in America, they must either, if they remain unmixed, form a subject race in the position, if not of slaves, yet of a class approaching to slaves; or if they mix they must form a bad hybrid. In either case, supposing the immigration to be large, immense social mischief must arise, and eventually social disorganization. The same thing will happen if there should be any considerable mixture of the European or American races with the Japanese.

The obvious misunderstanding of the term "Survival of the Fittest", which Darwin neither saw nor propagated in the "struggle for existence" described by him, shows the deep gap between Charles Darwin's scientific elaborations including evidence and the socio-philosophical "upper-lower-human-charac-terization" of a Herbert Spencer! How dangerous this kind of mono-causal thinking and acting of Spencer's coinage has rubbed off on later societies and their governments is something we Germans in particular can sing a song about. Reichholf (2018, p. 574) says in this regard:

The abuse (of preferred races in societies, i.e.) that happened with such terrible consequences should not, however, obscure the fact that it is precisely in the great diversity of human beings that our highest good lies, namely individuality (and thus also a high degree of biodiversity, which is fundamental for our further progress, i.e.).

3.2.3 Alarming Ignorance of Evolution

Conceptual clarification in the context of Darwin's theory of evolution is one thing. But as Reichholf (2018, p. 575) points out, there are more fundamental issues at stake. Not only in the USA do *creationists* (Bible-believing citizens) believe that life literally came into being according to the Bible's creation story, and they regard this as the basis for our existence. Also in Germany in 2006 – at the state minister level – creationists tried to influence the teaching of biology,[18] but after fierce protests without success. Reichholf sees it as follows (ibid., p. 575):

---

[18] http://www.spiegel.de/lebenundlernen/schule/hessische-schulen-kultusministerin-faellt-auf-kreation-isten-herein-a-445487.html (Accessed 01/20/2019).

It is certainly not only due to the language and the difficulties connected with translations [...] whether misunderstandings (of evolution, d. A.) reduce the acceptance or whether there are willful misinterpretations. Much more depends on the extent of the social impact of religious dogmas.

If, according to Reichholf, biology lessons today are largely dominated by biochemistry or environmental topics, it is high time that the evolutionary foundations of life elaborated by Darwin once again take their rightful place in biology lessons – also in view of the digitalisation or humanoidisation of our living environment.

3.2.4 Evolutionary Developmental Biology: Evo-Devo

While perusing the 2018 rewrite of Darwin's epochal work, On the Origin of Species by Means of Natural Selection, or the Preservation of Favoured Races in the Struggle for Life, 1972, one is struck by the enormous wealth of his findings, which in turn prompt new suggestions of their own. In fact, Reichholf writes (ibid., p. 576), this extensive work and its findings do not seem to have been nearly exhausted to this day. It goes on to say (ibid.):

> The outstanding example of how much Darwin thought ahead is offered by the currently (finally!) highly topical "evo-devo", which replaces the overly strict mechanical view of heredity as a biological one-way street. As the "New Synthesis," the view that hereditary traits, genes, do not interact with the environment was an unquestionable doctrine for more than half a century. There is only one path from genes to their products to organisms, but no backlash from them from their lives.
>
> Evolutionary developmental biology (evo-devo, the author) [...] is now challenging this doctrine with good findings.

The deliberately small number of evo-devo sources (Minelli 2019; Sanger and Rajakuma 2018; Sommer 2009; Carroll 2008; Müller 2007; Hoekstra and Coyne 2007) is not meant to obscure the enormous wealth of scientific publications on evo-devo. Figure 3.3 reflects how new questions for evolution arise from the concept of *evo-devo*.

Evo-devo sets in motion a new development that lies in the continuity of Darwin's thoughts, which do not contain incontrovertible truths as depicted by the immutable doctrine of the Bible. A comparison between the two approaches to evolution is nonsensical for this reason alone.

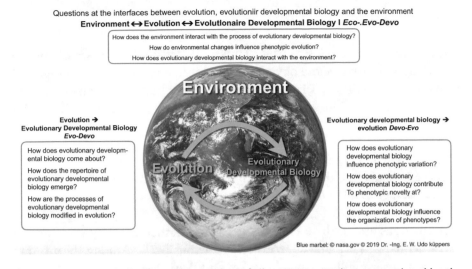

Questions at the interfaces between evolution, evolutioniir developmental biology and the environment
Environment ↔ Evolution ↔ Evolutlonaire Developmental Biology I Eco-.Evo-Devo

How does the environment interact with the process of evolutionary developmental biology?
How do environmental changes influence phenotypic evolution?
How does evolutionary developmental biology interact with the environment?

Environment

Evolution →
Evolutionary Developmental Biology
Evo-Devo

How does evolutionary developmental biology come about?

How does the repertoire of evolutionary developmental biology emerge?

How are the processes of evolutionary developmental biology modified in evolution?

Evolution   Evolutionary Developmental Biology

Evolutionary developmental biology →
evolution *Devo-Evo*

How does evolutionary developmental biology influence phenotypic variation?

How does evolutionary developmental biology contribute To phenotypic novelty at?

How does evolutionary developmental biology influence the organization of phenotypes?

Blue marble: © nasa.gov © 2019 Dr. -Ing. E. W. Udo küppers

Fig. 3.3 New questions on evo-devo (after Müller 2007, p. 944), text translated by the author. (©Graphic: the author, Image of the Earth: https://www.nasa.gov/content/blue-marble-image-of-the-earth-from-apollo-17. Accessed Jan 20, 2019)

The extension of the Darwinian evolutionary edifice, accompanied by scientific curiosity and scepticism, is precisely what makes biological research and its progress so interesting. However, in the age of the Anthropocene (Renn and Scherer 2015) and digitalisation, we are also experiencing evolutionary development paths that counteract biological evolutionary mechanisms through targeted genetic manipulations (e.g. using Crisper/CAS, genetic engineering tools for the targeted modification of DNA) and attempt to optimally control hereditary dispositions with specific goals.

It is not impossible that the self-proclaimed "crown of creation", evolutionary biological man will one day face competition from clones (artificially reproduced humans) and humanoids (mechanical or biomechanical human existences).

Whatever we humans, in whatever evolutionary, cloned or humanoid form of development, do to shape the future, it will affect all organisms and the inanimate environment equally.

After this small glimpse into a still nebulous future, we return to Darwin's theory of evolution. Reichholf writes about this (ibid., p. 578):

Why man has become the main factor of evolution in the present and near future is already clear as Darwin's *Origin of Species*. Because the book exposes the core pieces that make the process of evolution possible and keep it going.

3.2.5 Variation and Time

It is like many paths that lead us into unknown territory in the search for new knowledge: To perceive the obvious of an object, a process, even an organism or a species, requires special achievements. Reichholf (ibid.) takes a similar position when he addresses the biological mechanism of variation. He writes (ibid.):

> Biologically, it is by no means easy to distinguish species from one another. Because – and this is the crucial insight that Darwin used – they all vary locally, regionally, geographically and also in time.

Delimitations are on the one hand "[...] often rather arbitrary [...]" and on the other hand "[...] almost excessively generous [...]". Among the arbitrary demarcations, seagulls on the North Sea and Baltic Sea, in the Mediterranean and Black Sea are mentioned, which differ only in nuances, fine gradual differences in leg colour. To generous delimitations are mentioned dog breeds, "[...] all of which are said to be "wolf"." (ibid., p. 579). The perception about variations and how they are to be judged seems – according to Reichholf – to enjoy lively debates up to the present.

Darwin's special achievement was to have collected an overwhelming abundance of evidence showing that variations occur to both organisms always and everywhere, even when they are not readily visible externally. (ibid., p. 580).

Almost a century earlier, the famous Swedish naturalist Carl von Linné (1717–1778) had drafted a nomenclature, a directory of nature, in which everything has its immovable place. The all-encompassing world view of biblical creation was the only authoritative one in his time – until Darwin's insights culminated in his famous book after a 5-year voyage on HMS Beagle (a three-masted brig sloop used by the Royal Navy for surveying). On this voyage, Darwin delved into divergences, not focusing solely on type in "[...] describing and defining species" (ibid., p. 581).

He recognized how artificial and deliberate the boundaries were often when specialists described species. And the more he developed an eye for how species varied in place or across geographic spaces, the more time became central to his considerations; time in the evolutionary play on nature's stage and as the basis for the process of change. [...].

Darwin recognized that atolls as coral (ring) islands are formed by slow but apparently very long-term geological processes, that continents can rise and fall. [...] Darwin suspected that there must be some connection with the diversity of life, which, with all its change, did not vary quite arbitrarily.

Without describing an exact chronology of Darwin's life, we focus at this point on the breeding of new animal breeds, which was intensively pursued in the nineteenth century and which Darwin followed with interest. Darwin recognized a "[...] fundamental correspondence between the processes in nature and the action of breeders [...]." (ibid., p. 583).

Obviously there were species that varied (a lot), but also those that looked almost like they were made according to templates [...]. But what made the difference between breeding on the part of humans and the variations that became apparent when looking closely at the series, for example of butterflies, when they came from different areas? And across greater geographical distances, gave the impression that the ends of the spectrum looked (almost) as different as true species? (ibid., p. 585).

Does nature perhaps possess a mechanism comparable to that used in the breeding of animal breeds, which picks and chooses?

Darwin's *natural selection* differed from the selection of the breeders simply in the length of time, in the speed. What the latter produced in a few generations, Darwin's required thousands and thousands of generations or years.

While breeders proceed according to plan, with concrete ideas about what they want to achieve, and thus considerably shorten the start-finish time span of a breeding, evolution works without plan or aim. Nature is – if we may use technical terminology – her own "breeder". Due to natural selection, only a part of the offspring of a species survives within its habitat. The offspring that develop from this, with better conditions for survival, take a long time to establish themselves, precisely because no goal-oriented path – as with breeders – can be taken (ibid., p. 589).

That variation is available in nature is a necessary condition of evolution (Mayr 2003, p. 116). It goes on to say (ibid.):

On closer examination, it becomes clear that the variation concerns not only visible characteristics, but also physiological processes, behaviour, ecological aspects (for example, adaptation to climatic conditions) and molecular biological features. All of this underscores the recognition that each individual is unique in this or that respect. Only these constantly available differences account for natural selection.

Before we conclude Sect. 3.2 with today's view of evolution, we would like to take a more differentiated look at the term "Darwinian theory of evolution" and quote Ernst Mayr, who cites a total of five different evolutionary theories of Darwin, without having to deal with them in detail here (ibid., p. 114):

- Theory 1: Variability of species (the basic theory of evolution)
- Theory 2: Descent of all living beings from common ancestors (evolution by branching)
- Theory 3: Gradual course of evolution (gradualism, d. A.) (no jumps, no discontinuities)
- Theory 4: Reproduction of species (origin of biological diversity)
- Theory 5: Natural selection.

Regarding these five evolutionary theories of Darwin, it remains to be clarified what a species actually is as a central concept of evolution. Mayr explains it as follows (ibid., p. 204):

> Of course, one cannot deal with the gaps between species and their causes without first knowing what species actually are. However, naturalists found it excruciatingly difficult to reach agreement on this question. In the literature, the issue is known as the "species problem". Even today (2003, d A.), there is no unanimous opinion as far as the definition of the biological species is concerned. There are several reasons for the differences in opinion, two of which are particularly important. First, the term "species" is applied to two very different things: on the one hand, to the species as a concept, and on the other hand, to the species as a systematic group or taxon. The species term refers to the importance of species in nature and their function in the biological household. The taxon of the species is a zoological entity, a collection of populations that together meet the definition of the species concept. For example, the taxon *Homo sapiens* consists of numerous geographically widely distributed populations that can be collectively classified under a particular species concept [...].

In our context, we start – just like Mayr – from the biological concept of species, which applies exclusively to living beings that reproduce sexually.

A few years after the publication of Darwin's *Origin of Species,* Theory 1 and 2 had become generally accepted. Theories 3–5 could not come together until much later, when the so-called "synthesis of evolutionary research" took place. It "[...] led to general agreement, and the molecular biological revolution of the following years meant a further strengthening for the Darwinian doctrine [...]" (ibid., p. 11).

3.2.6 Today's Perspective

Darwin did not know much about the actual processes of heredity, although he had the publication of Gregor Mendel's[19] (Mendel 1866, on Mendel genetics see also Campbell et al. 2006, p. 293–317) crossbreeding experiments with peas, but did not pay further attention to them, according to Reichholf (2018, p. 597). When, around the turn of the century (1900), the rediscovery of Mendel's experimental results led to the emergence of the new research field of experimental genetics and, again half a century later, the two molecular biologists Watson and Crick (1953) discovered the gene structure of DNA (deoxyribonucleic acid), as the carrier of hereditary information, the material substructure of genes, a decisive building block was thus available that could be used to explain biological variations much better.

At the turn of the last millennium, the complete human DNA was sequenced for the first time by the international Human Genome Project[20] and published in Nature (International Human Genome Sequencing Consortium 2001; Baltimore 2001).

Moreover, genetics proves beyond doubt that all living things belong together (Reichholf 2018, p. 597). And this includes not only the relationship with apes (chimpanzees, bonobos), with which we agree in over 98.8% of our genes[21] (Vaki and Gagneux 2017), but also with all other organisms such as dolphin, sequoia, cheetah, snowdrop, etc..

3.2.7 Life from Non-Living Matter: Emergent Properties

From 1959 to 1961, the German physicist, philosopher and peace researcher Carl Friedrich von Weizsäcker gave the so-called "Gifford Lectures" at the University of Glasgow. In the first of two lecture series he also turned to the "evolution of life". At the time, a crucial question of life had not yet been solved: How does life arise? Can life arise from non-living matter? (Weizsäcker 1964, p. 135–153).

---

[19] Gregor Mendel (1822–1884) was a priest of the Augustinian order in Brno, Moravia. His crossbreeding experiments with peas in the monastery garden led to a new approach to the theory of heredity, which was not understood until 1900 by concordant scientific results and thus rediscovered. See this: https://de.wikipedia.org/wiki/Gregor_Mendel (Accessed Jan 24, 2019).

[20] The Human Genome Project Information Archive contains a range of data and information from the 13-year life of the project, https://web.ornl.gov/sci/techresources/Human_Genome/index.shtml (Accessed Jan. 25, 2019).

[21] http://humanorigins.si.edu/evidence/genetics, Smithsonian National Museum of Natural History, Washington, DC. (Accessed 25 Jan. 2019).

If we venture a long leap from these 1950s/1960s to the present and ask ourselves: Do necessary and/or sufficient answers to the previously posed questions about the origin of life exist today?

The leap from non-living processes of chemistry and physics into living biology is nowadays justified by *emergent* properties *of* matter, which in complex and highly complex states, e.g. by random cross-linking *of* single atoms to molecules, form so-called *"islands of order"* or *"islands of life"*, within an environment characterized by disorder (entropy). We ourselves are such *"islands of life"*, as, by the way, are all organisms, because an internal order (negentropy,[22] Schrödinger 1987) is maintained by our metabolic processes, which we feed with workable energy in the form of solid and liquid food and export their "waste products" as unusable energy for work into the environment.

Emergence is when new higher order structures or properties of a system emerge through the interaction of individual parts of the system, which cannot be predicted from the analysis of the individual parts of the system. This applies equally to all organisms and non-living systems in the environment.

Emergent properties seem to play a key role in this game of matter and living systems, because the interaction of special chemical structures, e.g. in living cells or organs, can lead to novel processes with properties such as self-organized or self-propagating functional processes. Whereby these new – emergent – properties cannot be derived from the previous elements that led to the new properties due to certain constellations.

In inorganic, non-living environments, something similar can be seen. Who can predict that the interaction of two hydrogen atoms with one oxygen atom will result in the elixir of life, water? Who can predict how the encounter of hundreds of cars and trucks will concretely affect the traffic situation of a city? All systems of this kind are inherently highly complex. Attempts to capture the functionality of this complex and dynamic interaction of biological and non-biological systems with conventional mathematical methods using linear analysis tools are doomed to failure. Only probability theory brings us closer to the processes occurring in nature.

---

[22] The term negentropy was coined by the Austrian physicist Erwin Schrödinger (1887–1961) in his book: What is Life? For Schrödinger, living systems were systems that absorb, store, and process negentropy or negative entropy. It is kept low by releasing entropy from the living system into the environment. See also bibliography.

3.2.8   Man and Humanoids

A living human being has consciousness – a machine, robot or humanoid does not. Who would not agree with this statement? In view of the previously described ability of non-living systems in nature to develop emergent properties, a second question arises – highly topical, in view of digitalization and "artificial intelligence" in our environment:

If we could consider machines as non-living systems, but program them to perform emergent properties with self-organized processes leading to a possibly higher emergent state and self-replicating, could we grant them consciousness?

I believe this will never be the case. Biological evolution and geological evolution, animate and inanimate nature form a dynamic, evolved unity over billions of years. Just as organisms depend, for example, on minerals for their own needs, the varied geological structures are a prerequisite for the high degree of differentiated adaptability for organisms. There seems to exist an unmanageable network between organisms and between animate and inanimate nature, which guarantees the progress of life.

In contrast, humans are the triggering moment for the development of non-biological artificial humanoids or robots, which still lack – and probably always will lack – the foundation of networked cooperation for self-preservation. Foresighted conscious planning in dynamic complex environments would be an equally difficult – if not unsolvable – task for artificial autonomous and self-organized machines, which cannot be ascribed consciousness – as we possess it.

While nature, with its inherent untargeted adaptation strategy, is unflinchingly working out its future, artificial machines, even if so-called artificial intelligence, which is always based on human intelligence, is in play, fail because of the smallest dynamics and complexity of all our environments. We humans have no less our problems with this, which we have also created for ourselves through the Anthropocene effects.

3.2.9   Final Look Ahead

Darwin's theory of evolution was based on his precise observation of the environment. As is well known, he was not yet familiar with the genetic changes in our genome, as a dominant, co-triggering feature of biological phenotypic changes.

Today, more than 150 years after his first publication of *"The Origin of Species"*, the genetic engineering method, so-called "gene editing", the targeted modification or switching on and off of genes that can have a damaging influence on hereditary dispositions, is state of the art. However, this is far from being so mature that genetic engineering interventions in the genome, whether in plants, animals and even more so in humans, are carried out without hesitation in such a way that every intervention is highly likely to be successful for the organism and its interaction with the environment.

The discussions about genetically modified seeds or food in general, e.g. maize (yield increase or herbicide resistance), or genetically modified animals, e.g. salmon (year-round instead of seasonal growth) or insects, such as mosquitoes (avoidance of infectivity) resistance and finally genetically modified humans (elimination of certain genes to avoid HIV infection) oscillate between rejection and hope. (See, among others: Arya 2015; Krimsky 2019; Jiankui et al. 2018; Paull 2018, 2015; Miotto et al. 2015; WHO 2014). Sources on genetically modified organisms number in the thousands, simultaneously underscoring their hopeful and dubious relevance to the artificially manipulated advancement of life.

In addition to "gene editing", the Crisper/CAS gene procedure, which was already mentioned at the beginning of Sect. 3.2, is increasingly coming to the fore in genetic manipulations by humans and on humans. The crossing of the ethical boundary by the Chinese He Jiankui (see source Jiankui et al. 2018), whereby the genetic material of twin sisters conceived by artificial insemination was altered by Crisper/CAS (Clustered Regularly Interspaced Short Palindromic Repeats) for the purpose of HIV immunisation, is currently viewed more critically than supportively.

However, genetic modifications – however they are carried out under what guidelines – will become an integral part of our lives, whether it concerns plants, animals or humans. What is crucial, in my opinion, are controlled boundaries in evolutionary genetic and genetically manipulated interventions, all of which interact with the complex dynamic environment and can therefore harm or benefit not only the organisms themselves, but also far more individuals or species.

Darwin's vehemently rejected competing evolutionary approach of a directed higher development of species through acquired characteristics, by Jean-Baptiste de Lamarck, (the classic example of the giraffe that gets its long neck by reaching higher and higher flights of a tree from generation to generation) is even somewhat strengthened by the findings of epigenetics (see also Sect. 3.7). Reichholf's words describe it this way (Reichholf 2018, p. 602):

Experiences that the organism has made can be passed on to the next generation because they have an effect on the genetic material. Evolutionary developmental biology has discovered astonishing things about this. The processes have been elucidated chemically and are not as strange biologically as was first assumed. The fact is that the hereditary material, the genome, communicates to a certain extent with the environment directly and with a lasting effect via repercussions. Which, in plain English, means the passing on of acquired characteristics. And for us, it means that lifestyle can have an impact on offspring. Consequently, we bear more responsibility than just for ourselves and for the future in general. Lifestyle affects our descendants very directly.

The aforementioned term *"feedback effects"* refers not least to the so-called *"feedbacks"* in the cybernetic sense. There, *negative* feedbacks stabilize a system, while *positive* feedbacks take risks up to certain limits. For complex organisms – let's just think of self-organized processes – this is similar to technical systems, which was already pointed out by Norbert Wiener, the "father" of cybernetics. However, to conclude from this in general that organisms are also cybernetically controlled in their evolutionary progress is completely absurd and profoundly contradicts the basic Darwinian idea! There is definitely no cybernetically controlled or regulated growth progress in nature, as was postulated in the 1980s by a "new theory of evolution" (Schmidt 1985).

3.3   Phenotype and Genotype

In developing his theory on the *origin of species,* Darwin's thoughts still revolved solely around the perceptible – visible – change characteristics of species in the environment. The appearance of organisms, their specific habitats with the resulting variability and behavior decisively supported Darwin's theory. Today we characterize these variabilities of species as change phenotypic traits. The dynamic phenotype is the visible sign of each species and an expression of its variable adaptations to variable environmental conditions.

The reason why in this section the phenotype is treated first before the genotype, from whose genetic mutations only the phenotypic appearances arise, is solely due to the historical development.

On the question of the causes of variability, Ernst Mayr (2003, p. 118) writes:

With this puzzle Darwin struggled all his life, but despite all his efforts, he never found the answer. The real nature of variability was understood only after 1900, after genetics and molecular biology have made great strides. How evolution occurs can be fully understood only when the basic facts of heredity are known, for they are the explanation of variation. This is why genetics is an indispensable part of evolutionary research. Only the hereditary part of variation plays a role in evolution.

Already a few years earlier Mayr (1998, p. 44 f.) writes about characteristic features of life:

Today [...] there seems to be agreement on the nature of living organisms. At the molecular level, all – and at the cellular level most – of their functions follow the laws of physics and chemistry. Nothing remains that needs independent vitalistic principles. Yet organisms are fundamentally different from inanimate matter. They represent hierarchically ordered systems with numerous emergent (to emerge E., d. A.) properties, which are not found in inanimate matter, and what is most important: Their activities are controlled by genetic programs that contain historically acquired information; again, something that inanimate nature lacks. [...].

Dualism in modern biology is thoroughly physicochemical and arises from the fact that organisms have a genotype and a phenotype. The genotype consists of nucleic acids [...]. The phenotype is formed on the basis of the information conveyed by the genotype and consists of proteins, lipids and other macromolecules [...].

Regarding the direct and indirect causes of effect on phenotype and genotype, Mayr writes (ibid., p. 165):

Direct causes affect the phenotype, i.e. morphology and behaviour; indirect causes help to explain the genotype and its history. Direct causes are mainly mechanical; indirect causes are probabilistic (calculable only in probabilities, d. A.). Immediate causes occur here and now, at a particular moment, a particular stage in the life cycle of an individual, in the course of its life; indirect causes have been operative over long periods of time, more specifically, in the evolutionary past of a species. Direct causes have to do with the unraveling of a genetic or somatic program; indirect causes are responsible for the emergence and changes in genetic programs. The determination of proximate causes is usually made possible by experimentation, that of indirect causes by inference from historical accounts.

Organisms thus internalise a dual functionality of genotypic and phenotypic variability, of genetic chance and phenological necessity, in short: organisms are the product of *"chance and necessity"*, as the molecular biologist and Nobel Prize winner Jacques Monod succinctly described it in 1970 in his book of the same name, in the original: *"Le hasard et la nécessité"* (Monod 1975).

In addition to the ability to evolve and the duality of genotype-phenotype changes, living systems are capable of other abilities that distinguish them from inanimate systems in the environment (cf. Mayr 1998, p. 47). We will discuss this in more detail in the following Sects. 3.4, 3.5, 3.6, 3.7, 3.8 and 3.9.

3.4 Self-Organisation, Equilibrium, Non-Equilibrium and Self-Regulation

3.4.1 Self-Organisation, Equilibrium and Non-Equilibrium

All life on our earth is connected with *self-organization*. However, self-organization is such a universal phenomenon that it also takes place in non-living environments. We will get to know examples of both spheres, whereby we concentrate primarily on nature and its organisms, which evolution has produced, in accordance with the title of the book.

At this point at the latest, I would like to draw the readers' attention to the following: Even though Ch. 3 takes up individual concepts or phenomena of nature in a structured way, it should be made clear that all the phenomena described in nature are interconnected in some way and that no stand-alone features or processes in nature characterize life.

The phenomenon of *self-organization* is the subject of much differentiated research. However, it cannot be the aim to collect all the results obtained from it, but to give you a series of examples in order to make clear the universality of the phenomenon and thus to understand its comprehensive effectiveness.

Let's start with some definitions or statements about *self-organization*.

- **Eugene P. Odum**, US ecologist (1913–2002), pioneer of the ecosystem concept:

  Biotic communities and ecosystems are not "superorganisms" but systems that are not in thermodynamic equilibrium and that are capable of *self-organization* […]. (Odum 1991, p. 199).

Odum further quotes Brooks and Wiley (D. R. and Wiley 1986) (ibid., p. 208) as stating:

[...] Self-organization (is) the result of the second law of thermodynamics[23] – with the consequence that living systems show an increasing complexity because of, and not despite, the expenditure of entropy. The "overall strategy" includes a decrease in entropy (disorder), an increase in information (order), an increasing ability of the ecosystem to survive disturbances unchanged (resistance stability), and an increasingly efficient utilization of energy and nutrients.

The fact that numerous ecologists in the 1980s were still hostile to Odum's hypothesis should not be concealed. Today, however, Odum's concept of ecology seems to be gaining acceptance.

Under the chapter on energy laws, Odum, a zoologist and ecologist, also explicitly discusses the First and Second Laws of Thermodynamics in connection with organisms and ecosystems (Odum 1991, p. 81):

The First Law states that energy can be converted from one form (such as light) to another (such as food), but can never be created or destroyed. According to the Second Law, processes involving energy conversion only occur when energy is converted from a more highly concentrated form (such as food or gasoline) to a less concentrated form (heat). Since a certain amount of energy is always lost to further use as heat energy, spontaneous (thermodynamically permissible) energy conversions such as the conversion of light into food can never occur with one hundred percent efficiency.

An example of energy conversion in flora explains the two main theorems in a practical way in Fig. 3.4.

Odum describes the link between *self-organization* and *equilibrium* or *non-equilibrium* as follows:

Organisms and ecosystems maintain their highly organized low-entropy [...] state by converting high-value energy into lower-value energy (for example, in the respiration of carbohydrates, (the majority of which consist of sugar molecules, d. A.)). Living systems and the entire biosphere are, as Ilya Prigogine

---

[23] The second law of thermodynamics (second HS) says something about direction and irreversibility (irreversibility) of a process. The state variable entropy can also be derived from the second HS. Since organisms are open systems and thus exchange energy (as well as matter and information) with the environment, entropy is also transported across the system boundary and produced inside the system. Therefore, the entropy change of a biological system consists of two parts. Accordingly, the entropy balance (Lucas 2007, p. 200) is: "The entropy of a system changes by inflow or outflow of entropy across the system boundary and by entropy production in its interior." Other sources on the second HS and entropy include Baehr and Kabelac (2016); Tipler et al. (2015); von Oppen and Melchert (2005), and many other textbooks and textbooks on thermodynamics.

Fig. 3.4 Energy conversion processes from point 1 via 2 to 3 using the example of a chestnut leaf as an energy conversion system. The first law of thermodynamics refers to the conversion of sunlight into food (sugar) by means of photosynthesis, in which a proportion of heat is also involved. The second law of thermodynamics states: The energy conversion product from photosynthesis (sugar) is always smaller than that of the irradiated energy of the sun on the energy conversion system chestnut leaf. (cf. Odum 1991, p. 81)

has called them, "far-from-equilibrium systems" with powerful "dissipative structures" that pump out the disorder. (Prigogine et al. 1972; Prigogine 1977, 1998; Nicolis and Prigogine 1977).

The *dissipative structures* mentioned above are connected with the phenomenon of forming dynamic, ordered self-organizing structures in nonlinear systems far from thermodynamic equilibrium, also known as ordered "islands in a chaotic environment", in which local entropy (disorder) is reduced, but through energy exchange with the environment its entropy increases. Odum writes about Prigogine that he succeeded in explaining "[…] better than anyone before how living systems seem to defy the Second Law of Thermodynamics by maintaining an open state far from equilibrium in self-organization." (Odum 1991, p. 82).

- **Manfred Eigen**, German bio- and physical chemist (1927–2019), Nobel laureate.

Rauchfuß (2005, p. 263 ff.) describes Eigen's achievements in the self-organization of matter as follows:

Already more than 30 years ago (today, almost 50 years ago, d. A.) Manfred Eigen presented a theory on the self-organization and evolution of living systems (Eigen 1971). Manfred Eigen, Nobellaureat for Chemistry 1967, from the Max Planck Institute for Biophysical Chemistry in Göttingen expanded Darwinian evolution with a mathematically based theory. It reaches back to the fundamental, molecular processes that could lead to biogenesis. [...] M. Eigen's theory describes the self-organisation of biological macromolecules on the basis of kinetic considerations and mathematical formulations, which ultimately find their theoretical justification in the thermodynamics of irreversible systems.

Evolutionary processes are irreversibly bound to a temporal sequence. Classical thermodynamics is no longer sufficient to describe them. One needs a thermodynamics extended to irreversible processes, which takes into account the "arrow of time" [...].

- **Erich Jantsch**, Austrian-American astrophysicist and engineer (1929–1980).

In his book on the self-organization of the universe (Jantsch 1992) he writes on *self-organization: the dynamics of natural systems* (ibid., p. 49):

Self-organization is the dynamic principle underlying the emergence of the rich world of forms of biological, ecological, social and cultural structures. But it does not begin with what we commonly call life. It characterizes one of the two basic classes of structures distinguished in physical reality, namely, the so-called dissipative structures, which are fundamentally different in this respect from equilibrium structures. Thus this kind of dynamics becomes the link between the animate and the inanimate. Life now no longer appears as a thin superstructure, but as a principle inherent in the dynamics of the universe. In the dissipative structures of chemical reaction systems, we now have the opportunity to study self-organization in "pure" form, as it were. The same conditions that occur at complex levels – openness, high disequilibrium, and self-amplification of fluctuations – are most clearly and easily seen here.

Examples of such dissipative structures are in animate nature all organisms; in inanimate nature river meander networks, dunes full of sand ripples; in chemistry so-called Bénard cells (pattern formation in liquid of a vessel when heated above a critical temperature gradient) or the famous Belousov-Zhabotisnky reaction (pattern formation by chemical reaction of certain substances); in physics, the candle flame or the constant height of water in a bathtub when water flows in and out at the same time; in the atmosphere and deep space, cloud structures and the granulation of the surface of the sun. These few examples of dissipative structures beyond a system equilibrium of different spheres could be continued at will. They show – often with special beauty of the structures – impressively the universality of occurrence in animate as well as inanimate nature, as indicated above.

Jentsch has compared in tabular form (see Table 3.2) both classes of systems, those of the structure-preserving (static and nearly static) and those of the structure-dynamic (evolving) systems, which once again compactly highlight the differences (Jantsch 1992, p. 67).

- Another source on the phenomenon of *self-organization – A System of Thought for Nature and Society* is Milos Vec et al. eds. (2006), including contributions on self-organization in foreign worlds (J. Eckert). Here, under *Culture in Change*, the absence of the concept of self-organization in ethnology is addressed; by E. Göbel, self-organization in companies is dealt with; M. Artzt and G. Gebhardt consider self-organization and state planability of social change.
- Finally, reference should be made to a small book from 1991 by Werner Ebeling, a German theoretical physicist, who compactly summarizes, among other things, the processes and principles of self-organization, its technical benefits, self-organization and information, and as a principle of future computer generations (Ebeling 1971).

Table 3.2 Structure-preserving and evolving systems

System aspect	Structure preserving systems		Evolving systems
Overall system-Dynamics	Static (no dynamic)	Conservative Self-organisation	Dissipative Self-organisation (Evolution)
Structure	Equilibrium structure permanent	Devolution towards equilibrium	Dissipative (far from equilibrium)
Function	No function or allopoiesis not self-reproducible	Reference to equilibrium state	Autopoiesis (self-referral)
Organization	Static fluctuations In reversible processes	Irreversible processes towards equilibrium	Cyclic (Hypercycle (M. Eigen)) irreversible rotation
Internal state	Balance	Near balance	Imbalance
Environmental relations	Closed or open (growth possible)	Closed or open (growth possible)	Open (constant balanced exchange)

A good 30 years ago, "artificial intelligence" or "swarm intelligence" was not yet known in connection with computer generations. Even today's computers are still based on the classical von Neumann architecture as 30 years ago. Even though a humanoid challenge (Küppers 2018a) has people experimenting with new communication and information systems between humans and machines and between machines and machines, the phenomenon of technical self-organization is still in its infancy. Biologically proven models of dynamic self-organized processes will – in the opinion of the author and from a holistic view of things – remain models for technical solutions of self-organized processes for a long indefinite time – if they ever reach the achievements of evolving self-organizing and self-replicating systems of nature.

The subject of self-organization is by far not sufficiently covered with the few explanations of this phenomenon, leaving unmentioned many important scientists and authors who dealt with the phenomenon. But it should have given you, dear reader, an insight into the world of animate – and non-animate – nature, which are full of examples of self-organizing processes in which we humans also participate.

3.4.2 Self-Regulation

Self-regulating systems are at the same time always self-organizing, if the functions of a system – for example a forest – maintain themselves or the system adapts to new changes. Closely linked to the phenomenon of self-regulation is the cybernetic process of *feedback*, in which disturbances to the system – for example, local short-term tree destruction caused by bark beetle infestations or gradual tree destruction caused by climatic influences that can last a human generation or longer – can be compensated for over time.

As an aside, our body's metabolism is also subject to clear controlled regulatory processes. Campbell et al. (2006, p. 119) describe it like this:

If all the metabolic processes of a cell were to occur simultaneously, chemical chaos would result. Imagine, for example, that a substance is synthesized by one metabolic pathway and immediately degraded by another. The wheels of cellular metabolism would go haywire. In fact, however, a cell tightly regulates its metabolic pathways by determining when and where its various enzymes (enzymes are proteins – protein building blocks – that serve as catalysts – amplifying a chemical reaction, ibid., p. 1508) become active. It does this either by switching on and off genes that encode certain enzymes or by regulating the activity of the enzymes after it has produced them.

If you further imagine that the total number of cells in an adult human being is in the trillions (a one with twelve zeros) and that each cell follows its own adapted regulatory program, one can only bow with awe before this achievement of a highly complex self-organizing and self-regulating individual.

Similar self-regulating processes can be seen in the flora and fauna, whose networked individuals and populations perform equally impressive feats. As already indicated above, the forest as a natural ecosystem is self-sustaining, which also means self-regulating processes.

Natural landscapes such as forests, meadows, steppes, rivers and mountains are sun-driven systems (Fig. 3.5), which also depend on other natural resources as indirect forms of solar energy. The movement of water is furthermore influenced by gravitation (cf. Odum 1991, p. 24).

Just as the rich flora of our earth makes use of the phenomenon of self-regulation in cooperation with other phenomena, this also happens in many, sometimes ingenious ways, in the fauna. Here, the small but extremely clever ants surprise us again and again.

Fig. 3.5 This figure shows an attempt to renaturalize a former managed forest towards a self-sustaining, self-regulating and self-sustaining forest ecosystem. (2020 E. W. Udo Küppers)

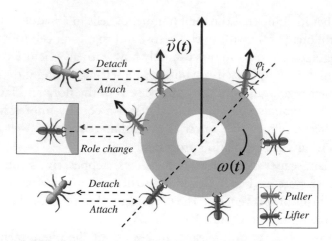

Fig. 3.6 Collective transport of an object by ants by means of self-organization and self-regulation (principle sketch). (With kind permission of Ofer Feinerman, Weizmann Institute of Science, Israel)

Israeli researchers led by Ofer Feinerman observed ants of the genus *P. longicornis* transporting an object to the nest together, which required non-trivial communication (Feinerman et al. 2018). Let us just think here of our own communication behavior in a group, for example, when transporting a bulky object, such as a sofa, through the hallway to the third floor. Figures 3.6 and 3.7 show the transport process of the ant collective, in which amazing things can be seen.

According to Fig. 3.6, the collective transport of a ring-shaped object by ants towards the nest leaves open several degrees of freedom of movement:

- Rotation and translation of the transport object
- Migration direction of the ants in alternating directions
- Restriction of the free direction of movement and the direction to the net is achieved by tying a thread to the transport object.
- Lifting (pressing if necessary) of the transport object of ants from changing directions
- Attachment and detachment of ants in alternation with other ants not directly involved in the transport, both on the pulling side and on the lifting side (possibly pressure side) as well as in alternation between pulling and lifting side (possibly pressure side) of the transport object.

The management of the rotational transport ring thus requires a sophisticated communication among the ants, so to speak an intensive self-organization or self-regulation with feedbacks, in order to transport the object of the transport ultimately goal-oriented to the network.

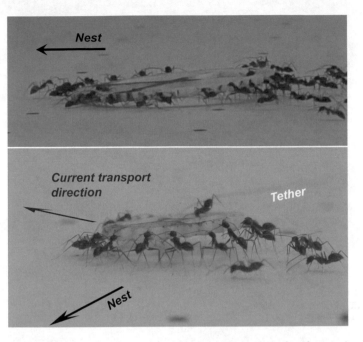

Fig. 3.7 Collective transport of an object by ants by means of self-organization and self-regulation (real image). (With kind permission of Ofer Feinerman, Weizmann Institute of Science, Israel)

What is sketchily visible in Fig. 3.6 is shown in reality in Fig. 3.7. Pure pulling movement of the transport ring would lead to a disproportionately high frictional resistance between the transport ring and the substrate. Whereas, by raising the transport ring significantly on the opposite pulling side, the frictional resistance can be reduced, thus increasing the effective transport speed. An ingenious move by the ants that cleverly links several coupled movement mechanisms of transport to save work and energy.

3.5 Self-Replication

From Sect. 3.2 we have already learned about deoxyribonucleic acid – DNA – as the carrier of hereditary material. DNA is a nucleic acid whose individual nucleotide building blocks form macromolecules. In all organisms, these contain the genetic code. The spatial structure of DNA (DNA deoxyribonucleic acid) corresponds to a double helix, a spiral of two strands wrapped around each other, which are held together by different *base pairs*. Base pairs are two

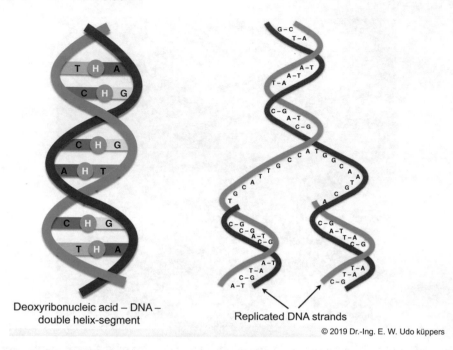

Deoxyribonucleic acid – DNA –
double helix-segment

Replicated DNA strands

© 2019 Dr.-Ing. E. W. Udo küppers

Fig. 3.8 DNA double helix structure with indicated base pairs (left), Principle process of DNA replication in which the original double helix of DNA is divided. At each strand, a new strand is synthesized in mirror image by complementary base connections. This results in two new DNA double helix strands (right)

complementary nucleic bases connected by hydrogen bonds. There are a total of four nucleic bases, A (adenine), T (tymin), C (cyanine) and G (guanine). A connects with T and C with G. Figure 3.8 shows the principle of the geometric structure of DNA in the simplest way.

On DNA (DNA) replication, Karp (2005, p. 679) notes:

A fundamental characteristic of the living is reproduction. Reproductive processes can be observed at all levels: Living things reproduce by sexual or asexual reproduction, genetic material duplicates itself by DNA replication. But the apparatus that duplicates DNA also goes into action in another process: repairing genetic material that has suffered damage. [...] The ability to self-duplicate may have been one of the first fundamental properties to develop in the evolution of the simplest life forms.

Schmidt (2017, p. 33) formulates on DNA replication:

> Replication duplicates DNA and is a fundamental process in all cells. It ensures that when the genetic material is passed on to daughter cells, both receive the complete genetic material and normally no information is lost. [...] Essentially, replication shows the following characteristics:
>
> It occurs semi-conservatively, one strand is retained, the second is post-synthesized. Several enzymes work in coordination and carry out replication together. The actual synthesis enzyme is always a DNA polymerase (an enzyme for controlling the biosynthesis of DNA, d. A.). Auxiliary proteins support the enzymes.
>
> Replication occurs in one direction only. [...] This results in a continuous synthesis of one strand and a discontinuous but simultaneous synthesis of the second. Replication is very accurate, but not error-free. The enzymes detect and correct many errors, but not all. There are fixed starting points for replication that are important for replication control.

A large number of other textbooks and reference books, in addition to the two mentioned, explain in detail to interested readers the chemical-biological processes associated with the replication of our carrier of genetic material, DNA, including Nordheim and Knippers (2018); Alberts et al. (2017); Plattner and Hentschel (2017); Graw (2015); Podbregar and Lohmann (2013).

The replication or reduplication of DNA is – as already mentioned above – a process in which errors occur! – They are the reason why in heredity, i.e. the passing on of hereditary dispositions from one generation to the next, a 100% identical living being, an identical image in all hereditary dispositions, can never arise as a parental offspring. In the genetic replication process, the foundation has already been laid, so to speak, for the sustainable further development of an environment rich in biodiversity, whereby many additional genetic and non-genetic influences reinforce this biodiversity of living beings.

> Genetically implied diversity acts upstream, so to speak, on the biodiversity of organisms, the very richness of life.

We will duly discuss the technospheric application of the natural principle of replication – as well as other natural principles – in Main Section II.

3.6   Genetically Directed Growth and Differentiation: Mutation and Selection

3.6.1   Ernst Mayr and the Clear Language of Science

The language of science is a double-edged sword for the great evolutionary scientist Ernst Mayr. With regard to the subject of evolution dealt with here, Mayr (1998, p. 90) writes:

> Each branch of science has its own terminology for the facts, processes and concepts of its field. If a technical term refers to an object or individual – mitochondria, chromosomes, nucleus, wolf, Japanese beetle, sequoia – it is usually unproblematic. But many technical terms refer to rather heterogeneous phenomena or processes; competition, evolution, species, adaptation, niche, hybridization, and diversity are some that one encounters in biology.

These heterogeneous phenomena from different disciplines, which are not understood in the same way, often lead to misunderstandings and controversial discussions or disputes. This is also the case with the term *mutation*. Mayr (1998, p. 91):

> A good example of this is Thomas Hunt Morgan's[24] application of Hugo de Vries'[25] concept of "mutation" to any sudden change in genetic material; de Vries understood mutation to be an evolutionary change that immediately gave rise to a new species. This concept was evolutionary rather than genetic. It took the non-geneticists 30 to 40 years to realize that Morgan's mutations were not the same as de Vries'.

The far-reaching effects that misleading language can have beyond the boundaries of disciplines can be seen in many social, political, ecological and economic discourses right up to the present time of great transformative activity, and is therefore not a phenomenon confined to science per se.

With this little introduction to the use and understanding of language, we move into the details of what mutation – and later selection – mean as a principle of evolution.

---

[24] Thomas Hunt Morgans (1866–1945) was a US zoologist and geneticist.
[25] Hugo Marie de Vries (1848–1935) was a Dutch biologist and was among the rediscoverers of Gregor Mendel's rules of inheritance

3.6.2 Mutation

Mutations are changes in the genetic material.

In this context, Thomas Hunt Morgan's concept of mutation referred exclusively to "[…] spontaneous changes in the genotype […]", in other words, to "[…] sudden changes in a gene." (Mayr 1998, p. 126). And further (ibid.):

> Gene mutations occur during cell division due to copying errors. The duplication (replication) of DNA molecules during cell division and the formation of germ cells is a remarkably precise process, but errors do occasionally occur. If one base pair (see Fig. 3.7, d. A.) takes the place of another, this is called a gene or point mutation. However, major changes in the genotype also occur, for example polyploidy or a rearrangement of the genes, as results from chromosome inversions. Such processes are called chromosomal mutations. However, mutations also include changes in the pathway from the DNA of the gene via messenger RNA and ribosomes to the amino acids or polypeptides of the phenotype. In addition, mutations can result from the insertion of a transposable element into a chromosome. Any mutation that results in an altered phenotype is favored or disfavored by natural selection.

In his detailed and exemplary chapter on "Changes in the genome: mutations", Graw (2015, p. 401–465) once again summarizes in great detail – here in sufficient brevity – the mutation varieties previously described *by* Ernst Mayr as a *classification of mutations:*

Theorem 1 on mutation
Any change in the genetic constitution of a cell that is not related to sexual reproduction is a mutation. Mutations can occur in the same way in germ cells and in somatic cells (ibid., p. 402).
Theorem 2 on mutation
The number of altered somatic cells or germ cells allows conclusions to be drawn about the time during development at which this mutation occurred (ibid., p. 402).
Theorem 3 on mutation
Mutations are a basic phenomenon of living systems. At the level of the individual, they often have negative consequences. For the evolution of organisms they are indispensable (ibid., p. 402).
Theorem 4 on mutation

Mutations, changes in DNA, can occur spontaneously and can be induced by radiation or chemicals. Mutations can be classified differently according to their size or the nature of their DNA or chromosomal changes (ibid., p. 404).

- **Chromosome mutations**

Chromosomal mutations play an important role both in nature and in experimental genetics because, unlike most genetic mutations, they can exert effects on the distribution mechanisms of chromosomes in meiosis and mitosis or can also stabilise certain allele combinations in the genome. (ibid., p. 404).

- **Spontaneous point mutations**

– *Errors in replication and recombination*

The DNA polymerase makes relatively frequent errors (approximately one in 10^4 nucleotides) in the incorporation of nucleotides due to base complementarity with the complementary strand in a growing DNA strand. Although the DNA polymerase complex can correct such errors immediately in the context of *replication* [...], nucleotide misincorporation still occurs. The error rate remaining despite the repair processes is still on the order of 10^{-6} to 10^{-11} nucleotides. This may seem low, but in a genome that contains 2.75 Å ~ 10^9 nucleotide pairs, as does the human genome, this misincorporation rate still results in a mutation frequency of one nucleotide in at least one in 100 replicating cells, or even in every replication cycle. Since as many as 10^6 cells replicate every second in a human individual, the total mutation frequency in any single individual is extraordinarily high.

(Another trigger for replication errors is the slippage of DNA polymerase over repetitive regions. According to Graw (ibid., p. 413), spontaneous base changes can also trigger mutations).

Another essential genetic process, namely *recombination* [...] is prone to disruption and the cause of many mutations. Here, too, repetitive regions are particularly at risk. Such sequence repeats can lead to mismatches during meiotic homolog pairing and, as a consequence, to recombination errors. This leads to duplications or deletions in the affected DNA region. These phenomena play a special role in the development of structural chromosomal aberrations.

During DNA replication, incorrect nucleotides are also incorporated at high rates of DNA polymerase error. However, most errors are corrected by a polymerase's own repair mechanism. (ibid., p. 413 f.)

- **Induced mutations**

In contrast to mutations caused by endogenous defects, which we discussed above, this section focuses on mutations caused by environmental influences. These include radiation (UV light, ionizing radiation) and chemical agents.

These mutagenic agents have always been present in different ways and to varying degrees or are dependent on human activities; therefore, for example, we do not distinguish here whether the UV light comes from the sun or from a solarium (ibid., p. 418).

• **Mutagenicity and mutation rate**

Considering the different relative mutation frequencies in mice after irradiation (with X-ray or γ-radiation, d. A.) or treatment with ENU (the major experimental mutagen, ethylnitrosourea, d. A.), it is understandable that each gene probably has its own mutation spectrum, which varies even more for different mutagens. Indeed, the mutation rates of different genes can differ by over two orders of magnitude from each other (10^{-5} to 10^{-7} mutations per cell generation) (ibid., p. 429).

• **Mutation rate and evolution**

About 80 years ago, John B. S. Haldane (1935) postulated that the male mutation rate in humans was significantly higher than the female mutation rate, since male germ cells experience significantly more cell divisions and thus DNA replication rounds per generation than is the case with female germ cells [...]. Although this hypothesis is now widely accepted, the magnitude of the ratio of male to female mutation rate (expressed as a factor α) has long been controversial. Knowledge of this magnitude is important in order to know whether the mutation rate is essentially caused by errors during DNA replication. (ibid., p. 435f).

• **Repair mechanisms**

Cells have various repair mechanisms that can remove damage to DNA. The mechanisms for this are also partly dependent on the type of damage, and they have different degrees of accuracy with which the repair is carried out: For example, while SOS repair [...] is very prone to failure, it allows the cell to survive at all despite extensive damage. In contrast, repair mechanisms are more accurate and tend to proceed via recombination mechanisms. The checking of DNA for possible damage and its successful repair is often also a prerequisite for the entry of DNA into di mitosis. Checking at appropriate "checkpoints" [...] can slow down progress in the cell cycle until repair has occurred. This checkpointing with cell cycle arrest and subsequent DNA repair is often referred to as DNA damage response (DDR).

In addition to the type of damage, the state of the cell in the cell cycle also determines the selection of the appropriate repair pathway. After recognition of the damage and the binding of specific proteins to the damage site, repair takes place over a longer period of time and with the aid of larger repair complexes. When the repair is complete, the complexes must be degraded again (ibid., p. 437 f.).

Nature always uses a variety of differentiated strategies to solve problems, including repair mechanisms on cells, which will be highlighted. For detailed insights into their functionality, please refer to the technical literature.

Repair processes can be characterized according to different criteria – once according to the underlying mechanisms, but also according to the different types of damage. We will consider the different mechanisms here (ibid., p. 437 f.):

- light-dependent repair of UV-induced DNA damage by photolyases (direct repair);
- Excision repair mechanisms (base excision repair, nucleotide excision repair);
- DNA mismatch repair;
- homologous recombination repair;
- non-homologous linking of free ends;
- post-replicative DNA repair.
- **Site-specific mutations**

In contrast to the random and undirected mutation events we have had up to now […], modern genetics has various possibilities to intervene specifically in the genome of model organisms. These genetically modified organisms are also referred to as "transgenic" (ibid., p. 446).

However, we will not elaborate on this group of directed mutations for our goals of using ingenious principles of nature in a sustainable and adapted manner in our biosphere and technosphere.

Because of the technical terms used in advance, interested readers are referred to the glossary at the end of the book, where technical terms are explained in an understandable way.

Mutation is therefore not equal to mutation! Nature not only works with ingenious principles, which she offers us to use for our purposes and goals; she also varies individual principles – such as that of mutation – in such a clever way that a wide range of changes result from it.

In the concrete case of mutation, selection, replication, etc., which we will bring to our purpose and goal of an "evolutionary optimization strategy" by means of a bionic application, we will go into more detail in Chap. 6.

For a detailed and exemplary account of the changes in the genome that we call mutations, see Nordheim and Knippers (2018); Graw (2015); Podbrega and Lohmann (2013).

3.6.3 Natural Selection

Let us look back again to the time in the middle of the nineteenth century when Darwin presented his evolutionary selection theory. From an evolutionary point of view, this period was characterized by the so-called type theory, which was recognized as the dominant way of thinking. The Encyclopedia of Biology[26] writes about this:

> "[...] widespread idea before the advent of the theory of evolution that the diversity of the forms of organisms can be traced back to a [...] morphological type (archetype) or a few types. Typology is the basis of pre-evolutionary taxonomy (C. von Linné) and of the morphological concept of species (Art), and especially of idealistic morphology and typological racial classifications (human races)." E. Mayr (1967) sharply contrasts typological thinking (essentialism) with thinking in terms of populations: "For the typologist the type (eidos) is real and variation an illusion, whereas for the population scientist the type (average) is an abstraction and only variation is real." (see Mayr. E.: Artbegriff und Evolution. Hamburg, Berlin 1967, d. A.).

Mayr himself puts it this way (2003, p. 147 f.):

> Natural selection, as conceived by Darwin and Wallace, was an entirely new and daring theory. It is based on five observations (facts) and three conclusions [...]. When one discusses natural selection and speaks of population, one usually means biological species that reproduce sexually, but it also runs in clones of asexual creatures. [...] It was a truly revolutionary idea [...]. Even today (at the beginning of the 21st century, and certainly at present, d. A.) many people find it difficult to understand the principle. However, if you use population thinking, everything becomes very simple.

Darwin's explanatory model for natural selection – according to Mayr 2003, p. 149 – is reproduced below in the outlined box.

---

[26] https://www.spektrum.de/lexikon/biologie/typologie/68190 (Accessed March 28, 2019).

Darwin's Explanatory Model for Natural Selection

Fact 1: All populations are so fertile that their size would increase exponentially without limitation. (Source Paley and Malthus).

Fact 2: The size of populations remains the same over time, apart from seasonal fluctuations (steady state). (Source: general observation).

Fact 3: Each species has only limited resources at its disposal. (Source: Observation, confirmed by Mathus).

Conclusion 1: There is strong competition (struggle for existence) between the members of a species. (Source: Malthus).

Fact 4: Two individuals in a population are never exactly alike (population thinking). (Source: Animal Breeders and Biological Systematists).

Conclusion 2: Individuals in a population differ in terms of their probability of survival (i.e. natural selection takes place). (Source: Darwin).

Fact 5: Many differences between individuals in a population are at least partially heritable. (Source: Animal Breeders).

Conclusion 3: Natural selection, continued over many generations, leads to evolution. (Source: Darwin)

Darwin in his time can not yet the influence of genetics on selection. He focused exclusively on the selection of the phenotype, on the "survival of the fittest" brought into play by Herbert Spencer.

Natural selection is – as we know today – a process that fundamentally takes place in two successive steps (cf. Mayr 2003, p. 152):

- *1st step* ➔ ***random*** *variations arise*

"Mutation of the zygote (diploid cell after fertilization, d. A.) from its formation to death; meiosis (maturation division, d. A.) with recombination by crossing-over (mutual exchange of corresponding chromosome segments, d. A.) in the first division and random migration of chromosomes in the second (the reduction division); any random elements in mate choice and fertilization."

- *2nd step* ➔ ***deterministic*** *aspects act on survival and reproduction*

"Greater success of certain phenotypes during their life cycle (selection by survival); non-random mate choice and all other aspects that cause the reproductive success of certain phenotypes to increase (sexual selection). In the second step, large-scale random elimination occurs simultaneously."

Chance and necessity (see also Jacques Monod 1975) are, in short, two concepts inherent in natural selection. Together they have an effect on the developing ability of organisms to camouflage themselves extremely perfectly – almost invisible to predators – on individuals that have extraordinarily good eyesight, on swimmers or flyers that are fast moving and yet have little flow, almost on everything that is connected with natural selection.

Three well-known processes of natural selection are exemplary for countless ones in nature:

- Example 1: Selection by hunter-prey coevolution
- Example 2: Selection by interspecific symbiosis
- Example 3: Selection by industrial melanism.
- **Selection by hunter-prey coevolution**

The interaction of two or more different species, whether within the animal or plant kingdom or between plants and animals, always leads to mutual selection pressure. This is shown particularly impressively in a predator-prey relationship when, for example, short-ranged arrow-fast predators such as cheetahs pursue wildebeest as prey, which in turn develop more effective escape mechanisms, to which the predator responds by new or extended attack techniques, and so on.

Similarly, in the following example of a predator-prey relationship between fox and hare, which is further complemented by a food source of the prey animal, see Fig. 3.9.

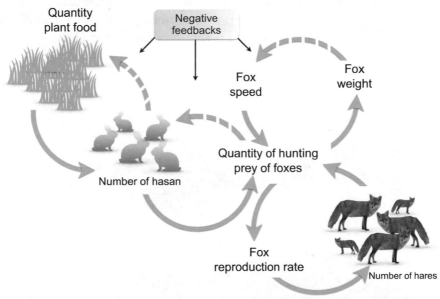

Original: F. Vester (1978) © 2019 Dr. -Ing. E. W. Udo küppers

Fig. 3.9 Selection pressure in a predator-prey relationship between foxes and hares. Solid circular lines act in the same direction and strengthen the relationship of … on, interrupted circular lines act in the opposite direction and weaken the relationship of … on. Example 1: The *greater* the weight of the foxes, the *lower* their speed. Example 2: The *greater* the number of hares, the *greater* the amount of prey hunted by foxes. Opposite effects stabilize a relationship – *negative feedback* –, Contemporaneous effects destabilize a relationship – *positive feedback*

The interlinked effect relationships of foxes and hares interpret the mutual selection pressure of population growth and population decline of both species. An increase in the fox population reduces the hare population over time. As a result of fox weight gain and consequent reduced hunting speed, the hare population can successively increase again and make abundant use of its food source. With a certain time delay, the foxes again succeed in building up a new population via the reproduction rate and hunt the hares more frequently. The population interplay of growth and contraction begins again, as outlined in Fig. 3.10.

- **Selection through interspecific symbiosis**
  The coexistence of two species for mutual benefit is called symbiotic behavior, the importance of which for evolution does not receive nearly the attention it deserves (cf. Ernst Mayr 2003, p. 259).
  This is due not least to the overwhelming number of symbioses between flowering plants and insects, which contribute to the mutual strengthening of plant and animal survival, in that on the one hand the plants provide nectar as food for the bees, for example, and on the other hand the bees ensure the spread of plant poles to other flowers of the same species, which contributes to the fertilization of plants and the life support of both organisms.

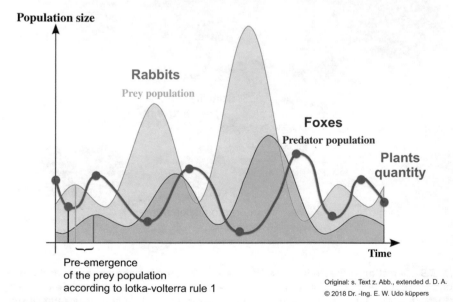

Fig. 3.10 Time course of the predator-prey relationship following the networked effect relationships in Fig. 3.9, outlined as coupled dynamic growth of the two populations. (Source: https://de.wikipedia.org/wiki/Räuber-Beute-Beziehung#Das_Lotka-Volterra-Modell. Accessed 29 Mar. 2019)

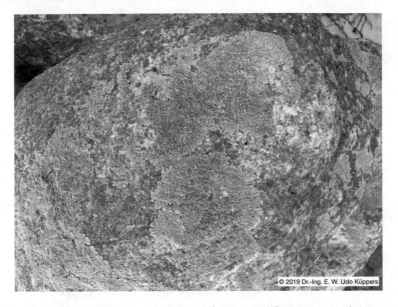

Fig. 3.11 Lichens, a symbiotic relationship of algae and fungi on a rock in a surf-rich region of the Baltic Sea

> ### A Small Anticipation of Chap. 6
>
> *The lack of attention to the natural principle of symbiosis that Ernst Mayr laments can also be mapped onto the technosphere (see Chap. 6) without any reservations – indeed, many times over. If technical-symbiotic mechanisms were to be used more consistently in sustainable material processing than has been the case to date, then it seems reasonable to assume that our produced mountain of high-loss materials and materials, some of which have persistent environmentally damaging properties, would be reduced much sooner than is currently the case, if at all. However, this would mean nothing other than a complete reversal of strategy from systematic maximisation of so-called "end results" of processes towards systemic optimisation of so-called "starting conditions" of processes.*

Figure 3.11 shows another symbiosis between bacteria and fungi, which we know as lichens and which settle on different substrates as pioneer organisms (cf. Straaß 1990, p. 81).

While the alga produces carbohydrates and proteins through photosynthesis, which serves as food for the fungus, the fungus provides water and minerals for the alga. As a common metabolic product, the so-called lichen acid is produced, which as colored crystals with the fungal filaments form an extraordinarily stable and resistant protection against environmental influences, which allows their settlement in extreme places, such as on bare rock in surf-blown coastal regions.

- **Selection by industrial melanism**

The *hypothetico-deductive method* is a scientific method from which hypotheses are formulated according to a question, which are intended to provide answers to question or problem formulations. Preliminary observations of natural processes or results from experiments or theories lead to scientific hypotheses with often general statements about the nature of the processes (cf. Hickman et al. 2008, p. 17). It goes on to say (ibid.):

> Darwin's hypothesis of natural selection, for example, explains the observation that many different kinds of living things reveal characteristics that make them appear to be adapted to their environment. Based on his hypothesis, the scientist may attempt to predict or estimate future observations or experimental results. [...]
>
> Useful hypotheses are those which are testable, from the derivation or continuation of which predictions result which can be tested in experiments and are thus falsifiable. If contradictions to the hypothesis arise in the experiment or in both observations to be measured, the hypothesis must be rejected – it is then said to be *falsified*. Confirmation, on the other hand, is considered *verification*.

With this brief introduction to scientific work, we move on to the phenomenon of industrial melanism. Darwin's hypothesis of natural selection was used to investigate the variation in appearance of moth populations (birch moths) in the coal mining area around Manchester, England, in the mid-nineteenth century. In this highly polluted environment.

> [...] many moth populations consisted predominantly of dark-colored (melanized) animals, whereas populations of the same species living in forests with clean air had animals that were light-colored with much greater frequency.
>
> The hypothesis is that moths survive best when they are adapted in appearance to their environment, making them almost invisible to birds looking for them to eat. Experimental studies have shown that, consistent with this hypothesis, birds are more easily able to detect and eat moths that differ in appearance from their environment. [...]

Another testable prediction, [...] that the number of lighter-coloured animals in the moth population should increase when air pollution decreases in the area under consideration, and exposure to soot and dust leads to a change in the environment [...] (has also been confirmed, d. A.). (ibid., p. 19)

Research on the genetic causes of the varying colour appearance of birch moths has clarified details of the so-called *carbonaria typica polymorphism.* According to this, the altered dark, anthracite-coloured wing surface can be attributed to a dominant gene mutation. In technical terms, this is referred to as a *transposome,* a limited DNA gene segment with the ability to change its local position, also known colloquially as a jumping gene. This transposome has thus contributed to the surface change of the phenotype by acting on a trait-shaping – colour-giving – gene (cf. van't Hof et al. 2011, 2016; Kronberg 2016; Majerus 2009; Kettlewell 1955).

Figure 3.12 shows two birch moths of different colouration on a birch bark. Figure 3.13 shows – on close inspection – an artificial birch moth that was part of a field experiment on industrial melanism in the Cornwall area, Great Britain (Walton and Stevens 2018).

Fig. 3.12   Birch moth of different coloration on typical birch bark. (Source: Martinowsky https://commons.wikimedia.org/wiki/File:Lichte_en_zwarte_versie_berkenspanner.jpg. Accessed 30 Mar. 2019)

Fig. 3.13 Artificial birch moth during an experiment on industrial melanism. (Courtesy of Olivia C Walton/University of Exeter, UK)

3.7 Epigenetically Directed Growth and Differentiation: Acquired Characteristics

First of all, let's take a look at the historical development associated with the concept of epigenesis.

The Austrian zoologist, physiologist and evolutionary biologist wrote in the chapter on "Evolution as an epigenetic process" in the book he edited, "The Evolution of the Theory of Evolution" (Wieser 1994, p. 35 f.):

> The fact that ontogenesis (development of the individual from the ovum to sexual maturity, d. A.) mirrors phylogenesis (phylogenetic development of all living beings, d. A.) was not first the idea of Ernst Haeckel[27] laid down in the biogenetic law. The correspondence between the development of an organism and historical processes that gave rise to this development, in an image that is already found in Aristotle and that was later used again and again in the search

---

[27] Ernst Heinrich Philipp August Haeckel (1834–1919) was a physician, zoologist, and philosopher. A contemporary of Darwin, he developed his theory of descent according to Darwin's ideas. https://de.wikipedia.org/wiki/Ernst_Haeckel (Accessed March 30, 2019)

for the ordering principles of the unfolding of the individual from the ovum. According to today's understanding, ontogenesis means the translation of the genotype into the phenotype (see also Sect. 3.3).

Although the program of this translation is fixed in the genome of the individual, the actual events can only be determined in the confrontation between the genes and their products – the transcribed (transmissible, i.e.) and processed RNA (ribonucleic acid, the essential task of RNA in the biological cell is the transformation of genetic information – gen. Code – into proteins, i. A.) as well as the expressed (expression of the genotype, genetic information, i. A.) proteins (biological macromolecule, consisting of amino acids, the building blocks of proteins, i. A.).

The genetic program is thus realized, so to speak, in the act of ontogenesis through constant communication between the controlling and the executing elements. In this process, the translation of a temporal sequence of genetic commands into the spatial shape of the organism plays a decisive role, as it could be made visible so impressively by means of molecular genetic methods in the development of the fruit fly *Drosophila* […].

Wieser attributes the concept of epigenesis – "after-creation" or the unfolding of new structures in the development of living things – to C. F. Wolff[28] (1759). In addition, Wieser points out that the importance of epigenetic interaction for the emergence of the biological individual occurred long before the great discoveries of molecular biology by developmental biology, referring to the physician and zoologist Waldemar Schleip (1929), cited in Sander (1990, p. 152).

On the information flows that take place between the environment and the organism, Wieser writes (ibid., p. 40):

An essential feature of ontogenesis (development of the individual, d. A.) are the interactions between cytoplasm (basic structure within the outer cell membrane, d. A.) and genome. Cytoplasmic factors control the expression of genes that produce and stabilize certain phenotypes. Now we know by numerous examples that the environment is capable of intervening in both the metabolism and the morphogenesis of organisms. This is the basis of any form of adaptation called *modificatory*, which consists in matching the phenotype with the changing conditions of the environment, for example: the pigmentation of the skin increases with the intensity of UV radiation; extreme environmental factors indicate the expression of specific stress proteins that protect the cellular environment from the damaging effects of these factors; the chemical composition of membranes change with the temperature of the medium; […].

---

[28] C. F. Wolff (1734–1794) German anatomist, physiologist and botanist. https://de.wikipedia.org/wiki/Caspar_Friedrich_Wolff (Accessed March 30, 2019).

In this way, information from the environment flows into the genetic programs of organisms – but only in such a way that the activity of existing program elements is altered, i.e. their expression in the phenotype is either promoted or inhibited.

A characteristic process for promoting the activity of existing genetic program elements are the multiple phenotypic routine cycles of incentive, routine and reward. They can be decisive for the fact that, for example, the strong urge to eat sugary sweets or to smoke becomes genetically consolidated and leads to the fact that offspring develop similar routine tendencies as their ancestors.

In addition, environmental factors can also influence the mutation frequency in cells, which is otherwise determined by endogenous processes. Like spontaneous mutations, induced mutations are also to be regarded as strictly non-directional (i.e. random, d. A.) (ibid., p. 40).

Based on this, Wieser describes two possible ways of information exchange between organism and environment (cf. ibid., p. 40):

1. *Phenotypic response*: Environmental factors alter the expression pattern of somatic cells Information flows from the environment into the somatic genomes of individuals and from these back out into the environment.
2. *Undirected genotypic changes in germ cells*: They are sorted by the environment and lead to changes in the genetic composition of populations. Information flow from the germ cells of a population into the environment and from the environment back into the population.

According to Wieser, *phenotypic flexibility* and *genotypic variability* stand for all mechanisms of adaptation of organisms to the environment. Thus, the most important motors of biological evolution are named.

After this description of interactions between environment and organism, which address genetic and epigenetic mechanisms, we go into more detail on the molecular processes that are genetically/epigenetically determined and contrast the concepts of *molecular genetics* and *epigenetics* (Graw 2015, p. 5, emphasis added):

Molecular genetics studies the biochemical basis of heredity. It wants to know how the hereditary material is molecularly structured and how it performs its function in a cell and in the organism as a whole.

Epigenetics [...] was originally understood as the interpretation of the genotype into a specific phenotype during embryonic development (Waddington

1940). Today, we understand it to mean permanent changes in gene activities that are maintained over generations of cells (or organisms) without altering the DNA itself.

The following description is intended to describe in a sufficiently concise form the molecular processes of epigenetics, which have become a current intensive research topic. It is not the task of this book to describe in detail the mechanisms of epigenetics, however valuable and interesting their effects on our organism and thus on our life in the environment may be. However, it is probably the partial aim to describe epigenetic effects in the light of Darwin's evolutionary selection theories, especially with regard to the biotechnological transfer processes discussed in Chap. 6.

Using Graw's overview of his Chap. 8 Epigenetics (2015, p. 292), we briefly describe the necessary processes associated with epigenetics. It will not be possible to avoid mentioning technical terms in the running text. However, interested readers will be helped by the comprehensible description of the technical terms in the glossary at the end of the book.

Epigenetics initially referred to genetic phenomena that could not be explained with the standard formal explanatory patterns of Mendelian[29] genetics (Mendel 1866) [...]. Over the years, "epigenetics" then became a kind of collective term under which everything was subsumed that was not quite comprehensible. Today, we understand epigenetics to mean stable changes in the regulation of gene expression that occur during development, cell differentiation and cell proliferation and are fixed and maintained over cell divisions without changing the DNA sequence. Accordingly, we can now describe the difference between hetero- and euchromatin molecularly by methylation of DNA and, most importantly, also methylation and acetylation of histones.

In this context, completely new mechanisms of gene regulation become apparent, which can be aptly summarized by the keyword "non-coding, regulatory RNA". In addition to the rRNA and tRNA genes discussed earlier, a completely new "RNA world" of small and large non-coding RNA molecules opens up to us. A long, non-coding RNA is also at the centre of dosage compensation in mammalian sex chromosomes: Quantitatively unbalanced gene constitutions (XX in female and XY in male) are usually not tolerated by the organism. In mammals, therefore, one of the two X chromosomes of the female sex is inactivated, resulting in a state that is functionally equivalent to the hemizygous X chromosome constitution of the male sex. Once inactivated, the chromosome

---

[29] https://de.wikipedia.org/wiki/Mendelsche_Regeln#cite_note-2 (Accessed Mar 30, 2019).

generally remains intact within an organism. The chromosome must therefore contain information that ensures that it remains inactive in all subsequent cell generations. Beyond this, however, other areas of the maternal and paternal genomes are obviously different. Before this phenomenon could be dealt with molecularly, the term "genetic imprinting" was introduced. We now know that imprinting is essentially based on methylation of imprinting centres and the inactivating effect of non-coding RNA. This imprinting is deleted in the early stages of embryonic development and later renewed in a sex-specific manner. Nutritional influences also play a major role in this process.

Do genes alone or also environment and experience shape us humans, animals and plants in the course of our life stage? Whether the specific diet of grandparents changes the life expectancy of grandchildren or we grow old like Methuselah or end our lives seemingly surprisingly at a young age? Genomes seem to be more than the sum of their cells from which organisms arise.

Doctrines persist, sometimes for decades and centuries, despite proven falsification of the theories. The economic sciences are an outstanding example of this with their theory of *homo oeconomicus,* to which they strictly adhere to this day, despite catastrophic misjudgements that have taken place.

For a long time, the purely genetic influence on heredity was advocated as the only mechanism in the evolutionary science space, despite the fact that numerous evidences on epigenetic inheritance have been presented across the organismal range (see, among others, Kegel 2018; Jablonka and Lamb 2008, 1995). However, it seems that the recognition of epigenetic influences on heredity is gaining ground, ushering in a new field of interdisciplinary research and development, more than 150 years after Darwin's theory of evolution, without compromising his brilliant idea in the process-which continues to endure! For the multitude of other sources on epigenetics, here is just a small selection: Mousavi et al. (2019); Ennis et al. (2018); Heil et al. (2016); Spork (2016, 2010); Deichmann (2016).

3.8 Metabolism: Ability to Bind and Release Energy

It is not surprising that the previous chapter on epigenetics also referred to the influences of nutrition, which play a major role. For all organisms, whether humans, animals or plants, are so-called *energy conversion systems,* the basis of which is metabolism! Campbell et al. (2006, p. 104) speak of the "[…] totality of all chemical processes of an organism […]", which they characterize as metabolism or metabolism.

Metabolism is an emergent property of life and results from the interactions between molecules within the ordered structures of a cell. (ibid.)

The chemistry of life is organized by a high complexity of cell metabolism, which is also called intermediary metabolism (intermediary = forming an intermediate product -). The figurative idea of intermediary metabolism or cellular metabolism resembles that of a highly complicated road network. Campbell et al. describe it thus (ibid., p. 105):

The reactions are arranged in branched metabolic pathways in which molecules are converted step by step. Enzymes (giant molecules, accelerate chemical reactions, d. A.) channel matter through the metabolic pathways by selectively accelerating each step. Analogous to the red, yellow and green of traffic lights that regulate the flow of traffic and prevent traffic chaos, enzyme regulatory mechanisms balance supply and demand within the metabolism, responding to a shortage or surplus of certain substances.

Overall, metabolism organizes the cell's material and energy resources. Some metabolic pathways release energy by breaking down complex molecules into simpler compounds. These breakdown processes are called catabolic metabolic pathways. A major route of catabolism is cellular respiration, in which glucose and other organic fuels are broken down, producing carbon dioxide and water. The energy stored in the organic molecules is now available to the cell to do work.

Anabolic metabolic pathways, on the other hand, expend energy to build complex molecules from simpler ones. An example of anabolism is the synthesis of a protein from amino acids.

The anabolic and catabolic pathways are the uphill and downhill paths of the metabolic map (Campbell et al., Fig. 6.1, p. 104 and Fig. 9.9, p. 184). They are interconnected in such a way that the energy released during the "downhill" reactions of catabolism can drive the "uphill" reactions of anabolic metabolic pathways. This energy transfer from catabolism to anabolism is called energy coupling.

Figure 3.14 shows a section of a cell metabolism (Campbell et al., p. 184).

According to Campbell et al. (2006, p. 104), several thousand metabolic reactions can take place in a (human) cell. This number is all the more impressive if we relate the size of a cell, ranging from a few single-digit micrometers (μm = 1 thousandth of a millimeter) to a three-digit μm size (in animal and plant cells), to the number of metabolic reactions. If one roughly multiplies an average number of a few hundred to one thousand metabolic reactions per human cell with the calculated total number of cells – approx. 10^{14} or one

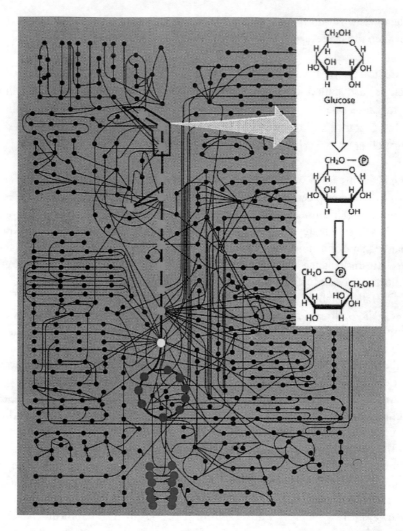

Fig. 3.14 Section of a cell metabolism (intermediate metabolism). The dots symbolize metabolites (metabolic products) and the lines represent the chemical reactions in which they are converted. The reactions gradually follow a so-called metabolic pathway in which each step is catalyzed by a specific enzyme. The entire cell metabolism is driven by the central energy-generating process of cellular respiration. This is divided into glycolysis (green), the citrate cycle (salmon/red) and the respiratory chain (blue). The end product of glycolysis (pyruvate) is one of the focal points of cellular metabolism. Another focal point is acetyl coenzyme A (yellow), which mediates between glycolysis and the citrate cycle. The citrate cycle, in turn, is the central hub of cellular metabolism. The section on the right in Fig. 3.13 shows the first two steps of glycolysis. Details of the chemical processes can be found in Campbell et al. Chap. 9. (Fig. 3.14 courtesy of Stark-Verlag, D-85399 Hallbergmoos, © Campbell, Neil A.; Reece, Jane B.: Biology, 6th edition, Pearson Studium, Munich 2006, p. 184)

hundred trillion – in the human organism,[30] the reverence, the respect for the extraordinary natural achievements and their principles becomes apparent for a further, but not for the last time.

3.9   Response to Environmental Stimuli

What distinguishes living organisms from inanimate systems? Everything that is presented and explained under this Chap. 3; so also reactions to environmental stimuli that we perceive through our sense organs, process and react to accordingly. Humans see only in a limited spectrum of light from about 400 nm (near ultraviolet) to 800 nm, (near infrared). In contrast, animals such as the rattlesnake detect their warm-blooded prey by infrared detection, beyond human optical perception. Bees flying to red flowers are guided by ultraviolet markers that we humans do not recognize. Elephants communicate and orient themselves by infrasound (frequencies below 16–20 Hz that humans cannot hear). Bats, on the other hand, hunt their prey with ultrasound (greater than 20 kilohertz to gigahertz) (1 billion hertz), far beyond human acoustic perception. Social insects such as ants, termites and bees communicate with a language cocktail of chemical signal substances.

> It is estimated that a single ant has about 30 pheromones at its disposal with which it can transmit messages to its comrades. If optical and acoustic signals can be ascribed symbolic character, why not chemical signals that convey messages such as "follow me", "get lost"? The pheromones of state-forming insects also include those mixtures of substances that make up the particular stick or family scent of a colony. (Müller et al. 2015, p. 692)

Even the following general description of ingenious sensory organ principles of nature can only hint at what qualitative and quantitative achievements across all organisms nature makes available to us for advantageous use for technospheric progress of human activities – on bionic development paths.

The adapted specialization of organisms by reaction to environmental stimuli is indescribably large and forms a rich treasure trove for bionic strategies and procedures, which we will discuss in detail later under main chapter II.

But for now, let's stick to general observations related to reactions to environmental stimuli.

---

[30] https://www.spektrum.de/frage/wie-viele-zellen-hat-der-mensch/620672 (Accessed 4/2/2019).

We have a total of five senses at our disposal with which we explore the environment (cf. ibid., p. 442 ff.):

1. *sense of sight* – optical sensory, see with your eyes -.
2. *sense of hearing* – acoustic sensory, hearing with hearing -.
3. *Sense of taste* – chemical sensing, tasting with taste buds on the tongue.
4. *Sense of smell* – chemical sensory system, smelling with in the olfactory sensors in the nose – 3rd/4th + trigeminal system, third chemical sensory system, extends over the mucous membrane of the mouth, nose and throat.
5. *Sense of touch* – mechanical sensing, feeling with mechanoreceptors of the skin.

Our five organic sensory "gates" to the environment can receive external stimuli together, quasi parallel multifunctionally, about which we decide whether they are of no essential importance for our existential progress or whether they are advantageous or disadvantageous. What we have problems with, however, is the temporally parallel, multifunctional reaction to external stimuli, in which we devote our full attention to each stimulus and activate subsequent reactions. In the context of the digitalization of our leisure and working worlds, the associated concept of *"multitasking"* or working in parallel on several tasks with the same intensity is more of a vision than a reality and ultimately runs the risk of ending up in a *"vicious circle" that builds* up risk (Küppers 2018a, b, p. 89–93).

Among the aforementioned five sensory modalities, human physiology, in conjunction with perceptual physiology, distinguishes different sensory qualities (Müller et al. 2015, p. 443):

Within the experiential world of seeing, RED is a different quality than BLUE. Within the modality of hearing, a low tone is a different quality than a high tone. These principles of classification are not too difficult to comprehend.

But what about sensations of heat, cold, pain? What about hunger and thirst? Is there a special sense of touch that is different from the sense of pressure? Is there a sense of vibration? What are sensations of rotation, force, and gravity? Is there a sense of the position of the body (horizontal, vertical, oblique) or the position of its limbs (angled versus extended)? Obviously, there are perceptions that are relatively clearly definable, but there are also perceptions that are lost in vague general sensations.

The sensory physiologist is also aware of many enteroreceptors that communicate states inside the body to the CNS without triggering any sensations. For humans, the rule of thumb is:

- **Exteroreceptors**, which receive information from the environment, convey more or less distinct sensations and perceptions. The more information-rich senses are, the more distinct the sensed and perceived world of experience.
- **Enteroreceptors** (interoreceptors, endoreceptors, *enteroreceptive senses*), which collect information inside the body, do not convey sensations. Such enteroreceptors are, for example, the blood pressure gauges and the sensors for the pO_2 and pCO_2 *of* the blood. Another group of receptors that are usually classified as enteroreceptors are the
- **Proprioceptors** They report whether and to what extent our muscles are tense and our limbs are bent. Since external forces also act on these receptors, they occupy a middle position between the enteroreceptors and the exteroreceptors. This functional middle position corresponds to a perception that can only be experienced faintly and indistinctly. Even with my eyes closed I can feel whether my arm is stretched or bent. However, as a rule I am not aware of the feeling of this. Everyday language has not named a special modality for this feeling and at best we complain about "heavy" limbs or the "heaviness of a load". The sense of weight, like the sense of position or space, occupies an intermediate position between exteroreceptors and enteroreceptors.
- **Pain receptors** occupy a special position. They react to a variety of stimuli that can lead to injuries or indicate injuries, and to internal alarm substances released during inflammations. It does not matter where such stimuli come from. Whether it is an external or internal stimulus: the pain intrusively announces itself with its alarming quality of experience.

After this somewhat in-depth description of human sensory systems, we will conclude with a few remarks on plant sensory systems, which will surprise us in every respect, although quite a bit has already been said in advance about the sensory performance of animals, and further and more detailed information will be provided in the topics of main chapter II.

Plants – for many fellow citizens certainly an unknown entity, although each of us cares for and nurtures plants at home or in the garden. Stefano Mancuso and Alexandra Viola have – among other "intelligent" characteristics of plants – also intensively studied the reactions of plants to environmental stimuli.

It is astonishing, to say it right away, that plants seem to possess another 15 senses for their continued existence in addition to the five sensory services for humans mentioned before – thus four times as much as we humans possess. (Mancuso and Viola 2015, p. 49–81).

And fifteen more senses! (some of which are reproduced, the author).
[…] As far as the senses are concerned, they (the plants, the author) are evidently in no way inferior to us, and one might almost think that plants strongly

resemble us in pucto senses. But far from it: they are much more sensitive, and have at least fifteen "senses" more than we have. Some have evolved for obvious reasons. Plants, for example, can reliably and precisely *measure* the *moisture content in the soil* and can even locate distant sources of water.

They have a kind of hygrometer (moisture meter) [...] that tells them how water-rich a soil is and where they can find the water. (ibid., p. 78, emphasis by the author)

As a stationary plant life form, this sensory ability is a necessary condition for its continued existence. Other extraordinary abilities of plants are:

(their) [...] *sense of gravity and electromagnetic fields* – by which their growth is influenced – and (they) can *determine* the *content of numerous chemical substances in the air and soil.* (ibid., emphasis added).

The local distribution of the sensory sensors of plants can be found in the roots, in the leaves and also in the whole organism. Most plants react to pollutants from industrial production, such as lead, cadmium or chromium in the soil and move towards the greatest possible distance from the pollutants.

Noteworthy in the context is the statement reproduced by Mancuso and Viola (2015) in an illustration of several (between pages 80 and 81) by Francis Darwin, professor of plant physiology at Cambridge, UK and son of Charles Darwin:

At the opening of the annual conference of the British Association for the Advancement of Science, 1908 he (Francis Darwin, the author) bluntly expressed his opinion that plants have cognitive abilities.

Another *technical* achievement of plants is, for example, their handling of the highly toxic trichloroethylene (TCE), a carcinogenic solvent used in the plastics industry, which practically does not decompose and whose long-lasting persistence can last thousands of years and destroy the fertility of soils.

Plants [...] absorb it (TCE, the author) without any problems and turn it into chlorine gas, carbon dioxide and water. In other words, they dissolve it.

Because plants have the extraordinary ability to render harmless some pollutants that are highly dangerous to humans – although most of these are produced by humans themselves – various technologies are attempting to detoxify soils and waters by means of so-called "*phytoremediation*". The economic and technological potential of biological soil remediation seems enormous, but its application is still in its infancy. (ibid., p. 81, emphasis added).

The final sentence concerns not only the plant kingdom – the flora – per se, but the entire biodiversity of our earth (ibid., p. 81):

> As we allow more and more plant species to become extinct, we are likely to lose countless as yet unexplored opportunities to rid our planet of pollutants effectively, cheaply and without side effects.

This concluding sentence and the aforementioned achievements of nature speak a clear language that must put the high risk of the continued existence of all organisms into a new future-oriented focus: A *"technical-economic carry on like this"* must be successively and consistently turned on its head in order to take the path for a *"sustainable, system-oriented development progress"* that makes use – not only but increasingly – of the highly qualified technical achievements of nature over billions of years.

Otherwise, we will remain trapped in a "vicious circle" of short-term, economically driven goals and, as a direct result, progressive destruction of nature. Do we really want that?

References

Alberts, B. et al. (2017) Molekularbiologie der Zelle. Wiley VCH, Weinheim

Arya, D. (2015) Genetically Modified Foods: Benefits and Risks. Massachusetts Medical Society, MA, USA

Assheuer, T. (2010) Leben 2.0. Was passiert, wenn es der Bio-Industrie gelingt, den menschlichen Körper neu zu programmieren. Die Zeit, 2. Juni, Nr. 23

Baehr H. D.; Kabelac, S. (2016) Thermodynamik. Springer Vieweg, Berlin

Baltimore, D.(2001) Our genome unveiled. Nature 409, 814–816, 15 February

Bartsch, G. M. (2016) Klimawandel und Sicherheit in der Arktis. Springer, Wiesbaden

Bateson, G. (1985) Ökologie des Geistes. Anthropologische, psychologische, biologische und epistemologische Perspektiven. Original 1972, Suhrkamp, Berlin

Begon, M. E.; Harper, J. L.; Townsend, C. R. (1998) Ökologie. Spektrum Akademischer Verlag, Heidelberg, Berlin

Blasberg, M.; Evers, M.; Glüssing, J.; Hecking, C. (2019). Schneise der Vernichtung. Brasilien. Der neue Präsident Jair Bolsonaro will indigene Schutzzonen im Amazonasgebiet für Bergbau, Rinderzucht und Farmer öffnen. Für das Weltklima eine fatale Entscheidung. Der Spiegel, 3/2019, 12. Jan., S. 84–88

BMU (1997) Agenda 21, Original Dokument in deutscher Übersetzung. Bundesministerium für Umwelt, Naturschutz und Reaktorsicherheit, Berlin

BMWE (2019) Was ist Industrie 4.0? Informationsschrift des Bundesministeriums für Wirtschaft und Energie

BUND (2018) Insektensterben 2018: Ursachen, Quellen, Studien & Untersuchungen – Gift, Neonicotinoide, Dünger, Monokulturen, Ferneintrag & Globalisierung. BUND: Bund für Umwelt und Naturschutz, mit vielen ergänzenden Quellenhinweisen, 43 S.

Brooks, D. R.; Wiley, E. O. (1986) Evolution as Entropy. University of Chicago Press, Chicago, aus: Odum, E. P. (1991) Prinzipien der Ökologie. Spektrum der Wissenschaft, Heidelberg

Calatayud-Vernich, P.; Andreu V.; Calatayud, F.; Simo, E.; Pico, Y. (2019) Influence of pesticides residues on acute honey bee mortality episodes. 20th EGU General Assembly, EGU2018, Proceedings from the conference held 4–13 April, 2018 in Vienna, Austria, S. 16584

Campbell, N. A.; Reece, J. B.; Markl, J. (Hrsg.) (2006) Biologie, 6. Aufl., Pearson, München

Carroll, J. B. (2008) Evo-Devo and an Expanding Evolutionary Synthesis: A Genetic Theory of Morphological Evolution. Cell, 134, July, S. 25–36

Darwin, C. (1859) On the Origin of Species by Means of Natural Selection or the Preservation of Favored Races in the Struggle of Life. Murray, London

Darwin, C. (2018) Der Ursprung der Arten durch natürliche Selektion oder Die Erhaltung begünstigter Rassen im Existenzkampf. Komplette deutsche Neuübersetzung seit 100 Jahren. Klett-Cotta, Stuttgart

Deichmann, U. (2016) Epigenetics: The origins and evolution of a fashionable topic. Developmental Biology 416 (2016) 249–254

Dobzhansky, T. (1973) Nothing in Biology makes Sence except in the Light of Evolution. The American Teacher, Vol. 35, No 3, S. 125–129

Dobzhansky, T. (1968) On Cartesian and Darwinian Aspects of Biology. Graduate Journal, 8, S. 99–117

Dobzhansky, T. (1939) Genetics and the Origin of Species. Columbia University Press, New York. In Deutsch: Die genetische Grundlage der Artbildung. Fischer, Jena

Dörner, D. (1989) Die Logik des Misslingens. Rowohlt, Reinbek b. Hamburg

Ebeling, W. (1971) Chaos – Ordnung Information. Deutsch, Thun, Frankfurt a. M.

Eigen, M. (1971) Selforganization of Matter and the Evolution of Biological Macromolecules. Naturwissenschaften, 58. Jahrg., H. 10, Oktober, S. 465–523

Eisch—Angus, K. (2019) Absurde Angst – Narrationen der Sicherheitsgesellschaft. Springer VS, Wiesbaden

Ennis, C.; Pugh, O. (2018) Epigenetik. Ein Sachcomic. Tibia Press, Mühlheim a. d. Ruhr

Feinerman O., Pinkoviezky I., Gelblum A., Fonio E. & Gov N. S. (2018). The physics of cooperative transport in groups of ants. Nature Physics. 2018 Jul, 14 (7):683–693

Fischer, A. C.; Hanemann, W. M. (1985) Option Value and the Extinction of Species. Beerkley (California Agricultural Experiment Station), 35. S., according to source from Norton (1992), original source not available

Forfert N.; Troxler A.; Retschnig G.; Gauthier L.; Straub L.; Moritz RFA. et al. (2017) Neonicotinoid pesticides can reduce honeybee colony genetic diversity. PLoS ONE 12(10): e0186109. https://doi.org/10.1371/journal.pone.0186109

Forrester, J. W. (1971) Der teuflische Regelkreis. Kann die Menschheit überleben? DVA, Stuttgart

Graw, J. (2015) Genetik. 6. Aufl., Springer Spektrum, Berlin, Heidelberg

Guillén, R. (2017) Dienstleistungstiere – die Ausbeutung der Bienen. In. Le Monde diplomatique, Dezember, S. 1 und 22

Habekuß, F. (2019) Der Wert der Vielfalt. "Menschen, kümmert euch darum!". Interview mit Edward Osborne Wilson und Antje Boetius, Die Zeit, 9. Januar, Nr. 3, S. 33

Heil, R. et al., Hrsg. (2016) Epigenetik. Ethische, rechtliche und soziale Aspekte. Springer VS, Wiesbaden

Hickman, C. P. et al. (2008) Zoologie, 13. Aufl., Pearson, München

Hoekstra, H. E.; Coyne, J. A. (2007) The Locus of Evolution: EVO DEVO and the Genetics of Adaption. Evolution, May, S. 995–1016

International Human Genome Sequencing Consortium (2001) Nature 409, 860–921, 15 February

Jablonka E.; Lamb, M. J. (2008) The Epigenome in Evolution. Beyond the Modern Synthesis. VOGis Herald 12, 242–254

Jablonka, E.; Lamb, M. J. (1995) epigenetic Inheritance and Evolution. The Lamarckian Dimension. Oxford University Press, Cambridge

Jantsch, E. (1992) Die Selbstorganisation des Universums. Vom Urknall zum menschlichen Geist. 2. erweiterte Neuauflage., Erstaufl. 1979, Hanser, München, Wien

Jiankui, et al. (2018) Draft Ethical Principles for Therapeutic Assisted Reproductive Technologies. The CRISPR Journal. https://doi.org/10.1089/crispr.2018.0051

Karp, G. (2005) Molekulare Zellbiologie. Springer, Berlin, Heidelberg

Kegel, B. (2018) Epigenetik. Wie unsere Erfahrungen vererbt werden. DuMont, Köln

Kettlewell, H. B. D. (1955) Selection Experiments on Industrial Melanism in the Lepidoptera. Heredity, 9, 323–342.

Kopit, A. M.; Pitts-singer, T. L. (2018) Routes of Pesticide Exposure in Solitary, Cavity-Nesting Bees. Environmental Entomology, 47(3), 2018, 499–510

Krupke, C. H.; Hunt, G. J.; Eitzer, B. D.; Andino, G.; Given, K. (2012) Multiple Routes of Pesticide Exposure for Honey Bees Living Near Agricultural Fields. PLoS ONE 7(1): e29268. https://doi.org/10.1371/journal.pone.0029268

Krimsky, S. (2019) Ten ways in which He Jiankui violated ethics. Nature Biotechnology Volume 37 Number 1, S. 19–20, January

Kronberg, I. (2016) Industriemelanismus: Die dunkle Form des Birkenspanners entsteht durch ein springendes Gen. Biologie in unserer Zeit. Heft 5, 276–277

Küppers, E. W. U. (2018a) Die humanoide Herausforderung. Springer Vieweg, Wiesbaden

Küppers, E. W. U. (2018b) Das Ende der Nachsichtigkeit. Springer, Wiesbaden

Lucas, K. (2007) Thermodynamik. Springer, Heidelberg, New York

Majerus, M. E. N. (2009) Industrial Melanism in the Peppered Moth, Biston betularia: An Excellent Teaching Example of Darwinian Evolution in Action. Evo Edu Outreach, 2:63–74

Mancuso, S.; Viola, A. (2015) Die Intelligenz der Pflanzen. Kunstmann, München

Mayr, E. (2001) Das ist Evolution. Bertelsmann, München

Mayr, E. (1998) Das ist Biologie. Spektrum Akademischer Verlag, Heidelberg, Berlin

Mayr, E.: (1967) Artbegriff und Evolution. Parey, Hamburg, Berlin

Mayr, E. (1954) Chance of Genetic Environment and Evolution. In Huxley, J.; Hardy, A. C.; Ford, E. B. (Hrsg.) Evolution as a Process. S. 157–180, Allen&Unwin, London

Mayr, E. (2003) Das ist Evolution. C. Bertelsmann, München

McKibben (2019) Der schrumpfende Planet. Blätter für deutsche und internationale Politik, 1/2019, S. 49–56

Meadows, D. H. (2010) Die Grenzen des Denkens. Wie wir sie mit Systemen erkennen und überwinden können. Oekom, München

Meadows, D. H; Meadows D.L.; Randers, J.; Behrens III, W. W. (1972) The Limits to Growth. A Report for THE CLUB OF ROME'S Project on the Predicament of Mankind. A Potomac Associates Book, Universe Book, New York

Mendel, G. (1866) Versuche über Pflanzen-Hybriden. Verhandlungen des Naturforschenden Vereines in Brünn 4 (1866), S. 3–47. http://www.deutschestextarchiv.de/book/view/mendel_pflanzenhybriden_1866?p=14 Accessed 25.1.2019

Minelli, A. (2019) An Evo-Devo Perspective on Analogy in Biology. Philisophies, 4,5, S. 1–9

Miotto, O. et al. (2015) Genetic architecture of artemisinin-resistant Plasmodium falciparu. Nature Genetics 47, 226–234

Monod, J. (1975) Zufall und Notwendigkeit. Philosophische Fragen der modernen Biologie. dtv, München

Mousavi, S. et al. (2019) Physiological, epigenetic and genetic regulation in some olive cultivars under salt stress. Scientific reports, 9(1): 1093, https://doi.org/10.1038/s41598-018-37496-5

Müller, G. B. (2007) Evo-devo: extending the evolutionary synthesis. Nature Reviews I Genetics, Dec., S. 943–949

Müller, W. A.; Frings, S.; Möhrlen, F. (2015) Tier- und Humanphysiologie. Eine Einführung. 5. Auflage, Springer Spektrum, Berlin, Heidelberg

Newbold, T. et al., (2016) Has land use pushed terrestrial biodiversity beyond the planetary boundary? A global assessment. Science, Vol. 353, Issue 6296, S. 288–291

Nicolis, G.; Prigogine, I. (1977): Self-Organization in Nonequilibrium Systems. Wiley-Interscience, New York

Nordheim, A.; Knippers, R. (2018) Molekulare Genetik. Thieme, Stuttgart

Norton, B. G. (1992) Waren, Annehmlichkeiten und Moral. Die Grenzen der Quantifizierung bei der Bewertung biologischer Vielfalt. In: Wilson E. O. (Hrsg.)

Ende der biologischen Vielfalt. S. 222–228, Spektrum Akademischer Verlag, Heidelberg, Berlin, New York

Norton, B. G. (1987) Why Preserve Nature Variety? Princeton, N. Y. (Princeton University Press) 1987, 281 S.

Odum, E. P. (1991) Prinzipien der Ökologie. Spektrum der Wissenschaft, Heidelberg

Paull, J. (2015). GMOs and organic agriculture: Six lessons from Australia. Agriculture & Forestry, 61(1), 7–14.

Paull, J. (2018) Genetically Modified Organisms (GMOs) as Invasive Species. Journal of Environment Protection and Sustainable Development Vol. 4, No. 3, S. 31–37

Plattner, H.; Hentschel, J. (2017) Zellbiologie. Thieme, Stuttgart

Podbregar N.; Lohmann, D. (2013) Im Fokus: Genetik. Dem Bauplan des Lebens auf der Spur. Springer Spektrum, Berlin, Heidelberg

Prigogine, I. (1977) Time, Structure and Fluctuations. Nobel Lecture, Dec., 8, 1977, Copyright © The Nobel Foundation 1977

Prigogine, I. (1998) die Gesetze des Chaos. Insel TB, Frankfurt a. M., Leipzig

Prigogine, I.; Nicoles, G.; Babloyantz, A. (1972) Thermodynamics and Evolution. In: Physics Today, 25/11 (1972) S. 23–38 und 25/12 (1972) S. 138–141

Randall, A. (1992) Was sagen die Wirtschaftswissenschaftler über den Wert der biologischen Vielfalt? In: Wilson E. O. (Hrsg.) Ende der biologischen Vielfalt. S. 240–247, Spektrum Akademischer Verlag, Heidelberg, Berlin, New York

Rauchfuß, H. (2005) Chemische Evolution und der Ursprung des Lebens. Springer, Berlin, Heidelberg, New York

Reichholf, J. H. (2018) Vom Sein zum Werden. Nachwort in: Darwin, C. (2018) Der Ursprung der Arten durch natürliche Selektion oder Die Erhaltung begünstigter Rassen im Existenzkampf. S. 571–603, Klett-Cotta, Stuttgart

Remsmeier, A. (2016) Fahrt ins Blaue – Russlands Traum von der Eroberung der Arktis. Skript des Beitrages des Deutschlandfunks am 6. Februar

Renn, O. (2014) Das Risikoparadox. Warum wir und vor dem Falschen fürchten. Fischer, Frankfurt, a. M.

Renn, J.; Scherer, B. (2015) Das Anthropozän. Matthes & Seitz, Berlin

Riedl, R. (2000) Strukturen der Komplexität. Springer, Berlin, Heidelberg, New York

Sander, K. (1990) Von der Keimbahntheorie zur synergetischen Musterbildung. – Einhundert Jahre entwicklungsbiologische Ideengeschichte. 83, 133–177. Deutsche Zool. Ges., München

Sanger, T. J.; Rajakuma, R. (2018) How a growing organismal perspective is adding new depth to integrative studies of morphological evolution. Biological Reviews, Vol 94, Issue 1, S. 184–198

Schellnhuber, H. J. (2015) Selbstverbrennung. Die fatale Dreiecksbeziehung zwischen Klima, Mensch und Kohlenstoff. C. Bertelsmann, München

Schleip, W. (1929) Die Determination der Primitiventwicklung eine Zusammen fassende Darstellung der Ergebnisse über das Determinationsgeschehen in den ersten Entwicklungsstadien der Tiere. Akademische Verlagsgesellschaft, Leipzig

Schmidt, F. (1985) Grundlagen der kybernetischen Evolution – Eine neue Evolutionstheorie. Goecke & Evers, Krefeld

Schmidt, O. (2017) Genetik und Molekularbiologie. Springer, Berlin, Heidelberg

Schmidt, R. F.; Lang, F.; Thews, G. (Hrsg.) (2005) Physiologie des Menschen mit Pathophysiologie. 29.Auf., Springer Medizin, Heidelberg

Schrödinger, E. (1987) Was ist Leben? Piper, München

Schweers, J. (2018) Die Vernichtung der grünen Lunge. Blätter für deutsche und internationale Politik, 12/2018, S. 65–68

Sommer, R. J. (2009) The future of evo – devo: model systems and evolutionary theory. Nature Reviews Genetics 10, S. 416–422

Spork, P. (2016) Geerbte Angst. Bild der Wissenschaft, 2, 16–19

Spork, P. (2010) Der zweite Code. Epigenetik oder: wie wir unsere Erbgut steuern können. Rowohlt, Reinbek b. Hamburg

Straaß, V. (1990) Spielregeln der Natur. BLV, München, Wien, Zürich

Tipler P. A.; Mosca, G.; Wagner, J. (Hrsg.) (2015) Physik. Springer Spektrum, Heidelberg

Vaki, A.; Gagneux, P. (2017) On Human Nature. Ch. 9: How Different Are Humans and "Great Apes"? A Matrix of Comparative Anthropogeny. Biology, Psychology, Ethics, Politics, and Religion, Pages 151–160

van't Hof, A. E. et al. (2016) The industrial melanism mutation in British peppered moths is a transposable element. Nature, Vol. 534, 2. June, 102–105

van't Hof, A. E. et al. (2011) Industrial Melanism in British Peppered Moths Has a Singular and Recent Mutational Origin. Science, Vol 332, 20. May, 958–960

Vec, M.; Hütt, M.-T.; Freund, A. M. (2006) Selbstorganisation. Ein Denksystem für Natur und Gesellschaft. Böhlau, Köln

von Oppen, G.; Melchert, F. (2005) Physik für Ingenieure. Pearson, München

Waddington C. H. (1940) Organisers and genes. Cambridge University Press, Cambridge

Walton, O. C.; Stevens, M. (2018) Avian vision models and field experiments determine the survival value of peppered moth camouflage. Communication Biology. S. 1–7, (2018)1:118 https://doi.org/10.1038/s42003-018-0126-3|www.nature.com/commsbio

Watson, D. J.; Crick, F. H. C. (1953) Molecular Structure of Nucleic Acids: A Structure for Deoxyribose Nucleic Acid. Nature, Band 171, S. 737–738, 25. April

Weizsäcker, C. F. (1964) Die Tragweite der Wissenschaft. S. Hirzel, Stuttgart

WHO (2014) Food safety Frequently asked questions on genetically modified foods, May

Wieser, W. (1994) Die Evolution der Evolutionsbiologie: Von Darwin zur DNA. Spektrum der Wissenschaft, Heidelberg

Wilson, E. O. (1995) Der Wert der Vielfalt. Die Bedrohung des Artenreichtums und das Überleben des Menschen. Piper, München, Zürich

Wilson, E. O. (Hrsg.) (1992) Ende der biologischen Vielfalt? Der Verlust der Arten, Genen und Lebensräume und die Chancen für eine Umkehr. Spektrum Akademischer Verlag, Heidelberg, Berlin, New York.

4

Biodiversity of Excellence in Flora and Fauna: A Selection with Two Extra Examples from Nature Measurement and BIOGEONICS

Abstract In this chapter I would like to take you into the biodiversity of nature, into the immeasurable wealth of plants and animals, at least to give you a first impression of the "technical" achievements with which plants and animals are ahead of us and that it is worthwhile to take a closer look at their principles. Towering redwood trees easily transport nutrients and liquids to heights of over 100 m, whereas we have to use enormous additional energies for the same transport processes.

Plants are ingenious in combining varied material-lightness with extraordinary material-breaking strength; we are still running a long way behind this technology, and this becomes even greater when we include our material-waste problems.

Producing bright colours without dyes, letting surfaces clean themselves with water only, gluing without using adhesives or constructing climate-friendly living spaces without using additional energy are just a few of nature's numerous practices that make us jam. If, like woodpeckers, our heads were subjected to such stress at tens of hundreds of times the acceleration due to gravity, they would be pulverized.

The techniques of organisms are not only ingeniously adapted to life in nature, they have also developed sophisticated tools, processes and organizations over long periods of time. We would be rather ill-advised – with a view to our own problem solving – not only not to take note of these but also not to use them for ourselves.

E. W. U. Küppers, *Ingenious Principles Of Nature*,
https://doi.org/10.1007/978-3-658-38099-1_4

In this fourth chapter, a few excellent natural achievements in partly extreme habitats are described. We can only marvel in awe at their functional abilities. However, the fact that their highly interconnected habitats are increasingly being destroyed and threatened by us humans is the sad flip side, which is decisively controlled by our short-sighted thinking and actions.

The Intergovernmental Science-Policy Platform on Biodiversity and Ecosystem Services (IPBES) report on the state of biodiversity or species diversity on Earth, released on May 6, 2019, describes an accelerated horror scenario[1] of species extinction on our unique, life-creating and protective Earth. Of the world's known 1.7 million species, nearly 25% are currently endangered. Extrapolations from data on well-studied species, e.g. birds and mammals, suggest an estimated 8.1 million species in total, of which about one million are threatened with extinction. According to this, the majority of all species are still unknown.[2] Instead of researching them as a matter of priority and deriving benefits for mankind from them, as the indigenous peoples – who are also threatened with extinction – do, individual politicians and entrepreneurs, as social decision-makers about our future, persistently continue to turn the big wheel of economics – environmental destruction included! The most recent example at the time of this writing of high-level arrogance and ignorance about the environment and nature is demonstrated by Brazilian President Bolsonaro.[3] He is cutting 95% of the funding for the Ministry of Environment Brazil, at the same time the clearing of the Brazilian rainforest as the largest CO_2-climate gas reservoir and O_2-oxygen producer on earth is increasing.

Protecting and preserving biodiversity is the indisputable cardinal task of all of us, especially of those who claim to want to lead societies into a "better" future worth living. Species conservation is therefore not least a highly political task. It is in our own survival interest to strengthen the diversity of species on earth. Those who permanently act short-sightedly against evolution lose in the long run, burden their own descendants with the burdens and consequences of the destruction of nature and deprive them of a self-determining future.

[1] https://www.ipbes.net/news/ipbes-global-assessment-summary-policymakers-pdf (accessed May 07, 2019).

[2] https://www.zeit.de/wissen/umwelt/2019-05/artensterben-ralf-seppelt-biodiversitaet-artenschutz--oekesysteme-ipbes (accessed May 07, 2019).

[3] https://www.zeit.de/wissen/umwelt/2019-05/klimawandel-brasilien-jair-bolsonaro-budgetkuerzung--umweltministerium-klimaschutzmassnahmen-rodungen (accessed May 07, 2019).

In the following, eleven plants and eleven animals are presented with their special abilities, which can hardly represent the biodiverse wealth and the associated ingenious principles of nature, but nevertheless allow small insights into their perfect "technical" survival achievements. In the background, the complex energetic, material and communicative interconnectedness should always be taken into account, without which the excellent organismic performances would not exist.

This selection is supplemented by two extra examples on plant metrology (Sect. 4.12) and on *BIOGEONICS* (Sect. 4.24).

Background Information

Vicious Circle into the Unknown

It is regrettable that we do not recognise and use the evolutionary achievements of nature consistently enough for our own protection of life and survival as practically adapted sustainability solutions. After all, the way we treat nature also has profound political and economic implications, and vice versa! Breaking this devastating vicious circle is not least an imperative in Anthropocene times, because we ourselves are an indisputable part of nature, but allow ourselves to increasingly intervene in the interconnected habitats of biodiverse wealth for selfish reasons, without suspecting the enormous damage we are thereby doing to nature and ourselves. This is not least due to the increasing flood of plastics, which are spreading over all parts of the world and are destroying nature with its biodiverse wealth in a recognisable and demonstrable way.

Those who work against nature, whether through direct or indirect human, ill-considered short-sighted interventions, will almost certainly get the receipt, if not us, then our grandchildren and great-grandchildren.

It is *now* in our hands to ensure that nature is not further destroyed by human life. Let us ask frankly which social prohibitions are still worth following at all, or whether new holistic thinking and action can lead to more promising paths of development than have been pursued so far. Last but not least, it is also a cautionary call for a survival strategy that can provide partial survival assistance through the use of digital algorithms (Küppers 2018), no matter how sophisticated, but will never replace nature. Despite all the warnings and facts that indisputably prove the far-reaching environmental catastrophes that destructive action against nature leads to, quite a few actors in analogue and digital *"anticipatory obedience"* and with *"frenzied standstill"* seem willing to consistently and ruthlessly pursue their misguided path to a future with less and less nature. Evolutionary nature will survive in any case and will certainly find new development paths – with or without us species humans.

Let us now begin with the natural examples, where the question inevitably arises:

Where to Start – Where to Stop?
There is such a large number of examples from bionics that it is no longer possible to keep track of them all. The most extensive part is certainly in product bionics. Individual principles and design features of natural organisms are recognised and technically applied by means of bionics. In process bionics, flow processes play an important role, whereby methodical information technology approaches are also to be subsumed under this. In the field of organisational bionics, structural and procedural processes in social systems are analysed. Of course, these formally chosen boundaries do not apply in a strict sense, they are rather fluid in their transitions.

If one looks at nature and its organisms from a holistic perspective, as is actually the realistic approach in the search for efficient, effective and exemplary solutions for technical applications, it becomes clear that biology alone – by definition – does present the dominant basis for bionic solutions. However, no biological development is possible without interconnectedness with the inanimate environment. When, for example, mountain goats riskily climb up dam walls over 80° just to supply themselves with the minerals they need to survive on the rocks, this is one of countless examples of the interplay between biology and geology in the environment.

An extended bionics draws equally exemplary solution approaches for technical applications from this. The corresponding term is called BIOGEONICS (Küppers 2008). It is also recognizable that in animate and inanimate nature parallel developments lead to similar forms and movement patterns, depending on boundary conditions. One example from the group of process bionics concerns the Meander Effect® (Küppers 2007).

> *BIOGEONIK, a composition of the syllables – BIO – from biology, – GEO – from geology and – NIK – from technology, means to use efficient physical-chemical processes in the bio- and geosphere for bionic purposes.*

It should be pointed out once again that technology, as a synonym for all bionics applications in the social sphere, does not remain concentrated on purely technical or engineering fields of action.

》》 4 A Eleven Evolutionary Principles from flora with an Extra Chapter on Plant Metrics

4.1 Communication among Plants: Warning and Protection

Plants are stationary and therefore place-bound creatures. They do not possess the ability of animals to run away from imminent danger, or at least to attempt to run away. Their protective mechanisms for the preservation of the species are scents, volatile substances that on the one hand attract insects for pollination, on the other hand also try to ward off predators and ultimately even warn neighbouring plants of the same species of approaching enemies through their mode of chemical communication. On biocybernetic communication among trees, see Wohlleben (2015, pp. 16–20), among others, from which an example of biocybernetic plant communication is explained:

> But often it does not even have to be a special call for help, which is necessary for insect defense. The animal world basically registers the chemical messages of the trees and then knows that some kind of attack is taking place there and attacking species must be afoot. Those with an appetite for such small organisms are irresistibly drawn to them. But trees can also defend themselves. Oaks, for instance, conduct bitter and poisonous tannins (such as tannins, the author) into their bark and leaves. They either kill gnawing insects or at least change the taste enough to transform it from delicious salad to biting bile. Willows produce salicin, which acts similarly, as a defense. [...] In addition to chemical signal transmission in the multiply networked cybernetic regulatory circuits, trees also help each other in parallel by means of the more reliable electrical signal transmission via the roots, which connect organisms largely independently of the weather. If the alarm signals have been spread, then all around all oak trees – the same is true for other species – pump tannins through their transport channels into bark and leaves.

4.2 Sequoia Trees: Giants in the Plant Kingdom, Nothing Grows Higher

They truly deserve the name giants in the plant kingdom – the *sequoias*.

As imposingly unique as they tower into the sky and as strong as their trunks make those of other trees look "tiny", they are not the sole rulers of the forest. Like all organisms, they are integrated into the network of nature. Marc Carwardine (2008, p. 187) describes the vibrancy of life in the 50th to 60th floor of the Coastal Sequoia (Sequoiadendron sempervirens):

One redwood tree that was examined had formed 209 saplings. Mostly small, but the largest was still 40 meters high and measured 2.60 meters in diameter. Even humus had accumulated at the top of the branch forks. Ferns, shrubs and other tree species grew on it. Insects, earthworms, snails and even a considerable salamander population live there.

Küppers and Tributsch (2002, p. 125–126) describe the achievements of plant giants as follows:

Probably the most interesting bark of all trees are the Sequoia species that grow on the west coast of California and in the Sierra Nevada Mountains. These trees are already remarkable because they are among the tallest (*Sequoiadendron sempervirens*, Coastal Se*quoia*) with a record height of about 114 m – other heights go up to 120 meters – and among the thickest (*Sequoiadendron giganteum*, Mountain Sequoia) tree species in the world with a circumference of 30 m.

A record redwood like the Big Tree in California's Sierra Nevada weighs in at over 5000 tons. Mountain redwoods apparently live up to 3500 years. What could the packaging of these largest of all living things, the bark, contribute to their survivability? When the forest burns to a blaze and the fire front reaches a Sequoia tree, it cannot destroy it, at most damaging it somewhat as neighboring logs lean and burn all ablaze. Its bark is remarkably resistant to fire.

For this reason, sequoias even benefit from forest fires because they create clearings. When they fall over in old age or due to storms, they don't just rot away. Rather, they remain for two or three centuries, protected by the bark. If fallen sequoias get caught in a forest fire, the trunks burn out, but the soft bark survives. Native Pomo and Sinkyone Indians built houses out of sequoia bark and also made clothing from crushed bark material.

More modern applications mainly involve building materials and products that require long lifetimes. For example, there are many products made from sequoia bark that date back to the nineteenth century. The cinnamon-colored bark of the mountain sequoia reaches a thickness of about 25 centimeters in a tree that is several hundred years old. In some places, however, it can be up to 70 centimetres thick. It has a strikingly fibrous structure that is reminiscent of asbestos and achieves almost analogous properties. Like asbestos, it conducts heat very poorly. There is probably no material in the plant kingdom that conducts heat worse. If lightning strikes, the electricity can only be dissipated insufficiently. The tree then literally explodes to get rid of the energy. Sequoia bark is also largely free of pitch, which other types of bark often contain. When fire is held to the sequoia bark, the outermost loosely attached scales usually burn away. But the thicker parts of the bark do not keep up the fire. It takes a long time of lighting the place with blazing wood to scorch the bark.

The bark of the mountain sequoia, which can still be found in California at an altitude of 2500 meters, still has the property that it protects the trunk from great cold. It also effectively protects against insects and fungi that cannot destroy the bark material. It is believed that tannin in the bark plays the role of insect repellent. What can we learn from the evolutionary experience of sequoias? Again, we will come to this further below, in chapters in which we discuss the technical – bionic – use of biological models.

Figure 4.1 shows a mature sequoia (sequoiadendron giganteum) of imposing size, probably in the range of 100 m, from Sequoia National Park in California, USA. It is not only the comparison to the size of an adult that makes us marvel at this plant organism.

Fig. 4.1 *Sequoia (Sequoiadendron giganteum).* (Source photo: Alberto Carrasco Casado – Sequoia National Park, California (2011), probably a few thousand years old)

Fig. 4.2 *Sequoia (Sequoiadendron giganteum)*, probably planted in Bremen, Germany, about 140–150 years ago

In Fig. 4.2, a relatively young giant squoia can be seen, with a trunk circumference of 6.90 m, measured at 1.5 m ground level. This gives a mean diameter of about 2.2 m, and a calculated height of about 25.00 m. The tree was probably planted 140–150 years ago in Bremen, Germany. The trunk circumference of just under 7 m, compared to the trunk circumference of a fully grown sequoia (25–30 m), still indicates its relatively "youthful" age.

Finally, Fig. 4.3 shows the layered structure of the very light, air-permeable fire-retardant bark of the young sequoia from Fig. 4.3. To the right, 3 immature green sequoia cones can be seen in the vicinity of sequoia needles. The cones are about 5 cm in diameter and about 6 cm long. For comparison, a 5 Eurocent coin is enclosed, with a diameter of 2.12 cm.

If the tree height and the trunk diameter of the sequoias are already extremely impressive natural features, the transport of water and nutrients

Fig. 4.3 Bark of young giant sequoia *(Sequoiadendron giganteum)*, with associated immature green cones

from the roots to the top floor of the trees is considered a supreme discipline. At sea level, the mean atmospheric pressure, which can be symbolized by the weight of a column of air, is 1013.25 hPa (Hekto Pascal) or 0.1 bar. This corresponds almost exactly to a 10 m high water column and is the height at which we can transport water upwards under the given conditions without additional energy input. In contrast, the sequoia uses various physical methods to transport water and nutrients to imposing heights that are more than 11–12 times the 10 m water column that occurs under mean atmospheric pressure. Root pressure and transpiration suction of the leaves are decisively involved in water transport at Sequoia tree heights.

It?s a fantastic performance of nature, which technically could only be done by enormous input of energy. Whether its practical application can also achieve the sustainable efficiency of the natural solution may be doubted.

In a dangerous current event (August 2020), the massive wildfires of the California wildfires, which also devastated the Redwood National Park with its giant redwoods, deprived many people of their belongings. However, it was also confirmed that most of the old redwoods had withstood the fire (including Mendoza, M. (2020) Redwoods survive wildfire at California's oldest state park., Associated Press, 25.8.2020, https://abcnews.go.com/US/wireStory/redwoods-survive-wildfire-californias-oldest-state-park-72580906, Accessed 23.10.2020).

4.3   Lotus Plant Leaves: Prime Example and Expert in Surface Self-Cleaning

No known organism in the field of flora possesses leaf surfaces that boast such an efficient self-cleaning effect as do the leaves of the lotus flower *(Nelumbo nucifera)*. This effect of hydrophobicity or low wettability of a surface was first discovered in the 1970s by Wilhelm Barthlott (Barthlott and Ehler 1977). Küppers and Tributsch (2002, p. 110, updated by the author write in this regard:

> The lotus plant is remarkable in several respects. It has had a stable evolutionary history of 160 million years, which shows that its adaptations have stood the test of time. Not only is it beautiful, but all parts of the plant can be eaten – the stems as salad, the roots baked and the seeds like almonds. From Egypt to all of Central Asia to Japan, the lotus plant is considered sacred. And it is considered a symbol of purity. For growing out of the muddiest pools and in the dustiest of environments, it remains pure. The background for this purity, or rather the ability to purify itself, has been scientifically researched since the 1970s. To understand it, it is useful to recall some structural properties of plant surfaces. Plants are bounded by an outer skin, the epidermis. This in turn is covered by a continuous thin layer, the cuticle. It is not made of plant cells but of a polymer layer that is impregnated with wax. This wax, which is composed of various long-chain hydrocarbons, has the property of crystallizing out in the form of various three-dimensional structures that spread three-dimensional patterns over the plant surface in characteristic dimensions of a few micrometers. Visually, they are occasionally visible as wipeable, light-colored dust, for example on the surfaces of grapes or plums.
>
> The prerequisite for the self-cleaning process known as the "lotus effect" (Barthlott and Neinhuis 1997) is a surface that repels water well and cannot be wetted by it. The angle of contact that a drop of water makes with the plant surface should therefore be as large as possible. Ideally, the contact angle between the droplet and the surface should be 180°. Superhydrophobic leaf surfaces achieve contact angles of up to 160 or 170°, which means that only less than 2–3% of the water droplet surface is in contact with the micro- and nanostructured leaf surface. However, if a surface now has hydrophobic wax structures 5–50 μm wide and 5–20 μm high, wetting interestingly decreases additionally (1 μm is equal to 10^{-6} m). Water beads off even more easily, which is what the lotus plant and many other biological structures have achieved with their microscopic wax structures. Self-cleaning now works in such a way that the water droplets rolling away move over the dirt particles lying around loosely, wetting them in the process. The dirt particles are drawn into the water and removed from the surface. Plants protect themselves in this way not only against ordinary

dust and fine sand, which could impair the function of their surfaces in respiration, light capture or water evaporation, but also against bacteria, viruses or fungal spores. They are simply transported away by the rolling water droplets.

Figure 4.4 shows the pearl effect on a lotus leaf and leaves of the lady's mantle plant *(Alchemilla vulgaris),* which, like the nasturtium, cabbage leaf and other plants, have hydrophobic, self-cleaning leaf surfaces, but whose efficiency is far behind that of the lotus leaf.

In Fig. 4.5 we dive into the depths of the micro- and nanostructure of green lotus leaves. This is not visible to the naked eye, but their physical effects of dirt-carrying rolling drops of water on the leaves are.

Fig. 4.4 Hydrophobic surfaces of lotus plant leaves and lady's mantle leaves with visible pearl effect of water droplets. The spherical water droplet at the lowest point of the lotus leaf will – like the smaller water droplets – roll off towards the edge of the leaf at the slightest instability of the leaf. This process is already in progress on the inclined leaves of the lady's mantle

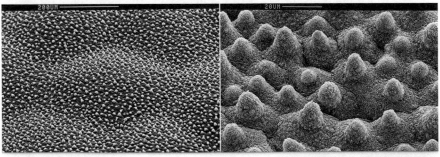

beide SEM-Fotos © 2013 W. Barthlott, lotus-salvinia.de

Fig. 4.5 Structured surface of a lotus leaf, photographed with a scanning electron microscope (SEM), magnification bars 200 μm, left and 20 μm, right. (http://www.lotus-salvinia.de/index.php/de-de/images256. Courtesy of W. Barthlott. Accessed 16 April 2019)

4.4   Giant Water Lily Victoria Amazonica: Ingenious Lightweight Supporting Structure on Water

The Victoria Amazonica, genus of the giant water lilies *(Victoria)*, family of the water lily family *(Nymphaeaceae)*, occurring in tropical regions and in temperate climates, is in many respects an imposing phenomenon in the plant kingdom. It has been handed down that in the nineteenth century the English botanist John Lindley named the genus Victoria with the species Victoria regia in honour of the just crowned English Queen Victoria (1819–1901).[4]

R. Ebbinghaus and B. Mory from the Botanical Garden Botanical Museum – BGBM – in Berlin describe the plant in their own way[5]:

The Victoria is certainly one of the most impressive aquatic plants [...]. It was discovered in 1801 by the botanist Haenke on a research trip to the Amazon region. In 1851 it bloomed in Berlin for the first time in Borsig's garden in Moabit "in an elegant glass temple made of delicate iron rods". [...]. Everything about the Victoria is extraordinary: its huge floating leaves, which in their tropical home reach a diameter of 3 m (here 2 m) and can carry up to 50 kg. The underside of the leaf is well equipped for this. Between the strong leaf veins, the underside of the leaf and the water surface, large air bubbles form on which the giant leaf floats as if on cushions of air. Additional protection against animal predators (fish and manatees) is provided by strong spines. The leaves have an approximately 6 cm high edge with two opposite incisions. This prevents the edge from tearing in the event of violent movements, the leaves from being pushed over each other and the water from flowing off easily in rain showers. In addition, the leaf tissue is criss-crossed in many places by vertical tubes (stomatodes), which drain off the water to the underside. The flower only shows its splendour on two days. In the early evening it blooms white, gives off a pineapple smell and is 10 °C warmer inside than its surroundings. This warmth, the flower colour and the scent attract beetles. In the morning, the flower closes and becomes a forced harborage. However, a starchy spongy tissue on the carpels provides plenty of food for the guests. It is not until the evening of the second day that the flower opens again, this time bright pink and scentless. By now the beetles are loaded with pollen, which they carry to the next white fragrant flower. The pollinated flower sinks to the bottom of the water. But life goes on, there the seeds develop. The ripe seeds are processed by the Indians into flour, from which pastries are made because of its good taste.

---

[4] https://de.wikipedia.org/wiki/John_Lindley_(botanist) (accessed Apr 16, 2019).
[5] http://www.victoria-adventure.org/victoria/victoria_history.html (accessed Apr 16, 2019).

Figure 4.6 shows a pond in Mauritius full of water lily leaves, occasionally with pink, scentless flowers in the evening (see text before). The two slit-like openings for the drainage of water from the inner part of the leaves can be clearly seen on the upstanding edge, which is about 10 cm in diameter. If the round leaf shells are turned upside down, a completely different picture emerges, as can be seen in Fig. 4.7. Here the water lily leaf shows itself to be a true master of natural constructive lightweight construction, ensuring a relatively high load-bearing capacity with optimum use of materials. Air chambers between the ribs also provide additional buoyancy and the spines on the rib skeletal struts efficiently ward off predators. All in all, a perfectly adapted natural solution – without ifs and buts.

Fig. 4.6 "Water lilies" by 2500, https://piqs.de/fotos/search/Victoria+amazonia/37610. html, is licensed under a Creative Commons Attribution 2.0 Germany License. (Accessed 16 April 2019)

Fig. 4.7 "leaf back side", left, "leaf back side detail", right, by Suguri F, https://commons.wikimedia.org/wiki/Victoria_amazonica#/media/File:Victoria_amazonica_back_side.jpg, and https://commons.wikimedia.org/wiki/Victoria_amazonica#/media/File:Victoria_amazonica_back_side2.jpg both licensed under a Creative Commons license: https://creativecommons.org/licenses/by-sa/3.0/ (Accessed 16 April 2019)

4.5   Bulrush: Large Length, Small Cross-Section and Yet Ingeniously Stable Construction

We all enjoy walks through nature unpolluted by humans, whether through forests or around lakes, with the towering colonies of reeds or cattails (*Typha colonies*) growing there, which the wind moves in all directions. On closer inspection, we are often amazed at the stability of the bending leaves and swaying cattails, because their stalk diameters of a few millimetres to centimetres seem relatively fragile compared to their length. But far from it.

What the water lily Victoria Amazonica described above achieves in the horizontal direction, applies to bulrushes and other long-stemmed plants in the vertical direction. For natural constructions there are no delimited building techniques, as engineers plan and execute them for concrete purposes. Nature combines what is expedient for the respective organism and ensures its survival (Patzelt 1972, p. 118):

> Trusses merge into shells, shells stiffen by internal pressure, tensile skins store solids and even solidify them. With all this, however, every plant and animal organism has a remarkable unity between smallest and largest elements.

Patzelt (ibid.) goes on to write about the "technical" achievements of the long-stemmed bulrush:

> The cattail stem is built up from curved compound shells. The sum of the leaves, laid close together, forms the stem. Each leaf consists of an upper and lower shell, which are connected by webs. The webs are partially stiffened with slightly spongy tissue. There are also folded webs. This folding offers the best resistance to forces in the direction of greatest stress. The most interesting thing about such a stalk, however, is that it is additionally stabilized by cell pressure.

The ingenious construction technique of the bulrush stem and its blades is unparalleled – by bioanalogical comparison – in technology. The twisted high protruding leaves cleverly give the wind coming from changing directions a minimal resistance surface. The danger of buckling under high bending loads is further reduced by the fact that the leaves themselves have the typical sandwich structure, which Patzelt refers to in advance as a lower and upper shell with intervening tissue, as can be seen in Fig. 4.8. Here, nature quite obviously developed optimal "carrier patterns" for any organism that could gain life advantages through a sandwich structure. The full sophistication and "natural intelligence" of this construction, however, can be seen in the cross-section at the lower end of the bulrush in Fig. 4.9. Although the circular

Fig. 4.8 Highly erect, torsionally stiff cattail leaves, left; cross-section through a cattail leaf with clearly visible stability structure, right

Fig. 4.9 Cross-section through a cattail stem with applied leaves. The ingenious composite construction of leaves and cob core gives the plant an optimal dynamic, because structure-growing resistance to breakage

cross-section of a vertically erected mast- or tower-like structure has the same load in all bending directions and would therefore be optimal in terms of the material-processing surface-to-volume ratio, the bulrush has chosen a different solution. Rightly so: because the organism is – in many cases in contrast to technical masts or towers – designed to grow. This is made clear in Fig. 4.9

by the right- and left-hand crescent-shaped cross-sections of the regrowing leaves. This greatly increases the mechanical stiffness of the elliptical cross-section of the piston core in these two growth directions against bending and fracture. The critical direction against bending is therefore approximately perpendicular to the growth direction of the sickle-shaped leaves. And it is in precisely this direction – in the white core of the trunk – that the green principal stress axes lie!

4.6 Onions: Perfectly Thermoregulated Universal Transparent Shell Composite System

We all eat the kitchen onion on many occasions and ways of preparation. However, we seldom or never think about the secrets that this plant has held for millions of years, which we use today – sometimes without knowing it – in many different ways.

The onion *(Allium cepa)* is – as one would expect – another prime example of a versatile biological packaging (see Fig. 4.10). Küppers and Tributsch (2002, p. 144–145) write in this regard:

> They not only appear as edible bulbs, but are also survival and reproduction facilities for countless plant species, among which flowers are the best-known representatives. They include lilies, tulips and hyacinths, as well as amaryllis, daffodils and snowdrops.
>
> The strategy of the onion is immensely mature in terms of packaging. Its round shape combines maximum content with minimum surface area, through which it must protect itself against the outside world. Striking features of this are multiple envelopes, which are translucent thin films that are easily peeled away from the onion (Fig. 4.10). Since they enclose air spaces and are very dry, it stands to reason that they are poor conductors of heat, as are the multiple paper wrappers of some round wasp nests. Onions thus compensate for environmental fluctuations in heat and cold, helped by their remarkable heat capacity, which results from their high water content inside. Certainly, during their evolution, onions have also learned to defend themselves chemically against bacterial attack. If onions are left cut open for a long time, fungal attack is rarely observed. As a rule, the plant only dries out completely. Because of its resistance to microorganisms, the onion can remain dormant under the ground for a long time. It is programmed so that it then begins to develop dynamic activity in response to a signal from surrounding temperature and moisture. It develops water-absorbing roots, energy-absorbing leaves and, for pollination purposes, a

Fig. 4.10 Onion. At the top, the whole onion can be seen with the protruding, wafer-thin, submillimeter-thick outermost skin. At the bottom left, the longitudinal section shows the crescent-shaped layered structure of the individual skin layers. At the bottom right, the cross-section provides an insight into the concentrically structured skin layers (white rings), which are connected to each other by cross-struts (also white) and give the onion a stable framework. Between them are air-filled chambers for climate regulation

stalk with a flower. Within its program of self-preservation, it also produces a replacement module that supplies it with energy and chemicals. Once this work is done, it decays to be recycled by the soil. The new bulb, however, encapsulates itself and waits, well packed, for the signals from the environment to initiate the new life cycle.

As with other ingenious biological masterpieces, we will discuss the broad relevance to packaging technology, with a view to our human packaging technology and the associated packaging materials, later (Fig. 4.10).

4.7 Fruit of the Coconut Palm: Nature's Incomparable all-Rounder

What ripens into a tasty fruit and quenches thirst with its sweet milk; what falls with a weight of more than one kilogram from a height of 10–15 m and lands without breaking; what drifts over a thousand kilometres through salty seawater without sinking or otherwise being damaged, in order – washed ashore again – to bring new life; what is finally used – almost completely – by humans for their own needs: the coconut of the coconut palm *(Cocos nucifera).*

It is a true multi-talent and its packaging helps it to do so. In the course of their optimised development, the packaging and the fruit have entered into a protective alliance that is second to none. Küppers and Tributsch (2002, p. 105–106), write in this regard:

When the young shoot of a coconut on a remote tropical beach breaks through the frayed, worn outer fibrous shell, the coconut may have already transported the young life thousands of miles and protected it from many deadly dangers. There was first the fall from the tall palm, perhaps in a storm. Then followed the long wait until a high tide washed the nut into the sea. Then came the endless trek across the sea, the constant attacks by busy small sea creatures and occasionally by large ones. This was followed by perhaps days of bombardment in the surf, for example against a lava coast with razor sharp rocks until a high wave finally deposited the precious package on a sandy beach. The technical performance of coconut packing is unparalleled. The coconut must not split open as a result of impact, and the lifeblood, the coconut milk, must not leak or mix with the salty seawater.

What sets the coconut apart are two important mechanical developments in particular. One is the shock-absorbing coconut fibre layer of cellulose, which brings protection against cracking. Measurements by the author (R&D project MANGO from 2004)[6] have shown that the spherical shell of the coconut can withstand a static pressure load of more than 10,000 Newton, which is about 1000 kg, without damage.

The other is a highly elastic, complex intermediate layer beneath the coconut shell, which compensates for injuries and thus prevents the loss of liquid. The shell of the actual coconut is pressure-resistant. But the coconut also has air-filled cavities in the pericarp that give it a high buoyancy in water. It is a veritable permanent preserve, using a variety of strategies to protect its cargo.

The coconut strategy already has parallels in technology, for example in glass bottles with a plastic coating. If the glass breaks, the liquid does not leak out. Could such a technical copy compete with the original biological model? Certainly not. If you think of the heat in the sand of tropical beaches a few meters away from the waves, you immediately recognize a problem. The coconut solves it by allowing the coconut fibre shell to absorb moisture, which evaporates slowly in the heat of the day, generating evaporative cooling. To achieve this effect, a bionic copy would have to become more complicated in many ways than just a plastic-protected wrapper.

Figure 4.11 shows two views of the coconut, which botanically is not a "nut" but a "drupe". On the left is the complete drupe with outer protective layer,

---

[6] R&D project funded by the Federal Ministry of Education and Research BMBF_PTJ FKZ 0311980.

© 2016 Dr.-Ing. E. W. Udo Küppers

Fig. 4.11 Coconut. On the upper left is a complete coconut, below which is a magnified surface section of the exocarp, clearly showing the resistance-reducing grooved structure for transport through water. On the right, the actual spherical protective packaging of life is evident as the inner hard pericarp, endocarp, with the flower side above and the stem side below

the exocarp, and on the right the flower and stem side of the inner hard protective layer, the endocarp, which has formed around the coconut fruit with the sweet-tasting fruit milk. This actual fruit packaging is only a few millimeters thick and consists of a break-resistant shell. Around it lies a fibrous fabric several centimeters thick with extraordinary damping properties against external pressure. The packaging composite is closed off by a smooth, furrowed outer layer, the aforementioned exocarp (Fig. 4.11, left). While the inner spherical fruit packing gives the growing life optimal space with the lowest material consumption, the outer shape of the packing seems to be characterized, among other things, by the fact that it assumes a more or less directionally and flow-oriented favorable shape and surface for transport through water. In addition to an optimal geometry of the packing, the coconut easily tolerates tropical high temperatures and ultraviolet-rich solar radiation. The water impermeability of the layered packaging (hydrophobic property) is an effective guarantee of life protection. The wrapped coconut, like many other wrapped organisms of nature, is unique and yet universal in terms of functionality.

As part of the BMBF research project "MANGO" (Küppers and Heyser 2004/2005), coconut shells were also subjected to a compression load test, in addition to other shells and casings of nuts. The fracture of the entire coconut shell only occurred at an axial compressive force of approximately 12,000 N!

Fig. 4.12 Section through a coconut, showing the outer protective layer adapted to water flow, the exocarp; a highly damping elastic intermediate layer, the mesocarp; the inner hard and spherical layer, the endocarp; and the fruit in the center of the coconut. (© Foto Tributsch, from Küppers and Tributsch (2002, p. 98)

(1 N = 1 kg m/s^2) and was by far the absolute front-runner among the fracture load values of the other nut shell pressure measurements (average fracture loads: peanut shell at approx. 10 N, hazelnut shell at approx. 500 N, pecan nut shell at approx. 650 N, Brazil nut shell at approx. 800 N and macadamia nut shell at approx. 2500 N). Details can be found in Sect. 7.13 Shells and Husks.

Figure 4.12 shows a coconut cut open, with all the composite layers visible, each perfectly adapted to its specific task of protecting life.

This incomparable natural talent contains multifunctional principles, which are summarized in Table 4.1.

4.8   Pitcher Plants and Tube Plants: The Most Slippery Traps in the Plant Kingdom

Diversity centres of today's pitcher plants *(Nepenthes)* are located in the mountainous regions of the Malayan islands of Sumatra, Borneo, New Guinea and the Philippines with about 50 species (Wittig and Nikisch 2014, p. 29).

In our temperate zone we find the plants mainly in tropical temperate aquariums of zoological gardens, but also offered for sale in nurseries.

The trapping mechanism of pitcher plant traps has evolved – like other plant insect traps – from an adaptation of their leaves for the purpose of

Table 4.1 Multifunctional optimization system coconut

Technical services of the coconut	Optimization parameters of the coconut	Boundary conditions of the coconut
Growth	Geometry variable	Temperature
Damping capability	Material composites	Humidity
Breaking strength	Material structures	Palm Heights
Material efficiency	Material quantity and material number	Feeders
Energy efficiency	Type of production	Parasites
Volume efficiency	Shape variable	Symbioses
Buoyancy	Density, aerodynamic drag shape	Water repellency
Hydrophobic property	Hydrophobic additives	Tropical radiation
Radiation protection	Surface structure	Residence time in saline seawater
Gas permeability	Membrane structure	
Long term protection of the contents	Structure complex	
	Interaction variable of envelopes	
	and content life	

survival. In a description of the Botanical Garden of the University of Bonn[7] on pitcher plants *(Nephentes)*, which belong to the carnivorous – carnivorous – plants, it says:

> Currently, about 120 species of *Nepenthes* are scientifically described and new species are still following. The old-world genus *Nepenthes* is the only genus of the family Nepenthaceae and has its center of diversity in SE Asia, especially on the islands of Borneo and Sumatra. Overall, the range extends from Madagascar to New Caledonia and from China to northern Australia.
>
> The leaves of the genus *Nepenthes* are composed of three sections:
>
> - a leaf-shaped part, which probably emerged from the base of the leaf,
> - a tendril, probably the petiole, that helps plants climb by entwining the surrounding vegetation,
> - and the leaf lamina converted to a pitcher trap.
>
> Each jug is in turn divided into three sections:
> The *attraction zone* includes the lid of the pitcher and the toothed rim of the pitcher, the peristome. Both are often brightly coloured and have nectar glands through which potential prey are attracted.
>
> The peristome (zoologically for environment of the "mouth", botanically in the figurative sense, the author), which becomes slippery with wetness, as well

---

[7] http://www.botgart.uni-bonn.de/o_samm/karni/nepenthes.php (accessed Apr 26, 2019).

as the upper pitcher-half, whose steep walls usually are covered by a slippery wax-layer, also belongs to the *Gleitzone*.

The *digestive zone* comprises the lower, liquid-filled half of the pitcher. In this area the pitcher walls are covered with glands, which on the one hand release digestive enzymes into the pitcher fluid and on the other hand absorb nutrients from the digested prey from the digestive fluid [...].

The *pitcher traps* are passive traps. On young pitchers, the lid is initially closed, but once it opens, it no longer performs active trapping movements. The digestive fluid inside the pitchers not only poses a lethal threat to some animals, but also provides a habitat for numerous aquatic animals. Thus, each *Nepenthes* pitcher becomes a small ecosystem.

Certain pitcher plants such as *Nepenthes bicalcarata*, native to Borneo, *have entered into a special symbiotic relationship with ants of the species Camponotus schmitzi. In particular, the ants consume mosquitoes and fly larvae as N_2-nutrient thieves of the plant*, with their own digestive remains returning valuable nitrogen to the plant,[8] see Bonhomme et al. (2011).

The pitcher plant as an insect-attracting organism for its own food intake and at the same time as a life-sustaining ecosystem for water-loving animals! An ideal interconnected system, as only nature could have devised, to counteract any waste of resources from the outset (Fig. 4.13).

Just as ingenious as the leaf surfaces of the lotus flower, the extremely smooth, steep gliding zone walls of the pitcher plant are not mirror-smooth and perfectly flat, but optimally structured for their purposes in the submillimetre range.

This repeatedly demonstrates the sophistication of nature, which is often diametrically opposed to our human notions of organismic functions and surfaces (Fig. 4.14).

In addition, a large number of other carnivorous species with sophisticated adaptive trapping mechanisms exist in the plant kingdom. In addition to gliding traps, these also include hinged traps, glue traps, suction traps and fish traps. It must be acknowledged without envy that the micro- and nanostructured manufacturing capabilities in flora, under the given functional circumstances and a sustainable loss-free material management, are far ahead of our comparable technology developments!

---

[8] https://www.spektrum.de/news/die-kleinen-helfer-der-kannenpflanze/1194978   (accessed   Apr 26, 2019).

beide Fotos © 2014 W. Barthlott, lotus-salvinia.de

Fig. 4.13 Pitcher plant *(Nepenthes madagascar)* left, longitudinal section through pitcher plant *(Nepenthes madagascar)* with organic food right. (http://www.lotus-salvinia.de/index.php/de-de/images256. Courtesy of W. Barthlott. Accessed 26 Apr. 2019)

4.9 Leaves of Rhubarb: Masters of the Folding Technique – Order Structures from Chaos and a Short Complementary Insight into the Spectrum of Natural Folding

A particularly unusual packaging phenomenon of nature can be discovered in a well-known leafy vegetable: rhubarb *(Rheum* rhabarbarum*).* From a spherical package in the soil, as if from nowhere and yet according to a sophisticated timetable, small leaf tips spring up whose surfaces look like furrowed and craggy mountains from ten thousand meters above sea level. The structural similarities between large and small things in nature are always startling. Could it be that the folds of a rhubarb leaf and those of a mountain range can be traced back to the same natural principle?

In seasonal growth, the plant's small unsightly leaf develops over several growth stages into a stately fan supported by just one sturdy stalk (Fig. 4.15 from top left to bottom right). The cross-section of this stalk (resembling the crescent moon) is again optimally matched to the bending stresses acting on

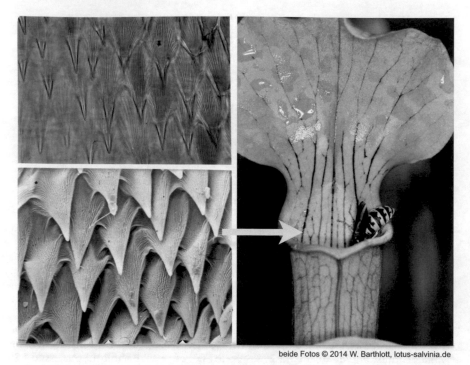

beide Fotos © 2014 W. Barthlott, lotus-salvinia.de

Fig. 4.14 Tubular plant *(Saracenia spec.)* Inner structured glide zone of tubular plant with SEM image on left, *Saracenia spec. on* right. (http://www.lotus-salvinia.de/index. php/de-de/images256. Courtesy of W. Barthlott. Accessed 26 April 2019)

the adult rhubarb leaf, which receives light for photosynthesis from above and shade from below. When unfolding, the leaves are stretched in one direction by the turgor pressure (pressure of the cell sap on the cell wall), causing the entire surface of the leaf to unfold.

In Fig. 4.15, top right and left, the complex mountainous structure of the rhubarb leaf can be seen particularly well, with its "mountain ridges and valley bottoms" merging into one another. The particular folding technique of the leaf is shown by the edges of a fold involving three "mountain edges" and one "valley edge". The enlargement in the middle of Fig. 4.15 highlights this folding specificity.

On closer inspection, the aesthetics of the *irregular regularity of* the characteristic folding geometry also reveal themselves to the viewer. At the same time, it is an example of how nature's own complex environment gives rise to relatively stable order structures that contribute to survival. The sophistication of nature cannot be overlooked in this example either.

> **Background Information**
>
> *Small experiment: Amazing order structures from chaotic paper crumpling.*
>   Take a smooth, wrinkle-free sheet in both hands and crumple it into a chaotic spherical shape. Then carefully pull it apart and discover regular irregularities of folds, including those described earlier with three mountain and one valley fold edge.

Fig. 4.15 Different developmental stages of folds on rhubarb leaves: top left shortly after emergence from the spherical bulbous sheath; top right in a middle extension with clearly recognizable "mountain structures"; bottom left extended rhubarb leaf as the plant's large-area "solar sail" for photosynthesis; bottom right a fully grown rhubarb bush. The circular section in the middle clearly highlights the typical "diamond folding" of the leaves – three "mountain ridges" in red, with a "valley bottom", in blue

- *Universal technique folds*
   The technique of folding in nature – like many other natural phenomena – is universal. To fully appreciate it alone is beyond the scope of this book. He who folds folds, interlaces, folds together, folds together, interlaces, bends over, folds over, interlaces, crimps, balls up, crumples or crinkles. Once folded, it can also be unfolded or pulled apart again. Folding leads to static forms, but also to dynamic processes. In almost all areas of life and on

all scales, living nature possesses sophisticated folding techniques that help organisms to ensure their survival. Folding is therefore one of several universal principles for shape or form change in nature.

For this reason, we find techniques of folding in the area of the smallest building blocks of life (proteins), in the macroscopic area in plant and animal organisms (e.g. leaves of trees, flowers, pressure-resistant patent fold construction in honeycombs of social insects as well as wing constructions in insects and mammals). Even in much larger forms of inanimate nature, in the structures of mountain massifs, specific folds can be recognized. Here only the keyword "folded mountains" may be mentioned.

Folded constructions are usually also stiffening constructions, which give a structure a higher resistance to compressive load and deflection. On the other hand, the pleated structure in dragonflies, for example, also leads to an aerodynamic advantage. But more on this later.

Hardly any surface of natural materials is – colloquially pronounced in the technical sense – smooth. Whether bird or insect wings, tree bark, cones, leaves, dolphin or whale skin, they all have more or less folds that curve straight or curved outwards, i.e. convex, or bumps that dent inwards, i.e. concave. Protagonists and antagonists, external folds and internal folds in interaction, is one of the secrets of biological, stable fold structures.

- *Submicrometer folds in the building blocks of life*

Even at the smallest dimensions of life, below one nanometer (1 nm corresponds to 10^{-9m}), spatial folding, in this case of amino acid sequences of deoxyribonucleic acid (DNA), helps to create, for example, the densest packing in a limited volume. The shape of our DNA molecule has the form of a spiral winding of two strands called a double helix, twisting around each other in a helical fashion. Both strands are held together by so-called base pairs. This shape can certainly be discussed as a model for a spiral, fold-like folding together or pulling apart, i.e. a kind of biological packaging by spiral arrangement of "packaged goods" (base pairs) in rotationally symmetrical "packages".

Another type of universal biological folding for the formation of ordered *scaffold* formations of proteins is the so-called parallel or antiparallel pleated sheet or β-structure. This involves more or less elongated peptide chains (strings of amino acids) connected by hydrogen bonds, in which two adjacent chains are either parallel or antiparallel. The *dynamic folding of* the interlocking proteins actin and myosin, the main components of muscle fibres and voluntary skeletal muscles respectively, is the cause of what makes us move. Here, folding is linked to repetitive motion. Folding is therefore part of the recurring cycle of stretching and contracting that is triggered by a variety of physical or chemical influences.

Of the multitude of macroscopic, biological materials with a folding technique in the submicrometer range, only wool fibers, as a representative of folding with a helical structure, and spider silk, as an example of folding sheets with parallel strands, should be mentioned here.

• *Wrinkles in the micrometer and macrometer range of life.*

In many cases, folding serves to stow away parts of an organism that are not needed at the moment in a space-saving way, i.e. also to protect them. One insect species even bears the term fold in its name, which makes this production technique even more obvious. This refers to the solitary and state-forming wrinkled wasps *(Vespidae),* which include the pill wasps, the paper wasps, the sand wasps and others. They get their name from the fact that they fold their wings lengthwise when at rest, which is not exactly common in insects.

Wafer-thin dragonfly wings would not be able to withstand the aerodynamic forces acting on the wings, let alone support the dragonfly's own body weight, without flexure-enhancing zigzag folding profiles. To ensure flight capability (with acrobatic, technically unattainable flight maneuvers) with minimal wing weight and relatively high bending stiffness of the wings could apparently only succeed evolutionarily by also incorporating two folds near the leading edge of the wing. This is a trick which – if applied correctly – could also prove to be energetically advantageous in the field of packaging transport, e.g. in the case of resistance- and energy-devouring, flexible tarpaulins of transport vehicles.

Beetles, peacocks, grasshoppers and many others are "animalistic" folding experts with extensive experience and always adapted optimization strategies to their lives. Flower petals, palm leaves, nutshells, tree folds and horsetail embody – with the highlighted example of the rhubarb leaf – ingenious plant folding techniques (see Küppers and Tributsch (2002, p. 75–76, 84, and others).

4.10   Window Plants: When It Gets Too Hot, ... Ingenious Survival Strategy in the Desert and Other Latitudes

Plants growing in dry – arid – climates such as those in southwestern Africa and Namibia use the same survival techniques as animals living there. Good water management ensures survival and many plants living there have developed sophisticated water storage techniques, whether in the roots, trunk and branches, leaves or special organs.

Other plants, such as the caddis tree *(Aloe dichotoma)*, reflect the sun's hot rays through bright bark. Still other plants radically reduce their surface exposed to the sun's rays. Among them are the so-called "living stones" *(Lithops) and* the window plants *(Fenestraria)*. Both are succulent, i.e. sap-rich plants of the genus Mittagsblumengewächse (Fig. 4.16). The entirety of the plant is underground (Fig. 4.17), except for one or two specialized leaves that flower at ground level. The succulent leaves are filled with water-retaining transparent tissue, which directs light to the rest of the plant, just like an open window to the sky. This is a masterpiece of survival by protecting against desiccation. Facing the sun, only the widened, flat ends of the leaves show, containing no chlorophyll, which is involved in photosynthesis and makes the leaves look green. It is only when the sun's attenuated rays penetrate into the deeper-lying area of the plant, where parenchyma cells, the thin-walled basic tissue of the plant, initiate the assimilation process – absorption of nutrients and conversion into endogenous substances – that photosynthesis becomes effective.

Lithops spec.   Fenestraria aurantiaca (South Africa)

Photo left 2014, photo right 2015 © W. Barthiotl, lotus.salvinia.de

Fig. 4.16 Lithops spec. and Fenestraria aurantiaca (Aizoaceae) (South Africa). (http://www.lotus-salvinia.de/index.php/de-de/images256. Courtesy of W. Barthlott. Accessed 26 April 2019)

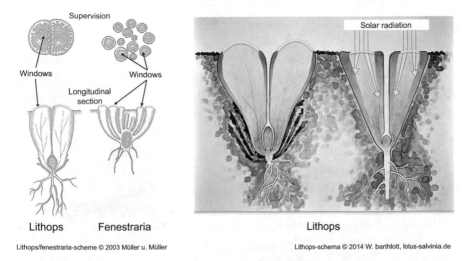

Fig. 4.17 Schematic sketch of sunlight uptake by window plants Lithops and Fenestraria. Left: Müller and Müller (2003), p. 83; right: (http://www.lotus-salvinia.de/index.php/de-de/images256. Courtesy of W. Barthlott. Sketches added by the author. Accessed 26 Apr. 2019)

Photochemical studies of light conditions inside succulent leaves were carried out as early as the 1930s (Schanderl 1935), and probably earlier. Further sources on the subject are Egbert and Martin (2002) and Krulik (1980); (Figs. 4.16 and 4.17).

4.11 Tree Growth Strategy: Perfect Stability Paired with Efficient Material Consumption

With communication among trees we have begun this small selection of examples of ingenious principles of nature in the plant kingdom; with their growth strategy we want to end for the time being, always with the thought in mind that we are describing only a drop from a vast sea of excellent natural principles.

Because of the special affinity my compatriots have with trees and forests, let us pay a little more attention to the final principle of plant ingenuity. But even more significant for our future dealings with nature and in view of the climate change taking place is the realization that the tree community of a forest not only provides us with the vital oxygen O_2, but is also a dominant guarantor for the elegant storage of the climate gas carbon dioxide CO_2. We humans have still not found a technically sustainable solution to reduce or store anthropogenic CO_2 – nature is far ahead of us with its solution strategies here too!

Trees are immovable, sometimes towering, sometimes sprawling organisms in our environment and must therefore also take special care of their very own survival, apart from the fact that they have developed many variations of dispersal techniques for their offspring.

A tree has a root system in the ground with which it seeks support against violent windstorms, snowstorms and snow loads, the effects of great heat, mechanical injuries and other influences that can damage the stability of the tree. The dangers are usually greater in solitary trees than in a climax community of the forest, which shows the final stage of a vegetation development.

The root system continues into a trunk with outer bark and inner water and nutrient transport vessels. Further up, branches of different thickness and length branch out in many different ways. Together with the leaves they finally form an imposing canopy, which again can take different shapes. As already described in Chap. 2, a tree or a forest is part of a rich community of plants and animals that we hardly notice at first glance. It is therefore all the more important that trees remain a stable part of this community throughout their life cycle. They therefore owe their ability to survive to a learned and skilfully implemented stability strategy, from the roots to the uppermost regions of the leaf canopy. Characteristic are the varied branching, their richness of form and spatial extension. Let us concentrate – with all the richness of more than 60,000 tree species worldwide[9] on the tree mechanics of the tree material.

If a tree grows symmetrically and vertically in height, similarly large forces act on it from all horizontal directions. The stresses in the wood are relatively uniform. Details follow below. The mechanical stresses on the tree change abruptly when branches break off, the tree is placed at an angle, the canopy snaps off, the tree trunk is otherwise subjected to strong mechanical influences on one side or is perhaps injured by fire.

How does a tree that has been thrown out of balance react to this? The tree tries to secure its continued existence by closing "wounds and injuries" through its ability to repair itself. The self-repair mechanism is inherent to every living organism. In addition, it tries again – under now changed external conditions and given instabilities – to regain a stable growth position. Ingenious manufacturing techniques of material application are involved in this, which increase stability by material accumulation where overturning threatens or fractures can tear further or phototrophic growth (light energy

---

[9] https://www.spektrum.de/news/wie-viele-baumarten-gibt-es-weltweit/1445973   (accessed   Apr 24, 2019).

use for metabolism) is hindered and material is saved where tensions are low. Claus Mattheck (1991, 1998, 2007, 2015) has intensively analyzed the adaptive growth, design and biomechanics of trees, creating the tree axiom[10] of constant stress (constant stress axiom).

In a healthy, largely injury-free normal state, trees therefore appear to live in a balanced mechanical tension relationship with their environment.

Figure 4.18 provides a view from the ground through a branch fork into the canopy of a stately, upright oak tree on the left. The load on the tree appears to be evenly distributed overall and extreme biomechanical stress peaks are avoided by the optimized curves between the branch forks. In contrast, the load distribution by the warped beech is clearly more asymmetrical, with high tensile forces acting on the left side of the tree, high compressive forces on the right and additionally high bending moments. The beech, which is also free-standing, has "all branches full" to accumulate material at the points where high tension forces act or where tension peaks occur in branch forks in order to reduce them.

Figure 4.19 shows on the left the root system of an upright growing pine tree, exposed by rain and evenly distributed around the trunk, suggesting a biomechanically balanced stress state of the tree. On the right in Fig. 4.18, a curvilinear and at the same time imposing shape growth of an oak tree can be

Fig. 4.18 Oak grown upright with a presumed holistic biomechanical uniform stress state, left; beech in leaning position avoiding the imminent fall of the tree by adaptive selective material rearrangement, right

---

[10] An axiom is a principle accepted without proof.

Fig. 4.19 On the left, a root system of an upright growing pine tree is visible, evenly distributed around the trunk and exposed by rain, indicating a holistic equal distribution of biomechanical stresses, without a unilaterally dominating bending stress on the trunk. On the right is an oak tree with multiple twisted branches. The rounded branch fork areas and windings are optimized in terms of material and stress

seen, whose material management provides the branch windings and branch bifurcations with material accumulations in a stress-optimal way. Pointed knot bifurcations are excluded because they considerably reduce the breaking strength. Well-formed curves, on the other hand, are the work of stress-equal distributions.

Figure 4.20 shows several material optimizations on tree trunks. On the upper left and right, optimal branch fork roundings prevent biomechanical tensile stress peaks. In the lower left, material accretions for compressive stability are clearly visible on the right side of the tree trunk, while what appears to be a face in the lower right has three features: The presumed "mouth" on the lower left shows the reinforced edge of what may have been a broken branch; the presumed "eye" also suggests a stabilizing thickening of a sore broken branch; and the presumed "thick cheek" is likewise a stabilizing feature following the removal of a branch.

Previous examples of trees in northern latitudes, such as oak, beech, chestnut, maple, pine and many more, have a similar structure when their tree cross-sections are observed and show the typical annual rings (Fig. 4.20 left). The self-repairs described and visualized before obey certain biomechanical laws, which can be summarized in the axiom of a *constant tension* (see Mattheck sources). However, not all trees have a biomechanical structure like the ones mentioned above. Exceptions are *endemic* plants that occur only in a narrowly defined area. One of these is the quiver tree *(Aloe dichotoma)* in the Namib

Fig. 4.20 Branch fork tension optimization and self-repaired wounds by wound closure on tree trunks

Desert of Namibia, South-West Africa, which the author researched there in 2001.

Heino Wolf (Sächsische Landesanstalt für Forsten, Pirna, Germany) describes the quiver tree in the Encyclopedia of Woody Plants (in ed. Chui, I. P. et al., 2002; Aloe dichotoma, p. III–IV, 1–5) as follows:

The quiver tree has an extremely attractive and striking appearance due to its characteristic appearance, light-coloured bark and bright yellow flowers. The species grows into a stocky tree up to 9 m tall with a diameter at the trunk-base of more than one meter, that colonizes the very hot and dry rock-landscapes of the desert Namib.

First described in 1685 by Simon van der Stel, Aloe *dichotoma* has no economic importance, but is a characteristic component of the desert vegetation of the north-western Cape and south-western Africa. The descriptive name "dichotoma" refers to the characteristic branching of the shoots. The colloquial name "quiver tree" is based on the fact that the Bushmen used the tree's branches as raw material for making quivers of arrows. [...]

In its natural range, *A. dichotoma* grows mainly on very hot and very dry sites located on rocky slopes. To thrive well, the species needs water-permeable soils; deep alluvial soils (alluvial soils near water, the author) or moist sites are avoided. [...].

In these areas at altitudes between 300 and 1300 m a.s.l. the average annual precipitation varies from 150 to 250 mm, falling mainly in the winter half-year. The driest areas of the Namib desert in the immediate vicinity of the coast are avoided [...]. In the hottest month of January, the average temperature is often above 33 °C. In winter the temperature can reach freezing point. In areas with summer rainfall, *A. dichotoma* survives only if it grows on extremely warm, sheltered and frost-free sites that have very good water permeability [...]. The quiver tree is an extremely drought-resistant tree species that can survive several successive dry periods and whose spread is limited by the occurrence of frost [...].

The wood of the quiver tree *Aloe dichotoma* (Fig. 4.21) is light, hard on the outside and has a fibrous, spongy inner tissue for storing water. The outer sheath forms a compression-resistant composite material with the fibrous tissue inside.

The yellow-gold coloured bark of the quiver tree is a highly reflective, layered structured protection against the intense solar radiation. The outstanding feature compared to tree branching of other deciduous trees and conifers is the always dichotomous, two-part branching of the branches (clearly visible in Fig. 4.21). The bifurcations between two branches described previously for oak and beech, which produce a stress-constant load, are not apparent in *Aloe dichotoma* – on the contrary. The curve between the dichotomous bifurcations of the branches always tapers to a point, which in the case of oak, beech or chestnut would biomechanically lead to considerable stress peaks and thus a high risk of breakage. The secret of the nevertheless high stability of the dichotomous branchings of *Aloe dichotoma* probably lies in the compression-resistant composite material and in the type of always identical symmetrical branching, from the uppermost end of the trunk to the top of the tree. Incidentally, the native San use the tree material as a quiver for their arrows, hence the German name for *Aloe dichotoma*.

Figure 4.22 shows the stem or branch cross-sections of native oak and beech and the fibrous-hardwood composite structure of the endemic quiver tree.

© 2001/2019 Dr.-Ing. E. W. Udo Küppers

Fig. 4.21 Quiver tree in Namibia, southwest Africa, with the typical bifurcation of branches in two, left and enlarged, right. The yellowish bark of *Aloe dichotoma*, reflecting the sun's rays, can be seen in the lower middle area

© 2001/2019 Dr. -Ing. E. W. Udo küppers

Fig. 4.22 Cross-sections through the trunks or branches of oak, beech and quiver tree

4.12   Excellence in Plant Metrology

The discoveries of physical-chemical performances of plants lead again and again to surprising results, which we note with awe. Plants are more than the organisms we perceive, which we use for our needs and technical purposes. Who today perceives plants as intelligent living beings? They use over 20 senses (Mancuso and Viola 2015, p. 9) to explore their environment and protect themselves and their species at an early stage through warning signals or specific protective mechanisms. Paturi (1974, p. 235–238) described the intelligent sensing techniques of plants in a marriage of bionic research and development. The fact that this happened more than 40 years ago compared to today hardly weakens the topicality of these natural achievements, which are evaluated in other time scales. Individual research results on plant measurement techniques (see Mancuso and Viola 2015) only reinforce Paturi's claims. Let us take a closer look at the metrological achievements of plants as described by Paturi (1974, p. 235) (emphasis and emphasis added):

> Adjusting to the environment requires first and foremost the ability to recognize one's environment. In addition to our own senses, we humans also make use of measurement technology. With their help, we also get a picture of things that we could perceive poorly or not at all without devices. Detecting toxic substances in the *air we breathe* or in drinking water, determining the correct illuminance for workplaces or the exact exposure values for sensitive films, precisely determining minute quantities of any substance or precisely measuring the exact moisture content of well-seasoned precious woods for making musical instruments – these are just a few of the infinite number of tasks that we would not be able to perform without our highly developed, sensitive measuring instruments.
>
> Living in an environmentally sound manner means recognizing the environment and acting appropriately according to this knowledge. Conversely, it can be said that an organism which behaves in an environmentally sound manner has certainly also correctly recognized its environment.
>
> Plants behave in an environmentally friendly way. They are in no way inferior to us humans in their adaptation to the habitat; on the contrary, they are often ahead of us in many respects. Does this mean that they can recognise their environment better than we can with all our technical aids? Generally speaking, the answer to this question is certainly not yes.

But on the other hand, it is indisputable that plants have also used their measurement techniques to communicate in a very efficient way and in a networked way across species, which significantly increases their chances of survival. Further (ibid., p. 235):

Man has succeeded in using radio telescopes to pick up electromagnetic waves from extremely distant or even long-exploded and disappeared stars (and today, 2019 to shoot photos of "black holes", d. A). Humans are able to register seismographic waves from an explosion in South America in the US many 1000s of miles away. Man can accurately determine the solar radiation on Mars or any other planet. But what does he gain by these extraordinary achievements?

His quality of life they do not improve. The plant has not learned *all* this. And what for; from the point of view of a system thoroughly attuned to expediency, the acquisition of such abilities must appear as an uneconomical and therefore not sensible digression from life itself. For the plant, grasping the environment does not include knowledge of any distant celestial bodies, but it does mean, for example, being able to register the migration of the moon precisely by means of measuring technology, in order to be able to derive appropriate behaviour in the habitat of the tidal zone from the measured data. Purpose measurements of this kind are, among others, the exact determination of light stimuli, gravity stimuli, humidity values, chemical substances or touch stimuli. They are directly related to the vital functions of the plants, and the plant masters them consistently as well as or better than man with his most elaborate precision measuring methods.

I have already mentioned how unimaginably sensitive the reaction of growing root tips or of unicellular organisms is to chemical substances: *0.000000028 milligrams (28 × 10⁹ mg)* of malic acid was what a sex cell of the fern still clearly detected and recognized as malic acid. That some bacteria are superior to all chemotechnical analyzers in detecting the smallest amounts of oxygen, I also said before. The conduction of excitation states inside a plant probably also works chemically. Extremely small amounts of active ingredient are also sufficient for this. An oxy-acid from the pressed juice of the mimosa evokes a clear reaction of the plant even in a *dilution of 1: 100 million.* The dilution corresponds to a quantity of 25 drops on a 1.5 m deep pool of 5 × 20 m size.

A small side calculation confirms the dilution ratio value:

(a) The volume of a drop: standardized by pharmacists as a measure in a metric system, i.e.: one drop equals 0.05 ml or 5×10^{-5} L, 25 drops then equals 25 × 0.05 ml or 1.25 ml × 10^{-3} = 0.00125 L.
(b) The volume of water used is 150 m³ or 150 × 10^3 = 1.5×10^5 L.
(c) 1.5×10^5 L water: 0.00125 L drops = 1.2×10^9: 1.

Further (ibid., p. 236):

Chemical analysis plants would be overtaxed here as well. The precision achievements of plants in the field of time measurement (I will speak of this later) would be sufficient to receive the predicate "chronometer" from the association of the world-famous Swiss watch industry as an award for particularly high-quality utility watches. Vital to climbers is the measurement of touch stimuli. If their tendrils, which are searching for a hold, bump into something during their circling search movements, they must detect this and immediately respond with a holding movement. The tactile sensitivity of such special clamp organs surpasses the human sense of touch many times over and leaves even chemists' scales and highly sensitive chemical laboratory weighing equipment behind. With a good chemical balance, quantities of 0.01 milligrams can still be estimated. A vine, on the other hand, reacts to a tickling stimulus, such as a wool thread of only 0.00025 milligrams, after only a few seconds and then bends so quickly that the movement can be followed with the naked eye. And with these tiny touch forces, the plant can still distinguish between different materials, in contrast to technical tactile measuring devices. Falling drops of water, which must not cause a climbing movement, or a glass rod, which would be too slippery to "hold on to", do not trigger any stimulus. Equally astonishing is the plant's ability to perceive the lowest levels of illumination. A 25-watt lamp in absolutely clear air and complete darkness could detect and locate the extremely light-sensitive tips of a vetch seedling (Vicia villosa) even at a distance of 30 km, a 100-watt lamp even at such a distance of 70 km! (For the light technician among the readers it should be said that this theoretical model case should explain an illuminance of 23×10^{-9} lux).

It is not too difficult to detect a 100-watt lamp 70 km away by technical means. With astronomical telescopes and suitable illumination measuring instruments [...] a candle at a distance of 28,000 km (which corresponds to a star of 23rd magnitude) could still be detected.

But these measuring methods assume that first a strong telescope or telescope magnifies the light source and thus its observed luminosity so much that it becomes detectable for good illumination measuring instruments. If such a weak light source is not point-like, the measuring technique fails. Under these circumstances, it can only with extreme difficulty detect illumination values which the vetch seedling still registers after prolonged exposure. However, an extremely sensitive measuring system suitable for this purpose will inevitably be destroyed on the spot if direct sunlight falls on it, because that is 4 trillion (in figures: *4,000,000,000,000*, 4×10^{12}) times as bright. The powerful optical measuring system of the plant, on the other hand, survives the enormous difference in brightness, which even the human eye, which is so adaptable, cannot cope with, unharmed.

These are impressive figures from the plant kingdom of nature. They come about not least because plants are stationary creatures. Once they have taken root, they remain there for life. All the more understandable in retrospect is their extensive differentiated sensory or measuring technology. Plants cannot run away from possible damage in their environment like animals. Therefore, they developed and continue to develop risk-preventing sensory systems, or at least sensory systems that ensure their survival. Even in the event of major natural disasters such as area-wide fires, plants have means to protect themselves against total loss.

>> 4 B Eleven evolutionary principles from fauna and an example from *BIOGEONICS*

4.13   Rhinoceros Beetles: True Powerhouses in Miniature

The size and force spectrum in nature's animal kingdom ranges from the mammalian blue whale *(Balaenoptera musculus), which is* over 30 m long and weighs nearly 20,000 kg – 20 tons – to the smallest organisms, smaller than 1 mm and 100 µg in weight, such as the horn mite *(Archegozetes longisetosus)* (Heethhoff and Koerner 2007). The physics of forces acting on organisms of such size differences was vividly described by Schlichting and Rodewald (1988, p. 4) in their paper "Von großen und kleinen Tieren":

> Many experiences suggest that small animals are comparatively much stronger than large ones. For example, it is well known that an ant can lift and carry a load many times its own body weight, but a human being can only lift and carry a load the size of his body weight (this refers to an average human being, not a competitive weightlifter). This fact again is a simple consequence of the surface-volume-relation. The number of muscle fibres and thus the cross-sectional area (e.g.) of the animal's legs are decisive for the muscle strength and thus for the maximum force to be exerted by an animal. With the size of the animal, the mass of the animal grows like the body volume, but the cross-sectional area of the legs grows only like the body area. The muscle power of the legs grows therefore unequally slower than the body mass. This has the following consequence: Small animals have a large body surface in relation to the body volume and thus relatively large muscle cross-sections. They can therefore develop great forces, which generally go far beyond "carrying" the relatively small body mass. In the case of very small animals – e.g. insects – it has even been found that nature does not use the potential reserves of strength at all. The cross-sections of the legs are thinner than one would expect on the basis of a well-proportionedness oriented towards larger animals […]. The strength of these tiny creatures maneuvering

on long "spider legs" sometimes fills us with great astonishment. This astonishment, however, is only an expression of the fact that we transfer the experiences made in the realm of human dimensions to other realms. What small animals have in abundance, namely "surface", the large ones lack.

Measurements of the horizontal holding force on a rough surface (mean roughness R_a = 30 µm) of the smallest organisms, such as horn mites, resulted in a tractive force of 1180 times their own body weight (100 µg = 0.1 g). On vertical rough surfaces (R_a = 30 µm), still 530 times the own body weight was measured as traction force, involving only two pairs of feet (Heethhoff and Koerner 2007, p. 3036). The muscles in the grasping claws of horn mites generate a load of up to 1170 kNm^{-2} (ibid.) (1 kN = 1 kN = 10^3 N, 1 kNm^{-2} = 1 Pa, Pascal).

On about the same order of magnitude as the vertical force of the horned mite just described is the maximum force of the rhinoceros beetle *(Oryctes nasicornis)* (Fig. 4.23). Various sources including National Geographic[11] give a carrying force of 850 times the rhinoceros beetle body weight of a male. If we calculate the average weight of a rhinoceros beetle about 20–40 mm in size to be 50 g (a Hercules beetle over 150 mm long with its long horn is about 100 g in body weight), we get a carrying force of 0.05 kg × 850 = 42.5 kg. In

Fig. 4.23   Male rhinoceros beetle *(Oryctes nasicornis)* with shiny medium brown surface.   (http://www.naturspektrum.de/ns1.htm.   Courtesy   of   Holger   Grötschl, Ganderkesee, Germany. Accessed 29 Apr. 2019)

---

a descriptive figurative sense, but without any realistic physical meaning or consideration of physical properties at different scales, a human of 80 kg would be able to lift a weight of 68,000 kg or 68 tons using a rhinoceros beetle as a model.

In Brandenburg and Berlin, in southern Germany and Austria, the small strong beetle, which is a protected species, is native. Its South American relatives in the tropical climates reach with 150 mm and more length the 5–8 times of their European relatives.

So let's take great care during our walks in the woods when we encounter the small, inconspicuous crawling powerhouses on the forest floor. Arthur Schopenhauer is attributed – unproven[12] – the quote, the truth of which, however, is indisputable:

> *Any stupid kid can stomp a bug,*
>   *but all the professors in the world can't make one.*

4.14   Morphofalter: Ingenious Trick of Colour Production Without Dyes

In the course of billions of years, nature has not wasted its time, as evidenced by the enormous biodiverse wealth and the associated principles of nature. So also the principle of color generation on body surfaces by light scattering.

White color is created in nature by light scattering, because the scattering particles are too large to produce blue light. So-called iridescent colours are produced in nature by superposition – interference – of different wavelengths of sunlight on highly structured optical layers. One of these structured layers at micrometer and nanometer sizes (10^{-6}–10^{-9} m) is exhibited by the magnificent, shimmering blue morph butterfly *(Morpho peleides),* with a wingspan of up to 120 mm. Morpho butterflies prefer the tropical regions of South and Central America, among others, and are also native to the West Indies, which are offshore from northern South America.

These tropical butterflies – as mentioned before – do not produce their bluish colour with the usual colouring substances – pigments – which also give organisms their typical colours or colour patterns. The complete avoidance of pigment substances for color production is the actual, very efficient trick of

---

[12] http://falschzitate.blogspot.com/2019/03/jeder-dumme-junge-kann-einen-kafer.html (accessed April 29, 2019).

saving material, which is also mastered by many other animals such as beetles, hummingbirds and pheasants.

In order to uncover the secret of coloration, we must delve deep into the hidden structures of the wings of morphofalters.

First, Fig. 4.24 shows two specimens and Fig. 4.25 one specimen from the family of the noble butterflies (Nymphalidae). Fig. 4.25 is supplemented by a schematic sketch showing how the scattering of incident sunlight strikes the nanometre-sized fir-like structures in the wings and, through multiple amplification, appears to us as the brilliant blue light of the moth's wing surface.

Morphidae

Morpho menelaus   Morpho cypris

© 2019 Dr. -Ing. E. W. Udo küppers

Fig. 4.24 Two morpho butterfly species, *Morpho menelaus* and *Morpho cypris*

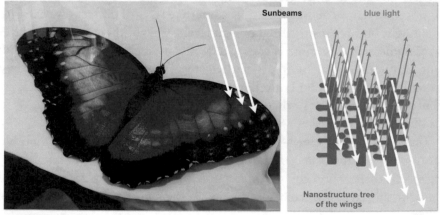

Morpho peleides

© 2009/2013 Dr. -Ing. E. W. Udo küppers

Fig. 4.25 A blue morpho butterfly *Morpho peleides* with principle sketch of blue light production

As part of my teaching at the HSU (Helmut Schmidt University, Hamburg, 2009), a trimester paper was written by Jörk Lackner, which describes the secret of colouring in detail and is reproduced here in excerpts (italics, fig. no. change and source insertion by the author):

Most morph butterflies have two types of scales, firstly the wings consist of the ground scales and secondly the covering scales. The covering scales are transparent and often serve as an *optical diffuser** for the incoming light. This has the effect that the appearing structural colours become visible at further angles [...]. The ground scales, on the other hand, have brown pigments at their bases. These pigments give many morphofalters a basic coloration and let them appear with brown wings if no light falls in. This basic colour can be seen well on the ventral side of the morph butterflies in question. In many species, however, the ground scale is the trigger for the structural colour (i.e. the blue reflected light). Only in the morpho Adonis is the ground scale responsible for the structural coloration (Kinoshita and Yoshioka 2005, p. 1448). Under the electron microscope, at a magnification of 20,000 to 30,000 times, it can be seen that the ground scales consist of a kind of tree structure (Kinoshita et al. 2002, p. 107). This structure is also described as a multilayer structure or lamellar structure. As can be seen in

Fig. 4.26a, b and d, each of these "trees" has about 6 to 10 lamellae. The special feature of these lamellae is that they consist of alternating, fine chitin layers and air layers, which have a thickness of only 0.055 nanometres (nm) for the chitin layers and 0.150 nm for the air layers, respectively.

In the case of the Morpho Adonis, it is not the ground scale (as already mentioned) that is responsible for the reflection, but the covering scale. Here, in contrast to the other specimens, differences in the basic structure of these reflective covering scales are worth mentioning. Thus, in Fig. 4.26c, it can be seen that the scales consist of only 2–3 skin layers, which are linked to another layer. Furthermore, another difference is that the individual lamellae interlock, whereas the "lamellar trees" of the other specimens each stand alone. The bright covering scales are lightly colored and contain the entire color spectrum, while the bottom scales have a dark blue tint.

A diffuser – the opposite of a nozzle – slows down the flow of gas or liquid while increasing the gas or liquid pressure (for those interested, see "Bernoulli's Law" of flow).

From the numerous recent R&D sources on morphofalters, some with new findings, only two are mentioned: Zhou et al. (2019) and Debat et al. (2018).

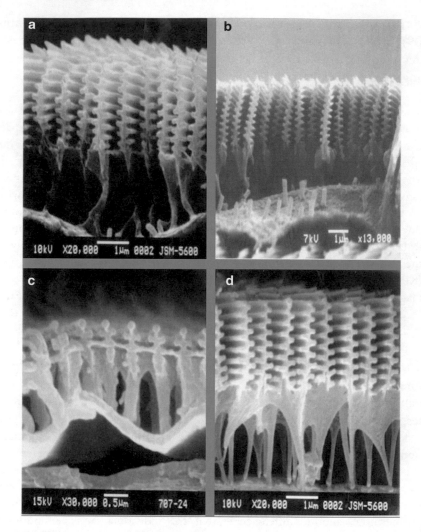

Fig. 4.26 SEM image of morpho butterfly wing cross sections. Original: Scanning electron microscope images of the cross sections of the iridescent scales of *Morpho* butterflies: **(a)** a ground scale of *M. didius,* **(b)** a scale of *M. rhetenor,* **(c)** a cover scale of *M. adonis* and **(d)** a scale of *M. sulkowsky.* (Source: Kinoshita, S. et al. (2002, p. 107). With kind permission from Shuichi Kinoshita)

4.15   Barnacles: Ultra-Strong Bonding in an Aqueous Environment and Other Natural Masters of Bonding

In the field of manufacturing joining, organisms have done preliminary work in a variety of ways for potential solutions by humans. Joining is defined as a permanent connection of two and more objects – in nature organisms plus objects. Often formless substances, joining substances or adhesives or adhesive secretions of organisms play a crucial role in such joints in nature, as in the case of barnacles *(Balanidae)*. Other marine organisms, such as the mussels *(Mytilus edulis)*, attach themselves to different substrates with their long byssus threads and cut this connection when they release their attachment to seek another location. Figure 4.27 shows en passant to the barnacles a group of mussels and the elastic byssus threads. In passing, a small barnacle can be seen as a permanent guest on the lower right mussel.

Barnacles, on the other hand, are – once attached to the watery or non-watery substrate – permanent guests and not easily detached. It is this particularly high adhesive strength and the bonding in an aqueous environment that these organisms have perfected over millions of years. They have therefore become organismic masters of permanent adhesion. Küppers and Tributsch (2002, p. 17–18) describe the organism as follows:

> Barnacles are at home on many beaches and rocks and in other marine biotopes of tropical and temperate latitudes, also on other marine animals such as mussels (Fig. 4.27, the author). They belong to the arthropods (arthropods, the author). Their way of life is strange because the mobile, swimming young larvae stick upside down to a suitable substrate and remain there until the end of their life – in total immobility. The biological adhesive of the barnacle is also called cement adhesive because of the extraordinarily high breaking strength of the

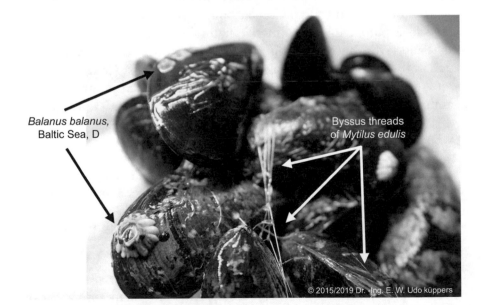

Balanus balanus, Baltic Sea, D

Byssus threads of *Mytilus edulis*

© 2015/2019 Dr. -Ing. E. W. Udo küppers

Fig. 4.27 Group of mussels with attached barnacles from the North Sea, Germany

adhesive bond with the substrate (more than 10 times higher breaking strength compared to technical, artificial epoxy adhesives). Investigations show that a multi-component adhesive consisting of different proteins, in combination with rinsing fluids enables the organisms to bond peripheral cracks occurring during their growth (quasi concentrically around the original bonding point) without fear of the adhesive solidifying in the feeding channels (Saroyan et al. 1970; Waite 1987; Abbott 1990). Especially under water, the superiority of this natural technique of a durable adhesive over artificial adhesives is evident. Particularly noteworthy is the automatism of self-healing in the event of injury, which, incidentally, is inherent in every living organism. Another advantage of natural adhesives in relation to their artificial counterparts should not be forgotten here, namely their environmentally friendly recycling.

Figure 4.28 shows the impressive difference in size of *Balanidae* from tropical and non-tropical biotopes.

Because of the universal nature of bonding, some specific subtleties will be briefly discussed. The elegant joining technique of bonding has found many applications in evolution. Some of them are listed in Table 4.2.

Küppers and Tributsch (2002, p. 21) state that nature saw no need to develop a universal adhesive. The following Table 4.3 shows some examples of this.

The production technique of bonding in nature – and technology – with regard to their raw materials is always a compromise to combine the physical properties of adhesion (intermolecular bonds) and cohesion (intramolecular bonds) in a purposeful way. Reactive natural adhesives are low molecular weight and favor adhesive bonding of surfaces. For this reason, reactive raw materials such as so-called prepolymers are advantageous. Cohesive requirements, on the other hand, are favored by high-molecular raw materials. Examples of both groups of raw materials are summarized in Table 4.4.

When it comes to the raw materials of natural adhesives, nature uses a relatively small number of starting materials for a wide range of specialized adhesives for demanding bonding processes. Only one type of adhesive technology does not seem to exist in nature: high-temperature bonding. However, one cannot be completely sure because biotopes with hot environments, such as black smokers in the deep sea and hot volcanic springs, have not been sufficiently studied with regard to adhesive joining techniques of organisms. (ibid.)

This statement from 2002 on high-temperature bonding in deep-sea biotopes, around hot hydrothermal vents, could be specified in the meantime. At this depth of 2500 m to 3000 m below sea level, water pressures of nearly 300 bar and water temperatures of 350–450 °C prevail around these hot

Fig. 4.28 A group of tropical barnacles, left, with a barnacle on a temperate mussel, right. The difference in size between the two barnacle species is impressive

hydrothermal vents.[13] A rich fauna has formed near the "black smokers", which includes Pompeii worms (Carwadine 2008, p. 64), spider crabs without eyes, Hoff crabs, Yeti crabs, beard worms, starfish clams and mussels. A subfamily of mussels, *Bathymodiolus Byssus* live in symbiosis with

[13] https://de.wikipedia.org/wiki/Raucher_(hydrothermal) (accessed May 02, 2019).

Table 4.2 Organisms and their bonding techniques – selection

Organisms	Principle of gluing
Barnacles	Permanent bonding
Mussels, orchids	Temporary bonding
Slugs	Fast bonding
Lacewing	Flash bonding
Venus flytrap	Sensor controlled joining
Termites	Gluing – Antique gluing
Abalone	Adaptive bonding
Potter birds	Non-porous bonding
Spiders	Energy-absorbing bonding
Plants	Temporary bonding
All the glue	Disposal-free bonding

Table 4.3 Basic building blocks of natural biological adhesives with biological application examples

Building blocks	Biological adhesive technology of organisms
Lipids	Termites (Nasutitermes, Trinervitermes)
	Enemies are immobilized by spraying a secretion into a kind of sticky package
	Cornifers (Gymnospermae)
	Packaging protection due to repellents containing terpene resins
Polysaccharides	East Asian gliders (Apodidae)
	Offspring protective packaging through nesting
	Wasps and hornets
	Nest building, in combination with polysaccharides (cellulose)
Polyphenols	Insects
	Multifunctional exoskeleton packaging, together with chitin and proteins
Proteins	Blue mussel (Mytilus edulis)
	Adhesion and release of the packaging tray to various substrates
	Liver fluke (Fasciola heptica)
	Hardening of the packing cover protecting the eggs

Table 4.4 Cohesive and adhesive biological adhesive raw materials

Cohesive-adhesive	Adhesive raw materials
Cohesive	Starch
High molecular weight	Cellulose
	Protein
	Beeswax
Adhesive	Rosin
Low molecular weight	Protein sequences, e.g.
	3,4-Dihydroxyphenylalanine (DOPA) in mussel proteins

Fig. 4.29   Hydrothermal fauna at the Mid-Atlantic Ridge at 3030 m depth. (With kind permission of ©MARUM – Center for Marine Environmental Sciences, University of Bremen)

chemosynthetic bacteria at hydrothermal vents (Thubaut et al. 2013; Kiel and Amano 2013). This adapted species *Bathymodiolus Byssus,* like its conspecifics, possesses byssal glands at the water surface, with the filaments of which it clings to the substrate (Brazee and Carrington 2006; Kádár and Azevedo 2006).

Despite the enormous differences in pressure and temperature between the two habitats of *Mytilus edulis* and *Bathymodiolus byssus,* the basic principle of adhesion has been preserved by both mussel species. An amazing feat of evolution.

We are certainly not entirely wrong, after these excursions into the realm of natural bonding techniques, in asserting that organisms, whether temporarily or permanently adhering to a substrate, have conquered the earth from the biosphere to the deep sea (Fig. 4.29).

4.16   Ostriches and Other Birds: Open Packing System Egg – Natural Product with Ingenious Performance Spectrum

Ostriches *(Struthio camelus)* are native to Namibia and South Africa. They belong to the running birds. Despite their weight of about 130 km and a height of almost 2.5 m of a male adult bird, they are the fast runners in the

Male adult bouquet on Molopo Farm, Namibia, South Africa   Female young pigeon in the Zoological Garden Berlin

© 2001/2019 Dr. -Ing. E. W. Udo küppers

Fig. 4.30 Adults and young ostriches

savannah. With top speeds of about 70 km/h, any attempt to escape is futile for humans. The particularly fine feathers of the ostriches have an antistatic effect and repel dust. They also have technologically advanced, adjustable structures for temperature control. Figure 4.30 shows two adult ostriches on Molopo Farm in southern Namibia, during a research visit by the author in 2001. The two young ostriches were encountered in the Zoologischer Garten Berlin in 2019.

Often it is the organisms of the fauna themselves that amaze us with their achievements and which, compared to man-made analogies, make them look very modest when we analyse the advantages and disadvantages of functions or techniques from a holistic point of view.

Occasionally, however, it is the "products" of organisms and their forms and structures that fascinate. So it is with the egg, or better: with the multitude of eggs, whose outer geometric shapes differ from each other in details, making no two eggs completely alike. This is also the reason why mathematicians have not yet succeeded in calculating the curve geometry of a universal egg. It simply does not exist in nature!

But how can we make the vast spectrum of eggs visible and comprehensible – an almost impossible task to solve?

In the German-speaking world, there was at least one person who dared to try and put together a handsome collection of eggs. It was the Thuringian surveyor and amateur researcher Max Schönwetter, whose egg collection included 19,206 at the number of 3839 bird species.[14] Today, 11,121 bird

---

[14] https://www.spiegel.de/wissenschaft/natur/sammlung-von-weltrang-eier-wir-brauchen-eier-a-954754.html (accessed May 03, 2019).

species are counted worldwide. This figure is the result of the updated Red List* by the NABU** umbrella organization Birdlife international 2016.[15]

Küppers and Tributsch (2002, p. 126–127) describe the egg as a biologically universal form of packaging in many details that give an idea of the value of this ingenious natural product (spelling has been adapted to the present, illustrations updated):

- **The eggshell – temperature**
  Probably the most conspicuous and impressive package in nature is the eggshell. It is a highly specialized mineralized structure that protects the bird embryo during its development (Burley and Vadehra 1980; Rahn et al. 1987; Tyler 1970). On the one hand, it must therefore ensure optimized gas exchange and adequate thermal conductivity. On the other hand, it is required to be highly resistant to breakage, although the chick's small beak must still be able to break open the shell. Bird eggshells are not always white, but are usually camouflaged by a variety of colours and patterns against predators who are after the high protein content of the eggs. Since the embryo can hardly survive a temperature of 39–40 °C, it is surprising how especially dark eggs in unprotected nests can withstand solar radiation. Research has shown that they are able to effectively reflect infrared sunlight, which accounts for the majority of the sun's radiated energy. They are helped in this by pigments embedded in the eggshell, particularly protoporphyrin and biliverdin. Sometimes, when the eggs are bluish or bluish-green in color, as in herons or ibises, the pigment bilin is involved. In the near infrared, a reflection of just over 92% has been observed in gull eggs (Larus heermanni). Practically, this means that unprotected eggs, which only reach a temperature of 28–30 °C in full sun, would heat up to 40–50 °C without infrared reflection, which has also been demonstrated experimentally. The infrared protection of the eggs was therefore a matter of survival, probably for a very long time.

---

[15] https://www.nabu.de/news/2016/12/21632.html (accessed May 03, 2019).

- **The eggshell – size, shape and weight**

  Eggs have very different sizes. The contents of an ostrich egg are about twenty-two times heavier than the contents of a domestic hen egg. Figure 4.31 shows an example of the differences in egg size.

  Again, there is an equally large difference between the egg of the domestic hen and those of the smallest birds, for example, the hummingbirds. Since the volume increases with the third power, but the surface only with the square, large eggs have a comparatively much reduced surface for the exchange of heat and gas with the outside world. This consideration already suggests that nature therefore had to adopt very flexible strategies with regard to the evolution of the special properties of eggshells, their thickness and pore density. But also the altitude of the clutch above sea level has an influence on the structure of the eggshell. At high altitude, water evaporates very easily, so eggshell porosity must be reduced to prevent desiccation. In fact, it has been found that the pore diameter changes with the barometric pressure around the clutch. The higher the clutch of eggs is laid, the smaller the cross-section of the pores. If you move birds to other altitudes, they need about two months to adapt the newly laid eggs to the changed environment. How they do this is largely unknown, but it is likely to work via blood composition, which is also adapted to altitude. Perhaps evolution experimented until such an automatic adjustment became established, which gave the birds the necessary range of movement in relation to

Fig. 4.31 Differences in egg size

altitude. This should convince us that we are really dealing with a highly developed packing technology.

But the shape of the eggs is also very variable. It ranges from round to elongated and eccentric shapes. Various factors control this, usually related to the birds' way of life. A striking example is the highly eccentric eggs of guillemots. They are seabirds which tend to lay their eggs on the bare ledges of vertically rising cliffs. Presumably, all reasonably normally shaped eggs rolled off sooner or later in the process. Only eccentrically shaped eggs survived. They have the property of rolling in a circular path. They then settle in such a way that they cannot roll any further on a slightly inclined plane.

- **The eggshell – structure and gas transport**

  We will now concentrate on the structure of the eggshell, which holds the microscopic secrets of this packaging miracle. A simplified picture (Fig. 4.32) of an eggshell, which consists of about 90% inorganic matter, shows its various structures. Characteristic is the thick, calcified, porous layer that we commonly know as the eggshell. Shown enlarged are the calcite columns, calcium carbonate crystallizing as a calcite mineral, arranged perpendicular to the eggshell surface, leaving open spaces for the pores of the shell. The pores are important for gas exchange. Usually the crystallites, which in domestic chickens reach a diameter of 20–30 μm, do not extend all the way to the outer surface of the crystallized layer, but are covered by

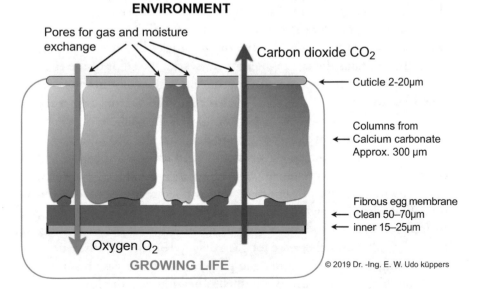

Fig. 4.32 Shell structure of a hen's egg with various pores and membranes

a thin, less crystallized, or powdery layer. Towards the interior, smaller crystals known as mammillae round out the crystallized columns. They are arranged around an organic core that serves to anchor the outer of the two inner shell membranes. The calcified layer also contains three percent of an organic matrix, part of which is protein. It also contains magnesium, which is arranged on the inner side, and sometimes additionally on the outer side, of the calcified layer.

Figure 4.32 also shows how the calcium layer is bounded on the outside by an eggshell skin, the cuticle, and on the inside by two membranes. The reason why freshly laid eggs are so shiny is the 2–20 μm thin outer cuticle layer. Of a 60 gram egg, it makes up about 18 milligrams in weight and is 90% protein. In addition, lipids and carbohydrates are found. Not all eggs have this outer membrane. It is thought that there is a connection with water vapor exchange or water repulsion. Perhaps this function is not important for certain birds, for example pigeons, which can do without this layer. Since the outer membrane remains sticky for a few minutes after the egg is laid, it could also help with proper positioning of the egg. This would be crucial when birds do not immediately place the egg in the planned position, but have to turn around and help it along with their beaks. It has also been suggested that the outer cuticle protects the egg from microbes. Some birds replace it with another layer. On booby eggs it consists of 0.4 μm thick grains of the calcium carbonate material Vaterite. It gives the eggs a chalky appearance. Perhaps this layer helps to camouflage them or to reflect the glare of sunlight. In any case, the covering layer must allow gas transport. Sometimes its structure resembles microporous glass, the kind used to filter liquids. The outer cuticle, like the outer calcified layer, also contains the egg pigments. This varies the color of the eggs over a wide spectrum. It ranges from a translucent white color in the European kingfisher to a greenish black color in the Australian emu. Egg colors are genetically controlled and vary, for example in the cuckoo, depending on the species of bird in which the eggs are deposited for brooding.

About 2–3% of the calcified layer is organic in nature. If the layer is decalcified, a fibrous structure remains. Its fibres are around 0.01 μm thick and up to 10 μm long. In addition, there are gas-filled corpuscles measuring 0.4 μm. While the fibres, whose network partly extends parallel to the eggshell surface, are thought to have a supporting function, it has been suggested that the small corpuscles serve for gas exchange or thermal insulation.

If you bring an egg under water and put it under excess gas pressure, many small bubbles escape from the eggshell. They indicate the function of the pores, whose shape and structure vary greatly among bird species. The pores control the intake of oxygen and the release of carbon dioxide as well as water vapor. A chicken egg has between 7000 and 17,000 pores. On the outside they have a diameter of 15–65 µm, on the inside a diameter of 6–23 µm. The pores are by no means always open. Rather, they are often blocked or covered by organic or inorganic structures. The density of the pores varies from 45 pores per square centimetre in the emperor penguin to 306 pores per square centimetre in the Japanese quail.

The egg's inner membranes serve primarily to confine the fluids inside the egg, while allowing gases to pass through. As a physical and chemical barrier – they contain antibacterial substances – they also keep bacteria out and serve as an anchor structure for embryonic development. In the domestic chicken egg, the outer of the inner membrane is about 60 µm thick, while the inner membrane is about 20 µm thick. The membranes are composed of dense networks of parallel, mutually anchored fibres. Their structure is reminiscent of the cellulose fibre meshwork of bacteriological filters. It has been measured that a chicken egg takes up 6 litres of oxygen during the incubation period of 21 days. The daily intake depends on the embryonic stage of development. It increases slowly at first and then rises dramatically between the 10th and 14th day to a limit value of approximately 0.6 litres per day. This is controlled by a membrane developed by the embryo, the chorioallantoic membrane, which covers the entire inner shell surface on the 12th day of incubation. An air chamber of 15 percent of the egg volume, which develops at the blunt end of the egg, also contributes to oxygen exchange. In addition, there is an exchange of 4.5 liters of carbon dioxide and 11 liters of water vapor released to the outside world during the breeding season. To achieve optimal gas exchange, the humidity of the nest must be readjusted, for example by regular ventilation. The biophysics of the eggshell and its auxiliary biological structures is therefore quite complicated and there is still a lot to learn in the direction of bionic oxigen generators, i.e. devices that provide oxygen for life functions.

The resistance of eggshells to compression is amazing. They can withstand up to 240 kg per square centimeter. Still, little chicks can crack eggshells because they do it right. With their small pointed beaks, they can locally penetrate the porous crystalline structure of the shell, which is also easier from the inside. The African egg snake, whose muscular jaws or body muscles could barely squeeze an egg, has also developed a trick. It guides an egg in swallowing along a saw-toothed palatal bone that literally rips it open. Then it can be squeezed effortlessly.

- **The Eggshell – Resume**

  All properties combined, an egg is a masterpiece of biological packaging that has stood the test of time through hundreds of millions of years. Already the dinosaurs have benefited from this amazing construction. Individual clutches have survived fossilized until our days.

  Not only birds use lime shells for packing purposes. Shells of mussels and snails also use the natural substance lime, which can be absorbed by the environment at any time and deposited there again (an ideal recycling material without the slightest impact on nature and the environment, the author).

What has been described above in all detail and appropriate to the ingenious natural product, we see – as a supplement to Fig. 4.32 and related to the ostrich egg – in Figs. 4.33 and 4.34, where some photo and SEM images of ostrich egg shell surface and cross-section show the principle of the supporting calcium columns, the internal protein membrane and the channels for gas and moisture exchange enlarged again.

The depressions to the gas channels, which still appear roundish in the upper image, become crater-shaped and recognizable as meander-like depressions with increasing magnification.

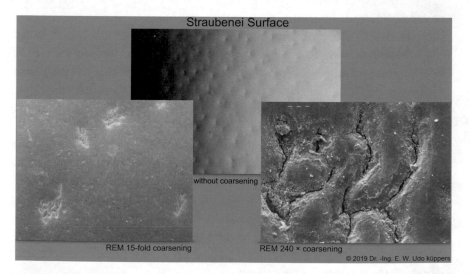

Fig. 4.33 Photographs showing scanning electron microscope – SEM – images of ostrich egg shell surfaces. (Sources: own photos, above, SEM-photos from 2004, with friendly support of W. Heyser, UFT, Uni-Bremen)

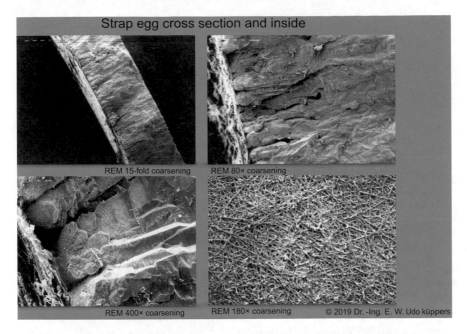

Fig. 4.34 Scanning electron microscope – SEM – images of ostrich egg shell structures

For all three images showing the cross-section of the eggshell, the inner space is on the left and the outer space of the shell is on the right. SEM images from 2004, with the kind support of W. Heyer, UFT, Uni-Bremen.

With increasing magnification from the top left to the bottom left, the calcium-like supporting columns of the eggshell become clear. At the bottom right, the finely branched protein tissue is visible as a closure to the interior, where life grows. It also ensures that no microorganisms can enter the interior and destroy life. This tissue layer is also visible in the three other SEM photos on the right.

After this very detailed description of one of nature's most ingenious and universally applied inventions, we now look into the dwelling-places of socially living insects and their biological achievements.

4.17 Termites: Social Insects, Outstanding Experts in Climate Regulation

State-forming insects have developed a sophisticated circulatory survival system in extended families. However, the indisputably greatest builders of nature are the state-forming bees, hornets, bumblebees, wasps, ants and termites living in social communities. Their constructions have a variety of

Fig. 4.35 Termite burrow in India. Image source: Bishnu Sarangi, 3571 image on pixabay, for free commercial use, https://pixabay.com/photos/termite-hill-termites-termite-mound-266587/ (Accessed 04 May 2019)

ingenious room designs with optimal air conditioning through natural air circulation, which – as with bees – is further supported by fanning with the wings. Termites, on the other hand, do not have wings. They must rely entirely on their constructive nest-building abilities. In concrete-hard structures with partly smooth, partly ribbed outer walls, more than 3 m high, in the case of *Macrotermes bellicosus* up to 7 m, there are neither doors nor windows (Fig. 4.35). Inside the nest are the fungal chambers. There, the air is heated by fermentation processes of the fungal cultures and by the respiration process of an army of working termites. The example described here involves the species *Macrotermes bellicosus,* which lives on the African Ivory Coast. The constant consumption of oxygen would inevitably lead to the death of the termites without a regulated supply of O_2. But a clever flow-circulation system in the nest causes the warm air rising from the mushroom chambers to be conveyed into a fine system of ducts within the outer ribbing of the nest structure. Here, gas exchange takes place, as it were, with the emitted carbon dioxide (CO_2) escaping to the outside and oxygen (O_2) being carried into the nest. In addition, the ribbed tubules have a cooling effect on the circulating air in the nest, which again provides the necessary supply of oxygen to the termites and fungi as the flow continues. Inside a nest housing millions of termites, a nearly constant temperature of about 28/29 °C exists day and night, while the external environmental temperature varies from about 10 °C at night to about 40 °C and more at the zenith of the sun (Fig. 4.36). See also on thermoregulation of

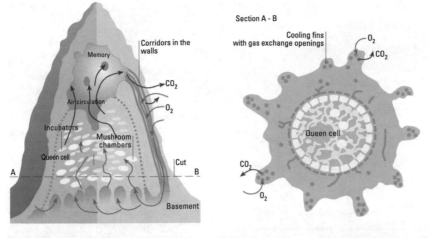

Fig. 4.36 Longitudinal section and cross-section A-B through an African termite burrow. Template by the author, Infographics with the kind support of Fr. Wulfken-Jung, Überseemuseum Bremen

termite burrows and on ecosystem services of termites in agriculture, among others, (Korb 2011, 2003; Ogoudetji et al. 2010).

Another termite species, the *Compass termites* (*Amitermites taurensis*), build their disc-like dwellings in the grassy savannas of Australia cleverly and consistently in a NORTH-SOUTH direction. The simple but ingenious reason for this is that their narrow upper and long sides offer little surface for the hot midday sun rays to attack, but the morning and evening sun pleasantly warm the broad sides – leaving the opposite one to cool – without overheating it. With constant temperatures around 30 °C inside and high humidity of nearly 100%, an ideal and regulatory climate has been created. Figure 4.37 shows constructions of Australian termites under sunlight conditions that vary with the time of day.

What if we were to consistently design our homes as thermoregulated innovative bodies with a feel-good effect and, in conjunction with new creative infrastructures, consistently align them optimally in terms of energy according to local solar radiation, as Australian termites show us in their biotope? More on this later in Section II.

The ingenious "architects of nature" live in a highly complex networked environment. They are masters of architecturally elegant constructions, functionally optimised specific use of materials, energy-efficient living environments and sophisticated communication systems at all scales. At one end of the size scale we have become acquainted with the filigree architecture of the

Fig. 4.37 Disc-like structures of Australian compass termite structures in Litchfield National Park, approximately 128 km south of the city of Darwin. (with kind permission of Hans Thiele, Kiel, Germany, https://www.hansthiele.de/australia-index.htm. Accessed May 04, 2019)

microorganism with a size of less than 1/1000 mm. At the other end is the largest living thing, the *sequoia* tree *Sequoiadendron giganteum,* more than 120 m tall and about 12 m in diameter at the base. The same building principles prevail in nature's small as in its large building system.

The discovery of ancient termite burrows in the DR Congo, Africa, determined to be over 2200 years old using the well-known radio-carbon method* (Erens et al. 2015), and a termite burrow extent in Brazil nearly the size of Great Britain (Martin et al. 2018), are recent evidence of the durability and immense areal extent of termites, which are tiny in proportion but huge in the confederation.

* The radio-carbon method,[14]C or C14 dating, uses electromagnetic radiation – radiometrically – to measure the age of carbonaceous organic materials, up to an age of about 60,000 years.

What distinguishes natural building could only come about through evolutionary principles that nature has perfected over millions of years. Some of these key principles are discussed in more detail below. They are also responsible for the incomparable aesthetics and functionality of biological buildings.

- **Nine key principles of natural building design**

The following principles of natural building show the whole breadth of organismic use in terms of sustainable survivability.

1. Synergy/Biocybernetics
   Nature's master builders often know how to design their structures in such a way that several substances or forces always create a mutually *beneficial* interaction. On the one hand, there is the actual mechanical protective function for the growing offspring. The colours of the nests, huts, hangings, tubes and other constructions camouflage. Structures are specifically placed in locations that enemies have difficulty reaching. They optimize the heat exchange with the environment or they minimize the flow resistance. The beaver, for example, shows us how various regenerative energy technologies can be used in parallel in a natural environment with its dwelling made of wood sticks, brushwood and other organic materials.

2. Choice of materials
   Nature focuses on a few basic organic materials for nesting and other structures. These include cellulose and keratin. By modifying their structure and by combining them with inorganic materials, new groups of building materials with new properties are created. For example, the rich variety of biological adhesives can be traced back to only four basic materials: Lipids (e.g. terpene resins of conifers or extremely sticky terpenes of termites of the species *Nasutitermes),* polyphenols (e.g. plant tannins, lignin), polysaccharides (e.g. gum arabic, basic substance of organic connective tissue, parts of wasp and hornet nests) and proteins (e.g. elastin, colagen, fibrin or the above mentioned keratin).
   It is always amazing to see how nature has created a rich palette of specific construction solutions with just a few material building blocks, whose materials are without exception perfectly recycled again after their life cycle, without leaving the slightest waste. What a difference to our technical building materials industry and the ever more advanced development of highly specialised artificial materials, whose natural disposal has not been solved at all!

3. Complete recyclability
   The basic materials used are fully recyclable through biological processes, even if the time required for this varies and depends on the demands made on the durability of the buildings. In most cases, the raw materials are recycled within a one-year cycle. It is not uncommon for materials to be recycled or reused immediately after being used by other living creatures. Although there are also long-term durable structural protective coverings,

such as the bark of 1000-year-old *sequoia* trees, they do not pose any environmental danger.

4. Energetic properties

Biological structures such as those of termites can provide food and life for millions of living creatures in their compact structure through intelligent, geometric construction of the structure in conjunction with a cycle-oriented energy-efficient heat exchange process. The continuing trend towards urbanisation and thus the housing of millions of people in a confined space in skyscrapers, combined with a prevailing trend in the industrialised nations to waste energy, is the human, building technology counterpart. Can we not, indeed must we not, fundamentally rethink our technical architecture if we increasingly put the sustainability of our actions, which is essential for survival, to the quality test?

5. Shape optimization

In many cases for the fulfilment of structural tasks, nature has developed spherical, sphere-like or elliptical, in any case rounded forms. In doing so, she realized numerous space-saving and material-saving forms from individual dwellings to the housing of entire peoples.

6. Structural optimization

Many natural structures exhibit a special design feature in their optimal spatial extent and on their surfaces, namely that of the fractal, i.e. self-similar structure. Fractal structures – combined with other features – provide optimal chemical and physical properties, for example high stability under static and dynamic load in conjunction with self-limited growth. The dwellings of bivalves (molluscs) are typical of this.

7. Dimensioning

Another trick of nature to excellently optimize structures according to their needs lies – as already indicated above – in the dimension of the dimensions. In the microscopic material range of micro- and nanometres ($1 \, \mu m = 10^{-6} \, m$, $1 \, nm = 10^{-9} \, m$), material composites of many organismic structures are models for improvements of technical material properties such as breaking strength, flexibility or stretchability.

8. Planned time program

Biological structures are often designed to perform different tasks over their life cycle, to which they adapt. In the process, they change their colour, their chemical composition or their mechanical structure. On the other hand, the lifespan of natural structures is limited precisely to the development period of one generation of descendants. Why shouldn't our technical buildings, which are erected for different purposes and needs in different places, also be given a function-adapted lifespan?

9. Evolutionary optimization

In nature's "building industry" everything is just right! Not a gram of building material is wasted, not a watt of usable energy is uselessly transformed into heat, rooms are optimized in shape, walls are multifunctional, the inhabitants construct pantries, corridors and living areas in a functional and efficient arrangement.

Biological structures, like everything else in living nature, are evolutionarily optimized, or more precisely: optimally adapted to their respective habitats, depending on their specific needs. The results of the technical construction achievements of plants and animals that we see today have proven to be sustainable. – One more reason to appreciate their technical achievements in our present time, which is driven by ecological social and economic changes, and to consider whether they might not help us, in our indisputably self-created mountains of problems, to find suitably fault-tolerant solutions for our future and that of our grandchildren and great-grandchildren (cf. also Küppers 2001, p. 88–103).

4.18 Woodpeckers: Hammering Without Headaches

Spring makes their hammering sound again on trees in forests and urban areas. Woodpeckers *(Picidae)* work the bark of trees for food to build nest cavities to raise their young. Woodpeckers are:

> Birds with highly specialized beaks. Two toes are directed forward, two backward. All cavity breeders. The most common woodpecker in Central Europe is the great spotted woodpecker *(Dendrocopus major)*; the largest native species is the black woodpecker *(Dendrocopus martius)*. In addition to the great spotted woodpecker, the green woodpecker *(Picus viridis)* also frequently encroaches into gardens. (Hickman et al. 2008, p. 891)

Figure 4.38 shows a great spotted woodpecker in a typical position on the tree. Besides the foot-claw attachment to the bark, the beak hammering against the tree bark is still used by the tail as a support element.

Studies showed that woodpeckers drum against the tree with a head velocity of 6–7 m/s and a deceleration of over 1000 g, (g = acceleration due to gravity of 9.81 m/s^2) without sustaining any head injury. During a period of drumming, the woodpecker hammers 10–20 blows in succession, with 1

Fig. 4.38 Great spotted woodpecker on tree bark. The static support posture of the great spotted woodpecker's tail, which facilitates hammering against the tree trunk, can be clearly seen. (©jLasWilson on pixabay.com, for free commercial use (https://pixa-bay.com/photos/woodpecker-great-spotted-woodpecker-1411082/.   Accessed   07 May 2019))

blow, or beak tip thrust, against the tree bark taking only about 50 milliseconds. Calculated over the course of a day, a woodpecker drums against the tree an average of 12,000 times (Wang et al. 2011, p. e26490). Other authors cite "beating frequencies" of 18–22 beats and a deceleration value of 1200 g, Farah et al. (2018) even 1200–1400 g, but this corresponds to the same order of magnitude as the previously mentioned values (Yoon and Park 2011, p. 1).

The mechanical load applied to a woodpecker head during hammering is damped by the interaction of various factors, such as skeletal structure of the beak, cranial bones, and hyoid bone (Yoon and Park 2011, p. 2–3). The extremely stable position of the avian brain and fast-acting muscles are part of the holistic load damping. Figure 4.39 shows the head morphology of the woodpecker *(Melanerpes aurifrons)*, with the hyoid bone highlighted in red.

According to Gibson (2016), there are three key events that allow woodpeckers to withstand the high delays:

1. their small size, this reduces the load on the brain for a given acceleration;
2. the short duration of the impact, which increases the tolerable acceleration; and

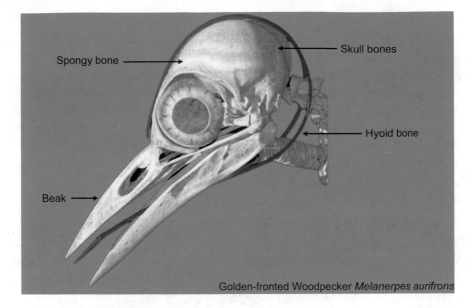

Fig. 4.39 Morphological structure of a woodpecker head of the species Golden-fronted Woodpecker *(Melanerpes aurifrons)*. (Picture with kind permission from Timothy Rowe, digimorph.org, A National Science Foundation Digital Library at The University of Texas at Austin http://digimorph.org/specimens/Melanerpes_aurifrons/)

3. the orientation of the brain within the skull, it increases the area of contact between the brain and the skull.

Smoliga and Wang (2018, Abstract) add to the previously mentioned data and information from woodpeckers as follows:

> Woodpeckers have several evolutionary adaptations to protect their brains. Computed tomography has confirmed that woodpeckers have numerous microstructural adaptations in the skull, including regionally specific changes in trabecular bone morphology (osseous tubercle-like structures, the author) that provide favorable mechanical properties. The unique tongue and lingual structure of the woodpecker dissipates shock. In addition, the woodpecker's beak is highly specialized, (in that the two beak tips, the outer tissue of the upper beak portion is about 1.6 millimeters longer than that of the lower, but in contrast, the bone structure of the upper beak portion is 1.2 millimeters shorter than that of the lower beak portion – causing both to impact with minimal delay). The impact load is thus distributed and the maximum value is reduced. (see Wang et al. 2011, p. e26490)

For us humans these are unimaginable, but for the small feathered birds evolutionary adaptive achievements of survival. Nature – and certainly not man with all his creativity for the development of technical systems – has developed what is probably the most efficient shock absorber of all with the head construction of woodpeckers. Whether technical solutions can already be derived from this, which humans can use better for the purposes of damping technology than has been the case to date, will be explored later in Section II, Sect. 7.22.

4.19   Geckos: Adhesion and Release Without Adhesive

What makes geckos run up steep glass facades at lightning speed or run upside down on ceilings without the slightest sign of falling? Their stick-and-release secret lies hidden in the soles of their feet.

Comparable to the ability of morphofalters to produce colours without dyes (colour pigments), described in point 4.13, small geckos *(Gekkonidae)*, which belong to the family of pangolins and are present worldwide with more than 1000 species, possess the unique ability to walk as described above.

The gecko has four legs, each of which splits into five toes. The underside of these toes has a tripartite, hierarchical structure. Visible to the naked human eye is the mesostructure shown in Fig. 4.40, left. This refers to the adhesive lamellae of the gecko. The lamellae are 400–600 µm long and consist of adhesive hairs arranged in rows, the so-called setae. These microstructures only become visible through an electron microscope. They consist of the protein creatine and have a length of 100 µm. The diameter of a seta is 6 µm, which corresponds to one tenth of the diameter of a human hair. Each toe of the gecko's foot has several hundred thousand setae. These split into even smaller adhesive hairs, which are referred to by the technical term spatulae. A seta has several hundred spatulae (Fig. 4.40, right). With a width and length of 200 nm and a diameter of 10–15 nm, these hairs are the smallest structure of a gecko foot. They are very elastic and consist of only four to five creatine molecules. The spatulae are in direct contact with the surface. Due to the fine ramifications of the adhesive hairs, the gecko is able to adapt to uneven surfaces (cf. Wengemeyer 2007, p. 2).

If there are no adhesives that create temporary chemical bonds between the gecko's feet and the surfaces it walks on, how do the little geckos hold on upside down without falling off? The Van der Waals interactions, named after

Fig. 4.40 Macroscopic feet of a gecko, left; scanning electron micrograph – SEM photo – of section of a gecko foot *(Hemidactylus frenatus)*, right. (©Dennis Kunkel, http://www.denniskunkel.com. Courtesy of Science Photo Library/Stock Food GmbH, Munich, Germany. Accessed 08 May 2019)

the Dutch physicist Johannes Diderik van der Waals (1837–1923), so-called covalent bonds (chemical bonds of atoms in molecularly structured compounds) play the main role. These are admittedly weaker intermolecular forces that become effective when atoms and molecules come very close to each other (cf. Kickelbick 2008, p. 98; Autumn et al. 2002).

Geckos therefore do not use chemical compounds for their hold but adhere purely physically. The feet of geckos are therefore perfectly developed for this special connection technique, as Fig. 4.40 impressively shows.

Autumn et al. (2000) made measurements of the adhesive force of a single specialized hairs *(seta)* on the foot of a tokay gecko *(Gekko gecko)*, which has nearly 5000 setae per 100 mm² of adhesive area. Thus, a gecko foot achieves an adhesive force of 10 N (N = Newton) per 100 mm², a single hairs thus an average force of 20 μN. Distinctive macroscopic orientation and biasing of the hairs increases the binding force by 600 times the force produced by a frictional force measurement of the material on which the gecko moves. Purposefully oriented hairs reduce the force for detachment from the adhesive base by simply lifting off through a critical angle. If all hairs are simultaneously and maximally adhesive, a single gecko foot is capable of developing 100 N of adhesive force (Autumn et al. 2000, p. 683). An adult gecko weighing 50 g is capable of briefly holding a weight of over 100 km with its four feet.

Other sources on exploring the mysteries of geckos include Lee et al. (2007); Higham et al. (2019) and Russell et al. (2019). The latter provides an overview on "Unified biology of gecko adhesion, history, current understanding and grand challenges".

However, more than 30 years of research with geckos have by no means uncovered all the secrets that nature has to offer: Geckos sprint at high speeds over difficult terrain, swing under leaves at lightning speed, right themselves by refueling with air just by swinging their tails back and forth, or cling upside down to slippery surfaces. Nirody et al. (2018) add to this range of capabilities the gecko's ability to run quadrupedally over water at speeds close to those they achieve on land. To do this, they use multiple techniques through surface flapping, undulating body and tail movements, and surface tension of the water, as well as their superhydrophobic – highly water repellent – skin surface.

As far as the ability of organisms to use highly water-repellent surfaces to their advantage is concerned, we can again look back to the realm of flora, where lotus leaves also produce this superhydrophobic surface effect.

Advantageous techniques are developed in nature not only once, but quite repeatedly and in parallel, as the examples of the lotus plant and the gecko prove.

4.20   Andean Condors: Majestic Gliders Without Precedent

As recognizable in the name of the bird, Andean condors *(vultur gryphus)* are New World vultures. The flight range of the birds of prey and scavengers in South America extends over about 8000 km along the Andes region from the southern tip of Chile (Tierra del Fuego) to Venezuela in the north.

With its approximately three meters long wingspans, they are outdone only by the pointed-winged sea-flyers, the albatrosses, with king-albatrosses, wingspans of almost 3.5 m were measured.

Figure 4.41 shows an Andean condor flying over Peru. Figure 4.42 shows the same condor with aerodynamic details. Despite weighing almost 15 kg, the bird soars through the air for hours – without any wing beat – circling, and this at an altitude of 5000 m and more. The extremely energy-saving gliding flight of the condor uses – like many organisms as well – several tricks of nature at the same time.

The bird of prey cleverly exploits the thermals present on the mountain slopes and in between, a warm updraft that rises to altitude as a convection current or vertical air current due to the warming of the ground by the sun. Its gliding direction inclined to the ground, combined with the direction of the vertical air current leads it to great heights, as already mentioned, almost without flapping its wings.

Fig. 4.41 Andean condor in quiet flight over Peru. (Image source: wrupcich/pixabay, for free commercial use https://pixabay.com/de/photos/kondor-flug-himmel-peru-fliegen-943300/. Accessed 16 May 2019)

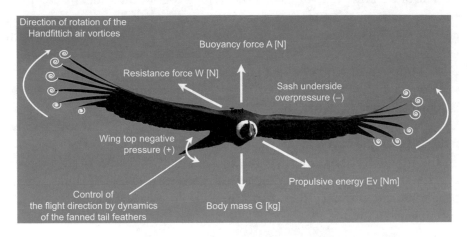

Fig. 4.42 Andean condor in quiet flight over Peru, as in Fig. 4.41, supplemented by aerodynamic details d. d. A

In the flight of the Andean condor – as with any other flying bird or insect – the air flows over the leading edge of the wings and is divided, into an area above and below the wing.

The profiles of the wings are convex on the upper side and concave underneath. This creates different pressure areas, above the wings there is overpressure, below negative pressure. The effort of the air to equalize this pressure difference takes place at the wing tips. If these were to end abruptly and with a smooth edge, the bird would probably have no chance of flying, the air resistance would be too strong, which would become noticeable there as induced air resistance or edge resistance. Large rotating air vortices would be created, which would be directed from the underside of the wing to the upper side of the wing and would "eat up" enormous propulsive energy of the bird.

The ingenious trick of the Andean condor and all birds, which have to generate relatively high lift due to their relatively large body weight through wide wings, is to minimize the energy-consuming edge resistance by the air pressure compensation zone at the end of the wing running out into several smaller hand-fits. This divides the theoretically large tip vortex into several smaller tip vortices, which in sum significantly reduce the tip vortex energy with the square of the vortex core diameter compared to one large tip vortex. Albatrosses, on the other hand, use their elongated, pointed wings to generate a minimum of boundary vortex energy.

Another aerodynamic trick of the bird is just as recognizable, in that the air flowing over the top of the wing, which acts as the frictional drag energy of the bird, is likewise reduced to a necessary minimum by the finely branched flexible bird feathers. Finally, along the trailing edge of the wing, further pressure equalization takes place with swirling and energy-draining drag acting against the bird's propulsive energy. The serrated edge along the entire length of the trailing edge of the wing is another trick of the Andean condor – or other birds – to fly with as little drag loss as possible for a given lift.

It was only in the first half of the twentieth century that the ingenious engineer Ludwig Prandtl (1875–1953) succeeded with his theory of lift in laying the foundation for calculable technical airfoil profiles, whose aerodynamics also apply to the flight of birds (cf. Sigloch 1996, p. 217).

The ingenuity of nature finds in the sublime flying skills of the Andean condor another example of aeronautical skill, with energetic and material properties and efficiency that man – despite all technical advances – is still running behind today.

4.21   Mantis Shrimps: Speed Is Not Magic

Mantis shrimps *(Stomatopoda)* have conquered the bottom of tropical seas and are distributed among approximately 400 species, including many solitary species of *Odontodactylus spec.* The name mantis shrimp is owed to the insect mantis shrimp or praying mantis, which in an upright position with bent forelimbs makes them spring forward at lightning speed after a prey.

A similar movement is performed by mantis shrimps, such as *Odontodactylus scyllarus,* which smash the chitinous shell of a prey animal with lightning speed and enormous force.

The colorful crab (Fig. 4.43) has two remarkable features:

1. his incredible lightning-fast punch when catching prey,
2. his extraordinary eyesight.

Fig. 4.43   Three perspectives of the mantis shrimp

1. **Powerful quick strikes secure the prey**[16]

Here the full force of the destruction can be seen.[17]

The following figures of blistering punching power are yet another proof of the unsurpassed patent wealth of nature, whose outstanding physical and chemical properties are based only on a handful of cleverly combined materials.

The following description of beating or flapping crustaceans, such as *Odontodactylus scyllarus* summarizes some numerical data on the movement process,[18] with the associated, scientifically sound source references inserted:

---

[16]To illustration: Female mantis shrimp *(Odontodactylus spec.)*, free use, https://reckenkrebse-weiblich-587725/, left.

Mantis shrimp *(Odontodactylus scyllarus)*, Andaman Sea, Thailand, photo Silke Baron, licensed under the Creative Commons Attribution CC (BY) 2.0, https://commons.wikimedia.org/w/index.php?curid = 8321651, right. The crab's two tense club-like "battering hammers" can be seen quite clearly. (both accessed May 18, 2019).

[17]About the illustration: a male clown mantis shrimp (Odontodactylus scyllarus) smashing the hard shell of a snail shell. © Roy Caldwell, with his kind permission, Department of Integrative Biology, University of California, Berkeley, Berkeley, CA, USA.

[18]https://de.wikipedia.org/wiki/Fangschreckenkrebse (accessed May 17, 2019).

Odontodactylus scyllarus hook parts of their exoskeleton for flapping, tense the strong muscles and then let the tentacles advance in an explosive movement (see also below: Patek 2016). The striking leg of the mantis shrimp develops a speed of 23 m/s (equivalent to 82.8 km/h) in this process; the accelerations that occur in this process are up to 8000 times the acceleration due to gravity (Patek and Caldwell 2005, even speak of acceleration values that exceed 10^5 m/s² (!)). A human blink takes about 40 times as long as this leg stroke. This blow is one of the fastest movements executed by an animal. The impact force is similar to that of a pistol bullet (Patek and Caldwell 2005; Patek et al. 2004).

However, the crab only develops its full power with the help of gas bubbles. When the hammer arms move through the water at high speed, they create a high negative pressure. Tiny gas bubbles form, which then implode, releasing an extreme amount of energy (cavitation). This creates a bang and sometimes even a flash of light. The victim is stunned by the blow (Patek and Caldwell 2005), [...].

Patek (2016, p. 24) contrasted "organisms with super-fast extremities" in her paper "The fastest movements of living things" in a Table 4.5, reproduced here.

Patek (ibid., p. 25) compares the ultra-fast movement of the club-like thickenings of the crab's tentacles to the technique of archery. Throwing an arrow with the bare hand results in moderate velocities and low throwing power. In contrast, the velocity of an arrow with a cocked bow and the force with which it strikes the target are many times greater. The mantis shrimp has therefore developed a sophisticated technique of club-like rapid movement involving joints, elastic structures, muscles, and tendons with bar-like devices. The diagram in Fig. 4.44 from Patek (ibid., p. 25) shows the moment of rest and release to strike.

The enormous force of the impact destroys even the hardest materials such as composite ceramic shells of crabs and mussels, which consist of biological structural polymers in combination with further organic and inorganic components. M. Vieweg (2018)[19] describes the experimental results of Grunenfelder et al. (2018) as follows:

> The researchers have already been able to show that the whimsical animals protect their attack tools from damage by means of a sophisticated composite structure. Special mineral compounds are used in combination with chitin – the material that also gives strength to the shells of insects and crustaceans. Unique structures and, above all, clever layering give the material additional stability,

---

[19] https://www.wissenschaft.de/technik-digitales/geheimnisse-tierischer-schlagkraft-gelueftet/ (accessed May 18, 2019).

Table 4.5 "Organisms with super-fast extremities, such as mantis shrimp and snapping-jaw ants, move their body parts far more rapidly than we move the lids of our eyes. Jellyfish with cnidocytes that explosively release a spiked cnidarian thread also produce record-fast movements." (Patek 2016, p. 24)

Duration 10^{-6} s or 1 μs	Speed (m/s)	Acceleration (m/s²)
0.7 Cnidoblast	67 Mandibles of snapping jaw ants, Termite mandibles	10^7 Cnidoblast
10 Fungal spores	58 Swoop of a hawk	10^6 Mandibles from Snapping pine ants
25 Termite mandibles	37 Cnidoblast	10^5 Butterfly strike of a mantis shrimp, Ballistospore
100–300 Mandibles of snapping jaw ants	31 Butterfly of a mantis shrimp	10^3 Spear thrust of a mantis shrimp
1000–6000 Blow of a mantis shrimp	25–26 Sprint of a cheetah, Water jet of a cracker	10^2 Fleeing fish, Leap of a grasshopper
10,000 Leap of a grasshopper	10 (9.58, Usain Bolt, 2009) Ø-speed of a 100 m sprinter 12.5 MAX speed of a 100 m sprinter	10 Sprint of a cheetah, Frog hop
100,000 Frog hop	2–3 Ballistospore, Froghopper, Grasshopper jump	
300,000 Blink of an eye		

the analyses showed. The exterior of the clubs, known as the impact area, has a hard, crack-resistant coating that ensures the prey receives the full force of the blow. The researchers' latest findings now concern the structures in the club that are responsible for shock absorption. According to the study, the inside of the club has two areas: an energy-absorbing structure that prevents cracks from forming along a series of long spiral fibers, and in addition, a special striped region. This striated region includes a highly aligned fiber structure that surrounds the entire lobe. As the researchers explain, this component prevents the club from expanding upon impact. "We believe that the role of the fiber-reinforced striped region in the club is similar to hand banding in boxers: This concept compresses the blow and prevents cracking. The combination of the

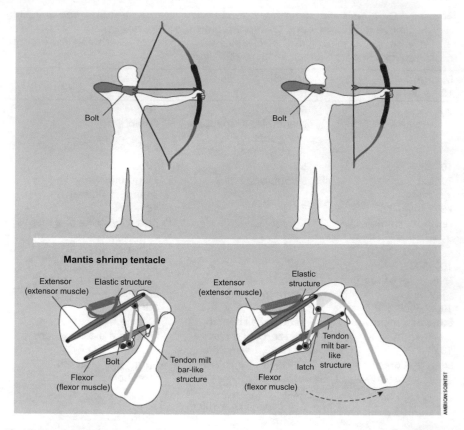

Fig. 4.44 Principle sketch of the rapid movement of the club-like thickening on the mantis shrimp's tentacles, with analogy to archery. Sketch from Patek (2016, p. 25). (Source: American Scientist)

different components creates a weapon of incredible strength, durability and impact resistance," sums up Kisailus (one of the researchers involved, the author).

Figure 4.45 shows the special architecture of the mantis shrimp club, which itself suffers little damage despite powerful high-velocity impact. "The regional substructural complexity of mantis shrimp lobes presents a model system for impact-tolerant biomineralization," according to Grunenfelder et al. (ibid.). The highly mineralized impact or contact zone of the club corresponds to a characteristic Bouligand[20] architecture, as shown in Fig. 4.45 on the right,

---

[20] A Bouligand structure is a layered and rotating microstructure similar to plywood. It exists of multiple, directional fibrous leaflets or layers that rotate in a stepwise helical or spiral fashion to their adjacent layers. This material structure increases biomechanical resistance to cracking while increasing strength in directionally independent isotropic – planes. See also: https://en.wikipedia.org/wiki/Bouligand_structure (accessed May 18, 2019).

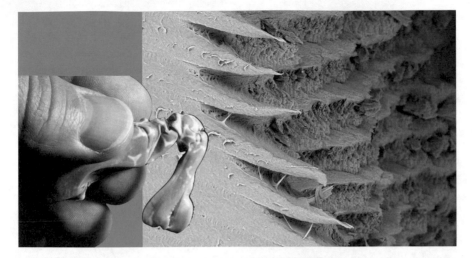

Fig. 4.45 Flapping club *Odontodactylus scyllarus, left,* Bouligand architecture as SEM image, right. (With kind permission from Pablo Zavattieri, Purdue University, West Lafayette, IN, USA)

which occurs in all arthropods. On Bouligand structure, see also Cölfen (2018) and Suksangpanya et al. (2017).

With such material-technical sophistication of nature, arthropods have lived and survived on earth for about 300 million years. What advantages from a material-technical engineering point of view could man draw from a systematic exploration of this highly efficient trick and other ingenious natural patents for his own sustainable development and environment, if he does not at the same time sacrifice too much and partly useless energy for destroying nature and thus also its known and still unknown secrets?

2. **Seeing with unimaginable variation**

Let us also look at the amazing ability of the mantis shrimp to see, which is far beyond human vision. Glaeser and Paulus (2014, p. 39) have written the following in "The Evolution of the Eye":

Particularly bizarre eyes are found in mantis shrimps (Stomatopoda). Their stalked eyes are either spherical or transversely oval (see Fig. 4.43, the author). They always possess a transverse band of about six rows of ommatidia in the middle, which divides the compound eye. This band is one of the most complicated sensors in the animal kingdom, analyzing not only over 100,000 colors but also ultraviolet and polarized light. This is because animals communicate via colored light signals. While the upper and lower halves are used to detect shapes and movement, the highly complex sensor with its six rows of ommatidia is responsible for color perception and detection of polarization. The field of view of this sensor is about 10–15 degrees. The independently moving eyes allow the crab to scan the shape with one eye and superimpose color detection and polarization with the other. The first four rows are specialized for color perception. Each row has eight different types of visual pigments, ranging from 400 nm (violet) to 550 nm (green). The next two rows allow the animal to detect linearly polarized light. This is achieved in part because the visual pigments in the individual eyes have a different orientation. This greatly improves contrast detection, which is important in the low-contrast underwater environment.

Glaeser and Paul (2014, p. 98) draw the following conclusion from this:

[…] the eyes of the mantis shrimp (are) almost phenomenally highly developed and far superior in complexity to those of other crustaceans, and probably even to the eyes of most other animals. In some species, the precision instruments may consist of up to 10,000 individual eyes (ommatidia). The central band can analyze not only over 100,000 colors, but also ultraviolet and polarized light (for review, see Chiou et al. 2008; Marshall et al. 1991, 1999).

With particular reference to the nature of seeing, Glaeser and Paul (2014, p. 99) note:

Spatial Vision with Only One Eye
The eye of a mantis shrimp (using *Odontodactylus scyllarus as* an example) is constructed quite similarly to that of other crabs with apposition eyes. The central area consisting of six rows of ommatidia divides the eye into an upper and lower hemisphere. The optical axes of the six rows are oriented exactly parallel to the front, while the first rows of the upper and lower spheres are slightly inclined inward and cross each other. Outer rows are parallel to the middle rows, and even more adjacent rows are inclined outward. This unique construction in the animal kingdom allows the crab to see a triple image with only one eye and thus to see spatially. While the upper and lower hemispheres are used to detect shapes and movement, the highly complex sensor with its six rows of ommatidia is responsible for color perception and detection of polarization. The field of view of this sensor is not particularly large and is only about 10–15 degrees. The independently movable eyes allow the crab to scan the shape with one eye and superimpose color detection and polarization with the other.

See and Interpret Polarized Light

Animals that have the special ability to perceive the alignment pattern of waves can, for example, recognize where the sun is even when the sky is overcast: depending on the position of the sun, the polarization pattern of the light in the atmosphere changes. Honey bees, for example, orient themselves according to this pattern. A previously unknown pattern of vision has been discovered in mantis shrimps *(Stomatopoda)*, namely to perceive circularly polarized light. Such light could be imagined as a kind of spiral in which the plane of polarization rotates in the direction of propagation of the light. The crayfish can even distinguish between left- and right-polarized light, which plays a role in sexual behavior and is not perceived by other animals. Technical polarization filters perform this essential function in DVD and CD players and to some extent in cameras. However, these artificial filters usually only work for one color of light, while those of the mantis shrimp eyes work almost perfectly across the entire visual spectrum, from near ultraviolet to infrared. Similar reflection patterns are known so far only from a Mexican green leafhopper *(Chrysina gloriosa)*.

Seeing across the entire visual spectrum, from near ultraviolet – UV –, with a wavelength of less than or equal to 390 nm (nm), to infrared – IR –, with a wavelength of greater than or equal to 780 nm, with the peculiarity of being able to provide this visual performance even in water, shows once again how far ahead of us nature is with its "technical achievements".

4.22   Radiolarians: World Champion Architecture at the Micrometre Scale

Barely visible to the naked eye, increasing magnification into the micrometer range of life opens up a whole new world of unimagined filigree constructions of nature. *Radiolaria* are unicellular organisms with radially protruding spine-like projections. The spherically or rotationally symmetrically constructed, skeleton-like structures reach a size of about 50–500 µm or 0.05–0.5 mm. Figure 4.46 shows the filigree beauty of these extraordinarily material-efficient and sturdily built little animals.

Hickman et al. (2008, p. 364–365) describe radiolarians as follows:

Fig. 4.46 Selection of radiolarians. (Plate on the left. Source: Public Domain Ernst Haeckel 1904, https://commons.wikimedia.org/wiki/Ernst_Haeckel#/media/File:Haeckel_Cyrtoidea.jpg. Panel in the middle. Source: Sijthoff, Leiden 1930, https://commons.wikimedia.org/wiki/Category:Radiolaria?uselang=de#/media/File:Radiolaria__Sijthoff_1930.jpg. Panel on the right. Source: WikiImages, Free download epub https://pixabay.com/photos/single-celled-organisms-radiolarians-63106/. All three downloads: Accessed 25 May 2019)

The term radiolarian refers to testate[21] amoebae of the sea with intricately built skeletons of great beauty. [...] Almost all radiolarians are pelagic (live in the free body of water). Most are constituents of the plankton (totality of organisms living in the water, whose swimming direction is without their own power, made only by currents of the water, the author) of shallow areas of the sea, although some also occur at great depths. The body is divided by a central capsule separating an inner and an outer cytoplasmic zone (cavity, the author). The central capsule, which may be spherical, ovoid, or branched, is perforated to provide continuity of cytoplasm. The skeleton is composed of silicate, strontium sulfate ($SrSO_4$), or a combination of silicate and organic constituents, and generally has a radial arrangement of spines extending outward – from the center of the body – through the capsule. At the surface, the capsule may be fused to the spines. Around the capsule is a foamy cytoplasmic area from which the axopodia (axial feet, the author) emerge. [...] Radiolaria may contain one or more nuclei.

---

[21] Testate amoebae form a group of protozoa* in which the cytoplasm of the amoeba is contained within a separate shell.

A protozoon, plural: protozoa, is a complete living organism whose entire life activity takes place within the boundaries of a single plasma membrane. Since their cytoplasm is not divided into (further) cells, they are unicellular life forms. These cells are very similar to those of multicellular animals (Hickman et al., 2008, p. 329).

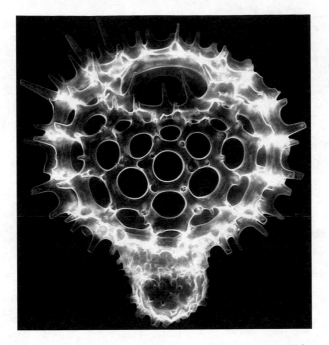

Fig. 4.47 Radiolarian. (Source: © Andreas Drews, 2016, CC BY 2.0. https://commons. wikimedia.org/wiki/Category:Radiolaria?uselang=de#/media/File:Calocycloma_sp._ Radiolarian_(32163186535).jpg. Accessed 25 May 2019)

Ernst Haeckel (1834–1919) was a German physician, zoologist and philosopher and not least known as a supporter and promoter of Darwin's theory (see Sect. 3.2.2). With a view to the underwater world, he developed undreamt-of drawing qualities, of which the right panel in Fig. 4.46 shows only a small excerpt (see also Haeckel 1998, original from 1904).

Figures 4.47 and 4.48 again show individual radiolarians, with corporeal skeletal structures that leave nothing to be desired in terms of constructive sophistication and beauty. In particular, the above-mentioned, hardly wasteful use of materials with high stability of the body structure of radiolarians has already inspired many architects to new building constructions. More on this in Practical Section II.

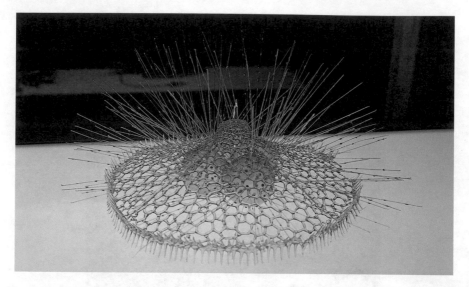

Fig. 4.48 Radiolarian. (Source: ©Blaschke-Lampromitra schultzei, 2018, CC BY 4.0, https://commons.wikimedia.org/wiki/File:Blaschka-Lampromitra_schultzei.jpg?uselang=de. Accessed May 25, 2019)

4.23   Polar Bears: Early Victims of the Anthropocene?

Polar bears *(Ursus maritimus)*, also called polar bears because they live in the icy environment of the Arctic in the northern regions of our earth, are predators of impressive size. Standing upright they reach more than 3 m and a stately weight of about 300 kg in adulthood. If we reflect the icy environment of polar bears as polar cold zone, these mammals as warm-blooded animals are able to survive temperatures of minus 70 degrees Celsius. Their dense fur, black skin and a layer of fat nearly 10 cm thick protect them from freezing to death.

But if we look at the effects of the existing climatic changes in the polar region, which, due to the melting of the ice in not too long time, will probably allow a free passage through the so-called "North-East Passage" all year round, this means a considerable restriction for the animals living there on the ice. No more and no less than an increasing decline of the habitat and presumably a drastic readaptation of the animals to the changed environment are the still unforeseeable consequences. The cause of this polar change can also be attributed to humans, who have played a large part in this global change through their massive interventions in nature, combined with their technical but not infrequently environmentally unfriendly advances.

It would be a naïve fallacy to hope that in times of increasing digitalisation *the* often postulated *power of algorithms will* find a way out of this dilemma.

But let's concentrate again on the technical achievements of the polar bear in an icy environment.

Figure 4.49 shows the underside of a rear polar bear paw next to adult polar bears.

The body of a polar bear is protected against the icy climate in three ways: First, by the densely hairy fur. The structure of the hair plays a special role in this. Secondly, by the black light-absorbing surface of the skin, comparable to the black areas of the paw in Fig. 4.49. Thirdly, by a layer of fat up to ten centimetres thick beneath the skin.

Polar bears are wanderers between the wavelengths of light. Their fur or hair appears white to us, but is transparent and comparable to hollow cylinders. They adapt perfectly to their white environment.

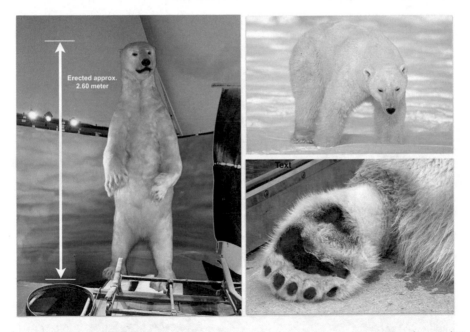

Fig. 4.49 Adult (stuffed) polar bear in upright position, Frammuseum Bygdøy, Oslo, own photo left, polar bear in natural environment, top right, own photo, hind paw of polar bear from Karlsruhe Zoo, bottom right, photo source: https://de.wikipedia.org/wiki/Benutzer:Srvban, Creative Commons license 3.0. https://de.wikipedia.org/wiki/Datei:Eisbaer_hintertatze-3.jpg (Accessed 25.05.2019)

When researchers wanted to photograph and count the warm-blooded polar bears with infrared (IR) cameras (infrared-sensitive films detect heat in the wavelength range around 770 nm and larger particularly well), they experienced a surprise: The developed films did not show a single polar bear, although they had seen them in nature with their own eyes! Only ultraviolet (UV) sensitive films (UV light of 390 nm and smaller has higher energy and shorter wavelength rays than IR light), with which the polar bears were photographed again, showed the polar bears as black dots in the white Arctic Ocean after film development (Lavigne and Øristland 1974). Figure 4.50 shows an IR image of a polar bear whose internal, heat-radiating metabolic temperature is well shielded from the colder external temperature by its insulated, dense, bluish-appearing pelt.

The fur of the polar bear, whose hair resembles small tubes as mentioned before, collects the energy-rich ultraviolet light at the north pole, conducts it with the help of various physical mechanisms through the dense tube-shaped hairs to the black body surface and is absorbed there as heat. Little of this heat is lost to the outside. For this reason, it was initially not possible for the cameramen to view the polar bears via heat radiation.

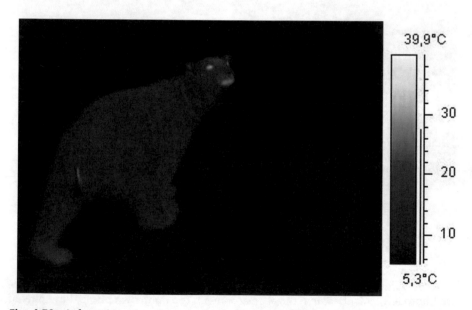

Fig. 4.50 Infrared image of a polar bear. (Source: Arno/Coen 2007. https://commons.wikimedia.org/wiki/File:Wiki_polarbear.jpg Public domain, for free use. Accessed 25 May 2019)

Experiments from the 1980s and 1990s attempted to get to the bottom of the physical phenomenon with concrete metrological processes, by answering the questions:

1. How does the polar bear manage to maintain a body temperature, like ours, of about 37 °C in an environmental atmosphere that can be −70 °C and even colder?
2. How does the polar bear orient itself in the fog of ice? Although they have an excellent sense of smell and hearing, they have similar vision to humans; surrounded by dense fog, we see – nothing!
3. **The pros and cons of scientific statements**

A number of experimental physical investigations into the presumed ability of polar bear fur to collect and transmit light took place in the 1980s and 1990s. Among other things, Grojean et al. (1980) investigated the thermal and thermoregulatory capabilities of polar bear fur and described a qualitative optical model of polar bear fur.

Tributsch et al. (1990) continued the idea of the polar bear as a "light collector" (heat influx) and investigated the question of the detection of solar radiation by the polar bear fur. The polar bear's fur, according to them, acts as a translucent – transparent – insulation, whereby diffuse light is absorbed by a combination of light scattering, and luminescence – physical luminous effect without temperature increase – and transmitted to the body surface.

The thesis of both groups of authors and many others was rejected by Koon (1998), taking into account other sources (Bohren and Sardie 1981; Bendit and Ross 1961), on the grounds that polar bear hairs have no fibre-optic ability to conduct light over typical hair lengths of 5 cm and more. The appearance of polar bears in UV light can be explained in terms of the known UV-absorbing properties of the protein keratin, which is what hair is made of. Keratin absorbs more light in UV instead of visible light.

Amstrup (2003), a wildlife biologist in Anchorage, Alaska (Unites States Geological Survey) supported the thesis of Koon (1998), as well as an earlier statement by Lavigne (1988) and states:

Lavigne (1988) and Koon (1998) established unequivocally that the hair of polar bears, although transparent in the visible spectrum, absorbs UV light. If the hair of polar bears absorbs UV light, it does not efficiently transmit UV light. As UV light moves down the shaft of the hair, its energy is absorbed, preventing significant energy from being transmitted to the skin.

He et al. (2011) again postulated that the distinctive labyrinth cavity structure of polar bear hairs play a significant role, as seen in the SEM images in Fig. 4.51 and indicated in the principle sketch in Fig. 4.52. The hairs not only avoid body temperature loss but are also able to absorb energy from the environment.

The special structure allows the hairs to absorb energy from the environment, including radiant energy, wind chill, and cold ice-free water. Each vacuum micro-maze is a good heat insulator and we suspect that at certain times the temperature inside the structural mazes of the hairs is equal to the outside temperature, for example –50 °C. However, this cannot be maintained for long because of the large temperature difference between the body temperature and the outside temperature, which is why the polar bear uses the warmest spots for it of 0 °C in ice-free water, according to the authors. After a short period of time, a smaller temperature difference is established, so that the hairs as thermal conductors conduct the energy of the cold water in the individual hair labyrinths and therefore a temperature of 0° of the ice-free water is established as. Now, due to the resulting low temperature difference between

Fig. 4.51 SEM images of polar bear hair, with kind permission from Ji-Huan He, Soochow University Suzhou, China

body temperature 37 °C and outside temperature 0 °C, the polar bear can again use energy for hunting. In this energy-rich phase of activities on the ice, a higher temperature difference (e.g. ΔT = 87 °C) between body heat and ambient cold is successively and gradually re-established in the hair labyrinths. The hairs can now absorb energy from the environment again if their temperature is higher than the lowest of wind, water and ice (ibid., p. 912–912).

This brief description can be understood in Fig. 4.52 on the right, by the different temperature indications in the labyrinth structures of the hairs.

But the authors also say that it is still a mystery how the polar bear maintains its body temperature of 37 °C against an environmental temperature of −50 °C and more – temperature difference (ΔT) of almost 90 °C.

Due to the objection that polar bear hairs cannot be defined as light-conducting and light-collecting transparent optical fibers, Khattab and Tributsch (2015) again started an experiment to resolve the contradiction – also to their own investigations by Tributsch et al. (1990). In summary, the following statements by Khattab and Tributsch (2015, p. 1):

> With the help of new spectroscopic, microscopic and laser-optical investigations, the contradiction in the statements for and against polar bear hairs as optical light collectors (heat influx) in the statements was attempted to be

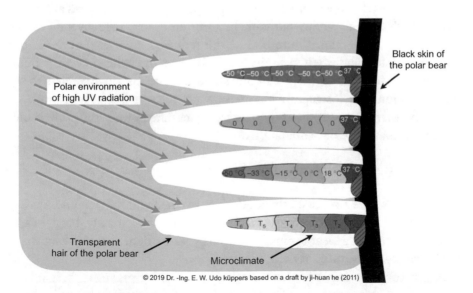

© 2019 Dr. -Ing. E. W. Udo küppers based on a draft by ji-huan he (2011)

Fig. 4.52 Principle sketch of polar bear hair with internal climate maze and adaptive temperature profiles, depending on the ambient temperature of the polar bear. Templates of the hair sketches with kind permission from Ji-Huan He, Soochow University Suzhou, China

resolved. The basic statement is: The light collecting mechanism can only be understood as a synergetic cooperation of many polar bear hairs. Light couples into the hairs by scattering processes over a short distance in order to couple out again by subsequent scattering processes, in order to couple in repeatedly into neighbouring hairs, etc., until the light changes into heat or is absorbed at the black hair surface of the polar bear. In between, a small percentage of the incident light is backscattered. The result is that the polar bear's transparent hairs appear white because they absorb most of the incident radiation.

Moreover, the solar optical technology includes a complementary strategy: IR radiation of body heat, between 8000 and 12,000 nm (nanometers) is effectively captured by an analogous mechanism. This is promoted by a high absorption capacity of the skin and the absence of any significant spectroscopic feature in the complete spectral region.

The final conclusion of the two authors is that the polar bear has developed an efficient optical nano-technology for light collection and energy conservation.

For some of you, the gentle reader, this enumeration of different research results on one and the same object, namely the dense hair fur of the polar bear, may bring little added value; likewise the following research results on the solar orientation of species such as polar bears and others.

On the other hand, research on polar bear skins over a period of almost 40 years shows that the sophisticated techniques used by organisms are not so easy to elicit from nature by humans. In addition, it may take several years or even decades to develop the results of the refined secrets of nature for our benefit and to apply their technical effectiveness and efficiency in a fault-tolerant and sustainable manner in our environment. Sustainable refers to the technical bionic microstructure materials, which are usually made of synthetic fibres and therefore (still) have to be critically evaluated in terms of their biological compatibility or degradability.

4. Orientation in visible and invisible space

Polar bears have a particularly good sense of smell and hearing and a visual ability that is comparable to human vision. This has already been briefly reported above. In a clear atmosphere, polar bears orient themselves visually to the geological features of their environment.

As for their hearing, they have a relatively sensitive organ in their watery environment. Measurement of acoustically evoked potentials for polar bears in air and estimates of their hearing ability based on environmental sounds concluded that polar bears have a wide frequency range, with a range of

11.2–22.5 kHz for the highest sensitivity (Nachtigall et al. 2007; Nummela and Yamato 2018).

Very well trained is their sense of smell, both in icy environment on the ice, as well as to detect prey below the ice surface. To this end, figures are cited that polar bears can still perceive their prey, seals, on the ice at a distance of 1000 m and more, and under thick layers of ice and snow of one meter and more.[22]

Concerning solar optical perception, Tributsch et al. (1990, p. 219) write, translated by the author:

> Because of the peculiarity of the transparent polar bear skin and the significant cooling of the surrounding tissue by the cooling atmospheric conditions, solar radiation could alter the subcutaneous (subcutaneous) temperature by more than 10 degrees Celsius. Therefore, it is likely that the polar bear's skin uses the temperature patterns created by light scattering, calibrated by wind chill against the body temperature controlled area of the large dorsal muscle, as a kind of sensing system. It may help the polar bear to approximately determine the position of the sun and in this way navigate under diffuse Arctic visibility, but also with local ice-free sea surfaces, with significantly reduced light scattering, comparable to ice surfaces.

Penacchio et al. (2015) noted on solar orientation of animals, considering their camouflage, that a large number of animals practice their orientation to the sun through proportions of thermoregulation and/or ultraviolet protection. Their results suggest that camouflage has been overlooked as a selective orientation pressure (ibid., p. 1164). Dawson et al. (2013) addressed the alleged conflict between thermal requirements and camouflage in mammals. They were able to show that although the polar bear *(Ursus maritimus)* and koala bear *(Phascolarctus cinereus)* have very different fur colorations, their heat gain from solar radiation is similar and it can be concluded that the lower the power of thermal insulation of the fur, the higher the influence of color on solar heating. It is worth noting that in some species, color in the visible light spectrum (400 nm, violet, to 750 nm, red, the author) does not correlate with spectral absorption of the body in the near infrared range (see Norris 1967 for this). So much for the ingenious techniques of the polar bear.

Polar Bear – current status: DANGEROUS!

---

[22] https://seaworld.org/animals/all-about/polar-bear/senses/ (accessed May 23, 2019).

This assessment of a species in danger of extinction could flip to status in not too long:

> Polar bear – deprived of its free habitat, therefore disappeared; only to be admired in zoological gardens or museums.

With the example of polar bears and their ingenious tricks of survival we close Chap. 4. The impression I wish to give you through the examples of the ingenious principles of organisms can only be a very superficial one; too numerous are the ingenious solutions we know or think we know. But far more numerous still are the tricks and subtleties of nature which remain hidden from us to this day. For nature does not let us look so easily into her clever cards, which reveal techniques from which we can learn a great deal of useful and valuable things for our own survival, above all sustainable further development.

In times of great upheavals, which threaten to "overthrow" our acquired *analogue* understanding of persons and things, perpetuated over millennia, an increasing look into a still often unknown treasure trove of nature's highest technical achievements can be helpful, probably more than we are currently able to suspect and reflect.

4.24 Biogeonic Meandering in Flow Processes: Flow Efficiency Through Entropy Minimization

Nothing in the biosphere happens in the airy space without taking into account the geological structures and forms. No organism develops in its biotope without taking into account the geographical and therefore also geological conditions in the non-living environment.

The fact that the living and non-living environment have often triggered parallel developments that have helped the associated organisms or processes to achieve optimal functionalities is illustrated by an example in which flow processes play a special role:

1. "Blood vessels as a *bionic* template for the optimal shaping of branched tube transport systems" (Zerbst 1987, p. 135–140).

2. "The Meander Effect®" of free-flowing water bodies as a *biogeonic* template for resistance-optimized flow and molding of engineered piping systems".

We will discuss the second example in more detail. Before that, Fig. 4.52 shows two meandering river courses, that of the Amazon in South America and that of the Green River in North America.

How Are River Meanders Formed?

Natural rivers rarely run straight for more than 10–20 × their width. At some point they begin to meander. They wind through the landscape in several bends in a sinusoidal pattern, as Fig. 4.53 shows. It is small changes in the structure of the river bed that alter the course of the flow so that a minimum of flow loss is maintained in this complex flow process.

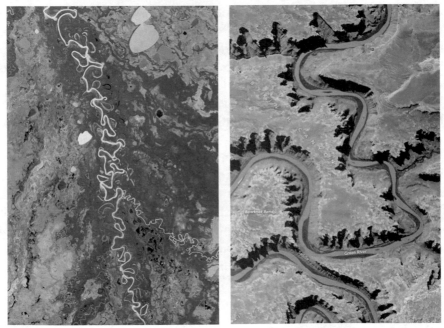

Meandering Amazon   Meandering Green River   © NASA.gov

Fig. 4.53 Amazon River in South America, left, Green River, in North America, right Source (Amazon photo): NASA photo 1985: meanders_oli_2014194 Credit: NASA Earth Observatory images by Jesse Allen, using Landsat data from the U.S. Geological Survey. Caption by Kathryn Hansen. https://earthobservatory.nasa.gov/images/84833/meandering-in-the-amazon. Accessed July 10, 2019. (Source (Green River photo): NASA photo 2018: Nasa photo iss055e031251 04/22/2018. https://earthobservatory.nasa.gov/images/92754/green-river-meanders. Accessed 10/07/2019)

Meanders, since they occur not only in free-flowing waters but also in glacial rivers or in the course of the Gulf Stream or – as mentioned above – in body currents of organisms, seem to be part of one or more universal laws of nature.

There are several dozen theories on the formation of river meanders (Mangelsdorf et al. 1990, p. 129 ff.), which will not be discussed in detail here. Finally, 3 interrelated criteria were worked out for the formation of river meanders:

1. great depth of water,
2. small flow velocity,
3. low turbulence.

Leopold and Langbein (1966) treated the formation of meanders as a probabilistic problem. They assume that meanders are *not* arbitrary natural phenomena but represent those channel forms with the *smallest work to be done by* the river – *entropy minimum principle* – when it bends. The mathematical description of these considerations by means of "random-walk technique" eventually led to the theory of "sinusoidally generated" curves. "The examination of typical river meander loops has well confirmed the analogy between the natural form and the mathematical model" (mutatis mutandis from Mangelsdorf et al. 1990, p. 137).

Different values are given for the individual coefficients and exponents in the empirically found equations depending on the author (see Mangelsdorf et al. 1990, p. 140). Leopold, in a remarkable paper from 1994, treated flux design or shaping by analogy with Darwin's selection theory. His conclusion with respect to meander shapes is that the river tends to minimize its energy loss through the characteristic and recurrent shapes of meanders.

Experimental work on flow optimization of pipe bends led Küppers (1997/1998) to the same thoughts as expressed by Leopold in 1994. In Küppers and Tributsch (2002, p. 162) it is therefore formulated that:

According to the laws of irreversible thermodynamics, the flow seeks – within the framework of given boundary conditions – a state of minimum entropy production. The conversion of free energy into unavailable energy (entropy × temperature), i.e. heat energy of the environment, strives towards a minimum.

The bionic energy efficiency approach is based on the current time window of geomorphological evolution of river meanders and posits:

Table 4.6 Comparison of characteristics of biological and technical flow systems

Features Biological flow systems	Features Engineering flow systems
Largely global networking	Largely local networking
System-optimized connections	Product-optimized linking
Long-term "input" optimization	Short-term "output" optimization
System is effective per se	Selective effectiveness control
Permanent energy efficiency control	Selective energy efficiency control
Adaptive long-term optimization	Short-term detail optimization
Intelligent loss-minimal flow redirection	Standardized, lossy Flow redirection
Geometrically intelligent design	Geometrically simple design
Dynamically variable flow parameters	Largely statically variable flow parameters
Adaptively optimized flow paths depending on variable Boundary conditions	Statically optimized flow sections
Adaptively optimized circuits of multiphase flows	Statically optimized flow characteristics of single-phase and multiphase flows
System-oriented, optimally structured surfaces of open and closed "flow beds" or active and passive bodies influenced By flows (including fish, birds, trees, leaves)	Technically "smooth" surfaces of flow guides (inner pipe walls, troughs, etc.)
Exclusively quality-optimised Flow systems	Predominantly quantity-optimized Flow systems

The current forms of free river meanders are the best possible forms of energy-efficient transport in their adaptive development within their water cycle within the open complex system of nature.

Table 4.6 shows thirteen efficiency characteristics of biological flow systems and, in anticipation of the practical application under main chapter II, the comparison with characteristics of technical flow systems.

References

Abbott, A. (1990) Bioadhesives: potential for exploitation. Science Progress Oxford, 74, 131–146

Amstrup, S. C. (2003) The Polar Bear – *Ursus maritimus* Biology, Management, and Conservation by Steven C. Amstrup, Ph.D. Published in 2003 as Chapter 27 in the second edition of Wild Mammals of North America. Edited by George A. Feldhamer et al.

Autumn, K. et al. (2002) Evidence for van der Waals adhesion in gecko setae. Proc. of the National Academy of Science, pub. online, https://doi.org/10.1073/pnas.192252799 (2002)

Autumn, K. et al. (2000) Adhesive force of a single gecko foot-hair. Nature, Vol 405, 8, 681–684

Barthlott, W.; Neinhuis, C. (1997) Purity of the sacred lotus, or escape from contamination in biological surfaces. In: Planta. Band 202, Nr. 1, S. 1–8

Barthlott, W.; Ehler, N. (1977) Rasterelektronenmikroskopie der Epidermisoberflächen von Spermatophyten. In: Tropische und subtropische Pflanzenwelt. Band 19, S. 367–467.

Bendit, E. G.; Ross, D. (1961) Techniques for obtaining the ultraviolet absorption spectrum of solid keratin. Appl. Spectrosc. 15, 103

Bohren, C. F.;. Sardie, J. M (1981) Utilization of solar radiation by polar animals: an optical model for pelts; an alternative explanation. Appl. Opt. 20, 1894–1896

Bonhomme, V.; et al. (2011) The plant-ant *Camponotus schmitzi* helps its carnivorous host-plant *Nepenthes bicalcarata* to catch its prey. J. of Tropical Ecology, Vol. 27, Issue 1, S. 15.24

Brazee, S. L.; Carrington, E. (2006) Interspecific Comparison of the Mechanical Properties of Mussel Byssus. Biological bulletin, Vol. 211, No. 3, S. 263–274

Burley, R.W.; Vadehra, D. V. (1980) The Avian Egg, Chemistry and Biology John Wiley, USA

Carwardine, M. (2008) Extreme der Natur. National Geographic Deutschland, Hamburg

Chiou, T.-H. et al. (2008) Circular Polarization Vision in a Stomatopod Crustacean Current Biology. 18, S. 429

Cölfen, H. (2018) Emerging artificial Bouligand-type structural materials. National Science Review, Volume 5, Issue 6, November 2018, Pages 786–787

Dawson, T.J.; Webster, K.N.; Maloney, S.K. (2013) The fur of mammals in exposed environments; do crypsis and thermal needs necessarily conflict? The polar bear and marsupial koala compared. Journal of Comparative Physiology B, 184, 273–284

Debat, V. et al. (2018) Why are Morpho Blue? In: Biodiversity and Evolution, 139–174

Egbert, K. J.; Martin, C. E. (2002) The Influence of Leaf Windows on the Utilization and Absorption of Radiant Energy in Seven Desert Succulents. Photosynthetica, Vol. 40, Issue 1, S. 35–39

Erens, H. et al. (2015) The age of large termite mounds – radiocarbon dating of *Macrotermes falciger* mounds of the Miombo woodland of Katanga, DR Congo. Palaeogeography, Palaeoclimatology, Palaeoecology, Volume 435, 1 October 2015, Pages 265–271

Farah G.; Siwek, D.; Cummings P. (2018) Tau accumulations in the brains of woodpeckers. PLoS ONE 13(2): e0191526. https://doi.org/https://doi.org/10.1371/journal.pone.0191526

Gibson, L. J. (2016) Woodpeckers pecking: how woodpeckers avoid brain injury. In: J. o. Zoology, Vol 270, Issue 3, 462–465

Glaeser, G.; Paulus, H. F. (2014) Die Evolution des Auges. Springer Spektrum, Berlin, Heidelberg

Grojean, R. E.; Sousa, J. A.; Henry, M. C. (1980) Utilization of solar radiation by polar animals: an optical model for pelts. Appl. Opt. 19, 339–46

Grunenfelder LK et al. (2018) Ecologically Driven Ultrastructural and Hydrodynamic Designs in Stomatopod Cuticles. In: Adv. Mater. 2018, 30, 1705295, 11 P.

Haeckel, E. (1998) Kunstformen der Natur. Prestel, München, New York, als Nachdruck des Originals aus 1904, Bibliographisches Institut, Leipzig und Wien

He, J.-H., et al. (2011) Can Polar Bear Hairs Absorb Environmental Energy? Thermal Science, Vol. 15, No. 3, S. 911–913

Heethhoff, M.; Koerner, L. (2007) Small but powerful: the oribatid mite *Archegozetes longisetosus* Aoki (Acari, Oribatida) produces disproportionately high forces. The Journal of Experimental Biology 210, 3036–3042

Hickman, C. P.; Roberts, L. S.; Larson, A.; l`Anson, H.; Eisenhour, D. J. (2008) Zoologie, 13. Akt. Aufl., Pearson, München

Higham, T. e. et al. (2019) The Ecomechanics of Gecko Adhesion: Natural Surface Topography Evolution, and Biomimetics. Integrative and Comparative Biology, icz013, 27. April, https://doi.org/10.1093/icb/icz013

Kádár, E.; Azevedo, C. (2006) Unidentified extracellular prokaryotes within the byssal threads of the deep-sea vent mussel Bathymodiolus azoricus. Parasitology (2006), 133, 509–513

Khattab, M. Q.; Tributsch, H. (2015) Fibre-Optical Light Scattering Technology in Polar Bear Hair: A Re-Evaluation and New Results. Journal of Advanced Biotechnology and Bioengineering, 2015, Vol. 3, No. 2, 1–14

Kickelbick, G. (2008) Chemie für Ingenieure. Pearson, München

Kiel, S.; Amano, K. (2013) The Earliest Bathymodiolin Mussels: An Evaluation of Eocene and Oligocene Taxa from Deep-Sea Methane Seep Deposits in Western Washington State, USA. Journal of Paleontology, 87(4):589–602, 2013

Kinoshita, S.; Yoshioka, S. (2005) Structural Colors in Nature: The Role of Regularity and Irregularity in the Structure. ChemPhysChem 6/2005, S. 1442–1459.

Kinoshita, S. et al. (2002) Photophysics of Structural Color in the Morpho Butterflies. Forma, 17, 103–121

Koon, D. W. (1998) Is polar bear hair fiber optic? Applied Optics, vol 37, Issue 15, S. 3198–3200

Korb, J. (2011) Termite mound architecture, from function to construction. In: D. E. Bignell, Y. Roisin, N. T. Lo (Hrsg.): Biology of Termites: a modern synthesis. S. 349–374, Springer, Berlin, Heidelberg, New York

Korb, J. (2003) Thermoregulation and ventilation of termite mounds. Naturwissenschaften (2003) 90:212–219

Klulik, G. A. (1980) Light transmission in window-leaved plants. Canadian Journal of Botany, 58(14), 1591–1600

Küppers, E. W. U. (2018) Die humanoide Herausforderung. Springer Vieweg, Wiesbaden

Küppers, U. (2008) Diskurs mit der Natur, Naturgesetze und Evolutionsprinzipien im Dienst nachhaltiger technisch-organisatorischer Prozesse. In: Reinauer, P. (Hrsg.) Bionik als Vorbild für die Gestaltung von Organisationsprozessen, 141–168. VDM Verlag Müller, Saarbrücken

Küppers, U. (2007) kleine Biegung, große Wirkung. Bionische Rohrbögen in der Lüftungsleitung. In: Chemie Technik, Sept. 2007, 24–26

Küppers, U.; Heyser, W. (2004) BMBF-Forschungsprojekt „Ideen-Wettbewerb Bionik", »MANGO«, PTJ-Bio/Fkz 0311980

Küppers, U.; Tributsch, H. (2002) Verpacktes Leben – Verpackte Technik. Bionik der Verpackung. Wiley-VCH, Weinheim

Küppers, E. W. U. (2001) Bionik und Bauen – Lernen von den Baumeistern der Natur. In: Becker, P.-R.; Braun, H. (2001) Nestwerk – Architektur und Lebewesen. S. 88–103, Isensee, Oldenburg

Küppers, U. (1998) Lehrmeisterin Natur. Bionik von Strömungssystemen. In: Chemie Technik, 27. Jg., Nr. 11, 84–89

Lavigne, D. M. (1988) Letter to the Editor. Scientific American, 258 (9), 8 (Sept. 1988)

Lavigne, D. M.; Øritsland, N. A. (1974) „Black Polar Bears", Nature 251, 218–9

Lee, H. et al. (2007) A reveresible wet/dry adhesive inspired by mussels and geckos. Nature, Vol 448, 338–341, July, 19.

Leopold, L. B. (1994) River morphology as an Analog to Darwin's Theory of Natural Selection. Proc. Of the American Philosophical Society, vol. 138, No. 1, S. 31–47

Leopold, L. B.; Langbein, W. B. (1966) River Meanders. Scientific American, Vol. 214, No. 6, 60–73

Mancuso, S.; Viola, A. (2015) Die Intelligenz der Pflanzen. Kunstmann, München

Mangelsdorf, J.; Scheurmann, K.; Weiss, F.-H. (1990) River Morphology. Springer Series in Physical Environment 7, Springer, Berlin, Heidelberg

Marshall, J. et al. (1999) Behavioural evidence for polarisation vision in stomatopods reveals a potential channel for communication. Current Biology 9 (14): 755–758

Marshall, J.; Land, M. F.; King, C. A.; Cronin, T. W. (1991) The compound eyes of mantis shrimps (Crustacea, Hoplocarida, Stomatopoda). In: Compound eye structure: The detection of polarized light Phil. Trans. R. Soc. Lond. B 334: 57–84.

Martin, S. J.; Funch, R. R.; Hanson, P. R.; Yoo, Eun-Hye (2018) A vast 4,000-year-old spatial pattern of termite mounds. Current biology 28, R1283–1295, Nov. 19

Mattheck, C. (2015) The Body Language of Trees. KIT, Karlsruhe

Mattheck, C. (2007) Secret design rules of nature. KIT, Karlsruhe

Mattheck, C. (1998) Design in nature – learning from trees, Springer, Heidelberg

Mattheck, C. (1991) Trees – The Mechanical Design. Springer, Heidelberg

Müller, G. K.; Müller, C. (2003) Geheimnisse der Pflanzenwelt. Manuscriptum, Leipzig

Nachtigall PE, et al. (2007) Polar bear *Ursus mariimus* hearing measured with auditory evoked potentials. Journal of Experimental Biology 2007 210: 1116–112

Nirody, J. A. et al. (2018) Geckos Race Across the Water's Surface Usind multiple Mechanisms. Current Biology 28, Dec. 2018, 4046–4051

Norris, K.S. (1967) Color adaptation in desert reptiles and its thermal relationships. Lizard Ecology: A Symposium (ed. W.W. Milstead), S. 162–229 University of Missouri Press, Columbia.

Nummela, S.; Yamato, M. (2018) Hearing Polar Bear and Sea Otter. In: Würsig et al. (Ed.), Encyclopedia of Marine Mammals (Third Edition), S. 462–470, Elsevier, Amsterdam, NL

Ogoudetji, G. P. C.; Nuppenau, E.-A.; Korb, J. (2010). The role of ecosystem services of termites *(Macrotermes bellicosus)* in agriculture in Pendjari region (Benin). Conference on InternationalResearch on Food Security, Natural Resource Management and Rural Development, ETH Zürich, Tropentag, 2010, 14.–16.9.

Patek, S. N. (2016) Die schnellsten Bewegungen von Lebewesen. In: Spektrum der Wissenschaft, Februar 2016, S. 22–27, ©American Scientist

Patek, S. N.; Caldwell, R. L. (2005) Extreme impact and cavitation forces of a biological hammer: strike forces of the peacock mantis shrimp *Odontodactylus scyllarus*. In: The Journal of Experimental Biology 208, 3655–3664. Published by The Company of Biologists 2005, https://doi.org/10.1242/jeb.01831

Patek, S. N.; Korff, W. L.; Caldwell, R. L. (2004) Deadly strike mechanism of a mantis shrimp. In: Nature, Vol. 428, 22 April, 819

Paturi, f. R. (1974) Geniale Ingenieure der Natur. Econ, Düsseldorf, Wien

Patzelt, O. (1972) Wachsen und Bauen. Konstruktionen in Natur und Technik.VEB Verlag für Bauwesen, Berlin

Penacchio, O. et al. (2015) Orientation to the sun by animals and its interaction with crypsis. Functional Ecology 2015, 29, 1165–1177

Rahn, R.; Paganelli, C.V.; Ar, A.J. (1987) Pores and gas exchange of avian eggs. A review, J. Exp. Zool. Suppl. I, 165–172

Russell, A. P. et al. (2019) The Integrative Biology of Gecko Adhesion: Historical Review, Current Understanding and Grand Challenges. Oxford Press, GB

Saroyan, J. R. et. al. (1970) Repair and reattachmant in the Balanidae as related to their cementing mechanism. Biological Bulletin, 139, 333–350

Schanderl, H. (1935) Untersuchungen über die Lichtverhältnisse im Innern von Hartlaub- und Sukkulentenblättern. Planta, Vol. 24, Issue 3, S. 454–469

Schlichting, H. J.; Rodewald, B. (1988) Von großen und kleinen Tieren. Praxis der Naturwissenschaften-Physik 37/5, 2, S. 1–7

Sigloch, H. (1996) Technische Fluidmechanik. 4. Aufl. Springer, Berlin, Heidelberg

Smoliga, J. M.; Wang, L. (2018) Woodpeckers don't play football: implications for novel brain protection devices using mild jugular compression. In: Sport Medicine, vol 53, Issue 20, Editorial

Suksangpanya, N. et al., (2017) Twisting cracks in Bouligand structures. Journal of the Mechanical Behavior of Biomedical Materials Volume 76, December 2017, Pages 38–57

Thubaut, J. et al. (2013). The contrasted evolutionary fates of deep-sea chemosynthetic mussels (Bivalvia, Bathymodiolinae). Ecology and Evolution 2013; 3(14): 4748–4766

Tributsch, H.; Goslowsky, H.; Küppers, U.; Wetzel, H. (1990) Light collection and solar sensing through the polar bear pelt. Solar Energy Materials 21, 219–236

Tyler, C. (1970) How an egg shell is made. Sci. Am., 222, 88–94

Vieweg, M. (2018) Geheimnisse tierischer Schlagkraft entlüftet. ©wissenschaft.de, 16. Januar 2018

Wang L, Cheung JT-M, Pu F, Li D, Zhang M, et al. (2011) Why Do Woodpeckers Resist Head Impact Injury: A Biomechanical Investigation. PLoS ONE 6(10):e26490. https://doi.org/https://doi.org/10.1371/journal.pone.0026490

Waite, J. C. (1987) Nature´s underwater adhesive specialist. International Journal of Adhasion and Adhasives, 7, 9–14

Wengemeyer R. (2007) Mit unbeschränkter Haftung – wie Gecko&Co die Materialforschung inspirieren, Techmax, München, Ausgabe 8 2007, Max Planck Gesellschaft, S. 1

Wittig, R.; Niekisch, M. (2014) Biodiversität: Grundlagen, Gefährdung, Schutz. Springer Spektrum, Berlin, Heidelberg

Wohlleben, P. (2015) Das geheime Leben der Bäume. Was sie fühlen wie sie kommunizieren – die Entdeckung einer verborgenen Welt. Ludwig, München

Wolf, H. (2002) Aloe dichotoma. III-4, S. 1–5 in: Schui I, P.; Weisgerbere, H.; Schuck, H. J.; Lang, U. M.; Roloff, A. (Hrsg.) (2002) Enzyklopädie der Holzgewächse 28, 7/02, Ecomed,?, aktuell: Wiley VCH, Weinheim

Yoon, S.-H.; Park, S. (2011) A mechanical analysis of woodpecker drumming and its application to shock-absorbing systems. Bioinsp. Biomim. 6 (2011) 016003 (12 pp)

Zerbst, E. W. (1987) Bionik. Biologische Funktionsprinzipien und ihre technischen Anwendungen. B. G. Teubner, Stuttgart

Zhou, L. et al. (2019) Butterfly Wing Hears Sound: Acoustic Detection Using Biophotonic Nanostructure. Nano Lett., 19, 4, 2627–2633

5

Operational Principles of Nature: Universal Development Tools of Long-Term Proven Biodiversity-Rich Management

Abstract This chapter describes universal developmental tools of nature that have been identified and are intended to show why nature's biodiversity has been able to unfold so richly in an environment whose complexity is almost impossible for us humans to penetrate, let alone fully grasp its effectiveness. Nevertheless, it seems to be easy for plants and animals in association to come to terms with the dynamics of the complex environment and to ensure their continued existence. Only one organism *steps* out of line: man. In his unfathomable overestimation of himself – to which the misinterpreted concept of the "survival of the fittest" still clings today – he exceeds the stress limits of natural and technical systems, which can only lead in one direction: system destruction.

Through individual sustainability strategies of plants and animals in their biological-geological biotopes – with tools that consist of self-organisation, self-regulation with stabilising feedback, symbiosis and much more – they maintain their ability to survive and to develop further through ingenious networking, despite occasional setbacks due to environmental disturbances, in which not least humans are involved or even the cause. Evolutionary nature is more stable than one might think.

There are things you have to be an expert not to understand.
  Hjalmar Söderberg (1869–1941)
  Swedish writer[1]

---

[1] Adopted from Egner et al. (2008, 9).

5.1 Understanding Complexity as a Solution: Not as a Problem

Nature is and remains the exemplary mother of all complexity on our earth. It regulates the interconnected relationships between organisms in animate nature and inanimate environment. For this, it does not need a goal-oriented planning strategy like humans do, because the advantageous adaptive progress crystallizes out of the foundation of evolutionary development. Complexity is, so to speak, the inherent law of organisms that accomplish highly specialized feats individually and collectively, but always under the maxim of holism and skillful interconnectedness. "A forest is more than the sum of its trees" describes complexity in a concise way.

The comparison between nature in general and humans in particular should be allowed at this point, because we humans often look for and realize different, rather more determined and straightforward ways of progress, although we know about or sense the dangers to which we are exposed by the real complex influences from the environment.

Excursus Man and Complexity

It is futile to keep pointing out that we live and work in a world of increasing complexity. But increasing complexity also means increasing uncertainty. Because we have enormous difficulties with our inbred causal analogue thinking to overview the totality of all influences – which can be a few but also hundreds and thousands – we create suitable aids, such as computers and model algorithms, which help us to bring a little order into the diffuse interrelationships. With a few dozen influencing parameters on a task to be solved, solutions usually succeed with good agreement and probability of the occurrence of a concrete expected goal. With thousands and thousands of influencing parameters, especially with different orders of magnitude and weighting, as well as the dynamic non-linear course of processes, as they are inherent to a complex system, it is already more difficult to generate useful solutions. Moreover, these have to be readapted again and again.

The increasing extinction of species on our planet is not least an expression of our inability firstly to recognise the interrelationships in nature, secondly to interpret them correctly and thirdly to deal with them sustainably.

Nature itself is a complex system of unimaginable size, with myriads of organisms and even more parameters influencing the individual and general goal of survival. We ourselves, as the most highly developed individual, with our neuronal network of 86 billion neurons (Herculano-Houzel 2009) and approximately

(continued)

(continued)

10,000 neuronal connections per neuron, hardly manage to remember half a dozen to a dozen different influencing parameters at once. We are too much led into dead ends by our causal partly monocausal approach to problems and their solutions. Human creativity is undoubtedly capable of *peak performance*. However, it has enormous difficulties in interpreting interrelationships correctly, i.e. in a fault-tolerant and sustainable manner, and in designing them in a practicable manner in our complex dynamic environment.

Up to this point we have used the term complexity several times without saying what complexity is in essence. Rasper (2000, p. 49) states already two decades ago that there are about fifty definitions of complexity:

> Every definition (of complexity, the author) captures only partial aspects, and none can claim general validity. Complexity has become its own symbol.

Ratter and Treiling (2008) also address the concept of complexity and, depending on the context, come up with several definitions of the term. However, complexity defies a unified definition, as we would certainly like it to be and what we could probably deal with most rationally, thanks to our causal linear thought processes. Two of many definitions of complexity (ibid., pp. 27–28) are:

> The geographer Leser, as one of the most prominent representatives of (landscape) ecological systems research, understands complexity as being composite: "The degree of complexity of a system is [thus] determined by the specification of the parts/elements as well as the interactions or relations between the parts/elements. These in turn can be expressed by the number of dependent and independent variables" (Leser 1997, p. 119).
>
> Environmental scientists Berkhoff, Karstens and Newig (2004) refer to complexity as a "state between order and disorder, which has a great structural diversity. Characteristic properties of complex systems are the existence of many, heterogeneous units, the dynamics of the system, the process of spontaneous self-organization and emergent behavior" (ibid., p. 101).

The author himself contributes two definitions of complexity (Küppers 2015, pp. 13–14):

> A complex system consists of many different influencing variables that are interconnected and change over time.

> Complexity is a multi-layered, time-influenced set of interrelated, equal and unequal elements (objects and subjects) at hierarchically ordered levels of varying quality **(emergence)**.[2]

(*continued*)

---

[2] **Emergence** characterizes the property (quality) of a – dynamic nonlinear – complex system, which its individual elements do not exhibit and which therefore cannot be derived from the analysis of the elements. Examples are: Neurons cannot think like a brain, genes cannot function like an organism, roots, leaves or barks cannot determine the functional quality of a tree, individual parts of a machine cannot determine its real working behaviour or in general: the whole is more than the sum of its parts.

(continued)

Understanding complexity as a solution rather than a problem has already been addressed by Baecker (1994, p. 112), who sees the rediscovery of complexity as an attempt to "[...] come to an understanding about the depth dimensions of problems." In doing so, Baecker brings economist Jürgen Hauschildt (1990) into play, quoting him as saying:

Jürgen Hauschildt sees one of the most important tasks of decision procedures in the assessment of this depth dimension, because decisions must not be too simple with regard to the complexity of the problem, but at the same time not too complex with regard to the accuracy and enforceability of the decision.

A crucial addition to the management of complexity follows (Baecker 1994, p. 113, emphasis and framing by the author):

The standard reflex of immediately inferring a need for simplification from complexity is understandable, but it is misleading. The problem one faces when dealing with complexity is not to be interpreted simply as unmanageability and lack of order.

Rather, the defining characteristic of complexity is that one is dealing with the necessity of choosing what is important to the detriment of what is unimportant, while at the same time knowing that what is unimportant today may be important tomorrow.

Figure 5.1 shows the variation of a vivid and compelling comparison for understanding complex relationships as we understand them (left in Fig. 5.1) and as we should understand them (right in Fig. 5.1).

In the left panel of Fig. 5.1, a gridded photograph can be seen up close that shows precisely many small, differently colored square patches that may be symbolic of specialized activities with progressively deeper views of detail. This fragmentary recognition says little about the overall information content of the photograph on the right of Fig. 5.1.

If you look at the grid image (left) from a few meters away, the precise boundaries between the squares that exist up close become blurred and the diffuse contour of the photo on the right appears. Already this view from some distance indicates how the holistic view of the multitude of detailed squares makes the image on the right of a Chinese and thus the overall information content of the photo – the actual target – increasingly recognizable. No matter how intensive detailed analysis of all the square content present only results in fragmentary information between which no relationship can even be established. The methodology for recognizing faces was first published by L. D. Harmon (1973).

Let us therefore learn to look at the nature surrounding us and our own and other people's activities and objectives again and again with new, reflected points of view, in order to develop solutions for tasks in this way, which allow us

(continued)

(continued)

© 2019 Dr.-Ing. E. W. Udo Küppers

Fig. 5.1 Linear isolated (left) and holistic (right) view of human subjects, objects and processes in complex environment

to process complexity in a goal-oriented way within suitable limits and to develop further in a fault-tolerant way.
- The appropriate management of complexity in the sense of sustainable developments compatible with nature and the environment is increasingly becoming the focus of human activity. It is becoming a central task of progress in view of the Anthropocene era and the intrusion of digital processes into the analogue development of humans that has evolved over five millennia.
- **PROposition** The adequate management of complexity becomes a survival task for humans.

This small foreshadowing of human approaches under complex situations will be encountered several times in Chapter II and the "Practical Chapters". Let us now continue to follow the exemplary principles of nature.

5.2 Nature's Production and Working Principles

Before we discuss in the following Sects. 5.3 and 5.4 central life flows and biocybernetic basic rules of nature, we devote ourselves to the special kind of production including the working principles, how organisms excellently master them.

5.2.1   Production Services of Nature

Given the biodiverse richness of nature, organisms have had sufficient time to perfectly adapt or advantageously evolve their production processes, including their species-specific services, to the highly complex dynamic environment. A core element of this adaptation is the consistent circular networked course of *producing – consuming – reducing,* as can be seen in Fig. 5.2.

The fact that in nature – by man's term – *nothing is superfluous is* also expressed by the ingenious interaction of natural principles, as they can also be seen in Fig. 5.2. We want to explain these briefly – after Fig. 5.2 from left to right – by means of examples, whereby we will encounter one or the other principle several times later (italic texts for quotations by the author):

1. **Cooperation** – *Within a species, cooperative behavior among individuals is evident, for example, in dolphin prey hunting, lion prey hunting, and other species.*

   *"In cross-species cooperation, the relationship that is essential for survival benefits of the two partners is either optional (cooperation) or obligatory **(symbiosis, mutualism)**"* (Odum 1991, p. 177). *Mutualism is a symbiosis of plants and animals, such as the pollination of plants by insects.*

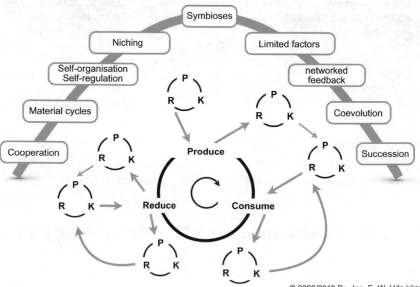

© 2008/2019 Dr. -Ing. E. W. Udo küppers

Fig. 5.2 Circular networked process flows in nature with a selection of natural principles

2. **Material cycles** – *"The functioning of any ecosystem is based on the one hand on the flow of energy, and on the other hand on material cycles. In contrast to the fact that the conversion process from solar energy to food always depends on the presence of solar energy, chemical substances – the elements and their compounds – can be used again and again without losing their usefulness. In a well-ordered ecosystem, many of these substances move in a cycle between biotic and abiotic components"* (Odum 1991, pp. 59–60). Cycles of this type are called biogeochemical cycles.

3. **Self-organisation and self-regulation** – Both principles are the prerequisite for order-creating, low-entropy structures in nature, see Sect. 5.3.

4. **Niching** – *This term points to the ecological role of an organism in its community and is closely related to the term* **habitat.** *"Habitat is, so to speak, the "address" (it indicates where an organism lives), and niche is the "occupation" (it indicates how it lives and its relationship to other organisms)"* (Odum 1991, p. 63).

5. **Symbioses** – *Symbiosis is the expression of a close coexistence of species for mutual benefit. According to* Odum (1991, p. 191), *symbiosis "[...] is in a sense the highest form of cooperation, (it) is widespread and extremely important."* Campbell et al. (2006, p. 1410) *speak of "[...] interspecific(s) interactions that are positive for both species involved are called symbioses or mutualism (q.v.). Such relationships sometimes require* **co-evolutionary adaptations** *(see point 8 above) in the partners, which means that changes in one species are likely to affect the survival and reproductive capabilities of the other."*

   In our business language, we like to speak of a WIN-WIN situation, which arises when corporate partners from different industries work together without either incurring losses of any kind that would jeopardize further cooperation.

6. **Limited Factors (Liebig's Law of the Minimum)** – *"The idea that organisms depend on the weakest link in the ecological chain of needs goes back to the time of Justus von Liebig more than 100 years ago"* Odum (1991, p. 144). *Liebig was the first to recognize the effect of inorganic chemical fertilizers in agriculture. His insight was to have found out "[...] that the growth of cultivated plants was often limited by that element which – in relation to the need – was in the minimum, regardless of whether the quantity required was small or large"* (ibid.).

   For the organisms in our context, Liebig's statement means that "[...] always that nutrient limits growth which is available in the smallest amount compared to the demand." It must be added that too much of a nutrient also limits growth. In this case one speaks of a **"limiting effect of the maximum"**.

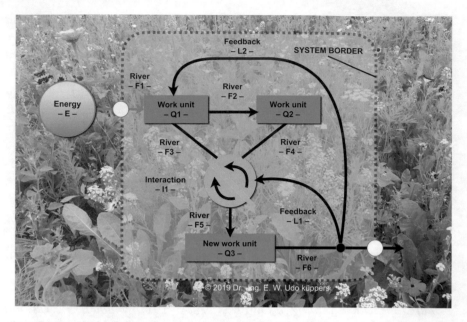

Fig. 5.3 Modeling the five basic components of a biological organized system. (After Odum 1991, p. 48 changed by the author)

7. **Networked feedback** – *feedback loops are formed by those process flows in which an output variable interacts with an "upstream" process variable. Nature is home to countless of these feedback loops or regulatory processes that ensure the growth of organisms in a coordinated (networked) manner. See, among others, Figs. 5.3, 5.9 and 5.10.*

Basically, there are two types of feedback that are subject to multiple linkages:

1. ***positive feedback*** *amplifies the effect of one argument (system element) on the other and leads in cascades to the limit being exceeded or to a system standstill (often marked with the (+) symbol).*
2. ***Negative feedback*** *reduces the effect of one argument (system element) on the other and also leads in cascades to system instability (often marked with the (-) symbol).*

Only a balanced combination of positive and negative feedbacks, in which negative feedbacks dominate, shows a system-stable development behaviour.

8. **Coevolution** – *"Coevolution refers to the emergence of traits (controlled by natural selection) in the course of long-term interactions between two or more groups of organisms that have a close ecological relationship but do not exchange genetic information with each other" (Odum 1991, p. 122). According to Campbell et al. (2006, p. 1411), coevolution refers to reciprocal evolutionary*

adaptation between two species, such as the relationship between flowers and pollinators (cherry blossom and bee), terrestrial plants and fungi (oak and oak firefly), parasite and host (mistletoe and birch). Coevolution occurs through change in one species that exerts selection pressure on the other species. "[...] This counter-adaptation in turn promotes evolutionary variation in the first species. This interplay of adaptation and counter-adaptation requires reciprocal genetic changes in the populations of both species involved" (ibid.).

9. ***Succession*** *– "The development of a biotic community (biocoenosis) within a short geological time span (1000 years or less) is commonly known as natural or ecological succession, but would perhaps be more accurately termed **ecosystem development**, since it is an active process involving changes in both organisms and the physical environment" (Odum 1991, p. 196). The change of a landscape including organisms from a grassland, to shrubs, to pine forest, to a deciduous forest rich in species are stages of a succession until finally a permanent community of life is formed, which is drawn as a **climax stage.** It is a stage that is "[...] in equilibrium with the regional climate as well as the local substrate, water and topographic conditions or is determined by their factors" (ibid.).*

We humans have only recently come to understand – if at all – what enormous advantages such natural principles hold for our own technospheric environment and our survival. But let us return to nature and its achievements. The contents of Table 5.1 summarize some of nature's essential achievements, which, in view of our increasing destruction of nature, make us realize what a wealth nature makes available to us, some of which we are carelessly destroying in full awareness.

5.2.2 Working Principles of Nature

Perfectly executable working principles in nature presuppose a functioning *structure and household of biological systems.* That both features persist at best and how they work, we will now try to describe.

Countless organisms populate the earth. They are inextricably linked to the inanimate (abiotic) environment. Organisms are shaped by their appearance, their food intake and processing, their relationships with other organisms, their habitat and much more. We humans are also part of this.

Ecology is a scientific discipline that attempts to better understand and describe the phenomena and mechanisms of biological systems and their organisms. Modelling is a common tool for this purpose.

Table 5.1 Selection of production services of networked nature

Production services of nature	
Sophisticated coupled food webs	Multifunctional nutrient cycles
Purification of air and water	Erosion protection of coastal areas
Mitigation of droughts and floods	Protection from ultraviolet radiation
Creation and preservation of fertile soils	Mitigation of weather extremes
Detoxification and breakdown of waste products	Species-rich productive economic management
Control of many agricultural pests by natural enemies	Aesthetic beauty and recreational value
Highly effective and efficient product technologies, processes and organisational principles for technospheric applications	Sophisticated optimization methods

Selection from various sources, including Campbell et al. (2006), Begon et al. (1998), Odum (1991) and own findings

For E. P. Odum, the functioning model of an ecological phenomenon, by which the structure and household of a biological system in the form of an organism, grassland, forest, desert, etc. can also be described and understood, consists of five basic parts (components), as explained in Table 5.2 and graphically sketched in a functional context in Fig. 5.3, on the background of a depicted grassland system.

The necessary condition for survivability (of an organism) and sufficient condition (practicing successful principles) for further development complement each other in nature in an ideal way. They are consistently implemented by all organisms. Figure 5.4 below symbolises the working principles of biological systems summarised in Table 5.3, whereby the physical quantities of energy, substances or matter and information are subordinate to the organisation and these in turn are available to the organism as an open system to the environment.

With these principles, nature provides us with a *toolbox* for value creation in the environment-society-economy network that has been tested over a long period of time and improved through permanent quality controls. What could be more meaningful than to apply these principles in a targeted manner – for example with the help of *systemic bionics* – for the sustainable survivability of companies? Activities of organisms and their biotic communities that are hidden behind working principles, including organizational principles, seem to work effectively in the biosphere. But what about *"ecological efficiency"*?

Table 5.2 Basic components of living (modelled) systems

Basic components (all components contain specific information)	Explanation (organism stands for biological systems, system analytical terms = italics)
1. **Properties – Q -**	*System and state variables* Structure of the organism, function of the individual components of the organism (e.g. heart as blood pump, liver as cleansing organ, lungs as respiratory organ)
2. **Forces – E -**	*Control and management variables* External energy sources or causal forces, which drives the organism (food intake)
3. **Rivers – F -**	*Nonlinear transfer functions* They indicate where system properties are connected to each other or to the forces by energy or mass transfer Entry: e.g. fluid intake Processing: e.g. food decomposition in the stomach discharge: e.g. heat release into the environment
4. **Interactions – I -**	*Interaction features* They show where forces and properties influence each other, and in this way fluxes Modify, amplify or regulate
5. **Feedback loops – L -**	*Feedback loops* One or more output variables in the system act – retroactively – on upstream system or state variables

After E. P. Odum (1991, p. 47), modified and supplemented by the author

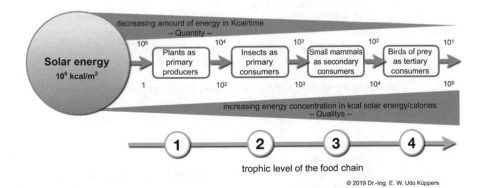

© 2019 Dr.-Ing. E. W. Udo Küppers

Fig. 5.4 Nature's trophic levels. Each stage of this idealized trophic food chain has a net production efficiency of 10%, with primary producers using only 1% of solar energy for their net production. (Adapted from Urry et al. 2019, p. 1670 and Odum 1999, p. 89)

Table 5.3 Selection of efficient working principles of biological systems

Working principles of biological systems	
Organism	Open system to the environment
	Dissipative structures
	Dynamic equilibrium, steady state
	Metabolism with cybernetic structures
	Optimal properties with regard to chemical-physical properties (lightweight construction, resistance, strength, floatability, etc.)
	coordinated energy-optimal time cycles for life processes
	Complete optimization of structure-form function
	Proven universal development tool "Evolution Strategy"
	Production of sustainable quality under low temperature conditions
	Cyclic succession (change in species structure and community formation)
	Isolation, niche formation, coevolution
	Alien association (symbioses)
Organization	Self-organizing autarkic process flows
	Dosed feedback processes in an open changeable network
	Pulse-controlled energy-saving initiation or modification of process sequences by closed-loop control instead of complex control systems
	Optimisation of the development capacity of work networks
	Use of the optimal interaction of different forces (synergy) also for the development of new "product qualities" (emergences)
	Small-scale optimization of cross-species energy and material networks
	Limited factors principle
Energy	Efficient (minimum loss) energy conversion chains
	Direct and indirect energy use
	UV radiation protection techniques
Fabrics	Use of a few types of raw materials for a broad application of high-quality specialized material properties
	Micro- and nanostructured composites
	Multifunctional properties
	Preferred efficiency optimisation of an "entry management" system
	Complete material recycling while retaining the same initial quality
	Regenerable self-repairing functions
	Multiple use
	"Intelligent" composites with opposing physical properties of the solitary substances
Communication	Exchange by time-triggered programmed chemical-physical signals or processes
	Highly effective stand-alone transmitter-receiver systems (sensor-actuator systems)
	"Intelligent" positioning systems
	System-oriented, system-integrated communication

Selection from various sources, including Campbell et al. (2006), Begon et al. (1998), Odum (1991) and own findings

5.2.3 Ecological Efficiency

Ecological efficiency should indeed be understood here as efficiency that relates to ecology and should not be confused with the term "eco-efficiency", which is often applied to technical systems.

With his description: "... the whole science of the relations of the organism to the surrounding external world,..." Ernst Haeckel coined the term ecology in 1866 (Begon et al. 1998, p. XVII). If we generally consider ecological systems as energy and material transforming systems, it makes sense to group species of a community into trophic levels of a feeding relationship. It is the interconnected material cycles that help to ensure that materials of high "quality" or low entropy, depending on the duration of life cycles, are continually recycled for new generations of organisms. The solar energy source, through transformative (dissipative) energy cascades, controls these cycles through all trophic levels, from producers to destructors (decomposers). These are connected to all other levels of an ecosystem and thus hold a key position in the material cycle, through:

1. decompose (decompose) the organic material and
2. converting the chemical building blocks into an inorganic form,

which are mineral components of soil, air and water, are used by producers as feedstocks for new biomass.

The conversion of energy through the individual stages of the food chain is not 100% efficient, as is well known, because ecosystems are also subject to the basic laws of thermodynamics – in particular the law of entropy. With each conversion step, energy is lost in the form of heat, similar to technical systems. The efficiency of the ecological energy transition by individual organisms or through all trophic levels can therefore in principle be measured in exactly the same way as the efficiency of technical systems (Campbell et al. 2006, p. 1432).

With a preview of the practical chapters under main chapter II, however, caution is advised to directly compare the efficiency of one – natural – system with the efficiency of the other – technical – system, because the preconditions are too different after all! From a technical point of view, seemingly low efficiencies of natural systems, if their sustainable complex interdependencies are taken into account, can take on a completely new meaning for technical systems, which may well lead to an emergent leap.

Models (including compartmental models, see; Campbell et al. 2006; Begon et al. 1998; Odum 1991) for capturing the different pathways of energy in organisms and communities consider different types of efficiencies. Along an energy pathway, starting from primary production in ecosystems. Campbell (2006, p. 1443 ff.) describes primary production in ecosystems as follows:

> The total amount of chemical energy (in the form of organic compounds) produced by the autotrophs of an ecosystem within a given period of time using light energy is called *primary production*. This photosynthetic output is the basis for the study of metabolic and energetic turnover within an ecosystem.

Autotrophic organisms (especially plants) carry out photosynthesis, i.e. they use light as a source of energy. In Sects. 5.3.1 and 5.3.2 we will talk in detail about energy and material in nature. Continue with Campbell, ibid.:

> The total primary production is called **gross primary production (BPP)** – the amount of light that is converted into chemical energy by photosynthesis per unit time. Not all of the molecules produced are stored as organic matter, however, because primary producers use some of it as fuel for their own cellular respiration. **Net primary production (NPP)** is therefore equal to gross primary production minus the amount of energy consumed by the producers during respiration (**R of Respiration**):
> **NPP = BPP – R**
> Of interest to us is net primary production, which indicates how much stored chemical energy is available to consumers of an ecosystem. Net primary production may be only a quarter of total production. This is the case in forests, for example, because trees consist largely of photosynthetically inactive structures – such as trunks, branches, and roots – whose metabolic activity must be maintained by respiration. [...].
> Of the visible light reaching photoautotrophic organisms, *only about 1 percent* is *converted to chemical energy by photosynthesis,* and photosynthetic efficiency varies with organism species, light intensity, and other factors (italics added and emphasis mine [...]).
> (Looking ahead to Sect. 5.3.2, it is worth noting that although) the fraction of *incident solar energy (1022 J (J) of radiant energy,* Campbell et al. 2006, p. 1434, Urry et al. 2019, p. 1664) that is eventually captured by photosynthesis is very small, the Earth's primary producers collectively generate about *170 billion tonnes of organic matter per year* – an impressive amount. (ibid., emphasis mine).

The 170 billion tonnes of organic matter or biomass that nature leaves behind each year are remarkable for another – highly topical – reason. Because not a fraction of this biological material that accumulates year after year remains unused! Our handling of raw materials and materials is – despite some trends towards material recycling – stone-age in comparison. According to the World Bank report "What a Waste 2.0" (World Bank Group 2018, p. 3), a total of 2.01 million tonnes of municipal solid waste was generated on our planet in 2016, of which 33%, 0.663 million tonnes, was – extremely conservatively – processed in an environmentally damaging manner.

> Nature recycles 170 tonnes of biowaste every year without the slightest loss of material. This is approximately 85 times the amount of urban solid waste generated annually in 2016, of which 33% still has environmentally harmful effects. "Without restrictive measures, global waste (and thus very likely environmentally harmful recycling problems, the author) will increase by 70 percent by 2050 compared to current levels." (ibid., p. 17)

Here, too, the principles of nature's highly complex food webs demonstrate their undisputed leadership in the efficient use of materials. In view of the expected increase in material waste in our habitats, our societies would be more than well advised to pay more attention to the – holistically speaking – highly economical processes of nature and to use its principles instead of carelessly destroying them – and thus, in the broadest sense, ourselves as food consumers.

With regard to the focused "ecological efficiency" of individual organisms within a trophic level on the one hand, and through several trophic levels of food webs on the other hand, secondary production in ecosystems plays the decisive role.

Organisms with the same mode of nutrition are grouped together at trophic levels or trophic levels, which form a pyramid. The primary producers or autotrophic organisms (see above) are classified on the basic level. They are followed by consumers of different order (primary, secondary, tertiary consumers etc.) on the second, third to rarely more than fifth level.

Primary producers convert only about 1% of the available energy into net primary production. The higher level of this food pyramid utilizes only a fraction of the biomass of previous trophic levels. See Fig. 5.4.

Starting from the low concentrated form of energy of the sun, a successive decrease of energy takes place over each of the trophic levels of the food chain. As mentioned above, only 10% of the previous amount of energy is processed. In contrast, the energy concentration (energy efficiency) increases proportionally per trophic level, in Fig. 5.4 from solar energy via the primary producers to the tertiary consumers. Odum writes in this regard (1999, p. 89):

(The reduction of) the amount of energy within a food chain [...] (and the increase of) the energy quality [...] from trophic level to trophic level [...] is shown, for example, by the higher energy content of animal compared to herbal biomass. 1 g of animal matter corresponds to 23.4 kJ (kilojoules, an energy equivalent, the author), 1 g of plant matter to only 19.2 kJ. A suitable measure (quality factor) of energy quality can be the amount of calories of sunlight consumed to produce one calorie of a qualitatively higher form of energy (e.g. food or wood (from which coal can be produced in the next stage and electricity in the stage after next, i.e.)).

According to this, 10,000 kcal of sunlight are required to produce 1 kcal of bird of prey. In anticipation of the later practical chapters under Chapter II, we can follow Odum's (ibid.) comparison as follows:

When comparing energy sources (meaning energy conversion systems, since the sun is the only energy source available to us, n.d.) for direct human consumption, one should compare both the quality (exergy, the part of the energy that is converted into work, n.d.) and the quantity available, and if possible match the quality of the energy source with the quality of the use.

Before looking at the production efficiency of an example organism, we first look at secondary production in a little more detail. On this, Urry et al. (2019, p. 1669) write:

The quantities of organic matter taken up by heterotrophic organisms in an ecosystem and used to build up their own body substance within a certain period of time are referred to as secondary production. What is the pathway of organic matter from primary producers to primary consumers? In most ecosystems, herbivores consume only a very small fraction of the total phytomass produced. In addition, they also cannot fully digest the plant material they consume, but often excrete large amounts of undecomposed plant material. Accordingly, a large part of the primary production is not used by the consumers. This aspect of energy transfer is discussed in more detail below.

As an example organism for secondary production with subsequent production efficiency, we take a caterpillar.

If it feeds on a leaf, only about 33 out of 200 J (48 cal), i.e. one sixth of the total energy of the leaf, is used for secondary production and thus for the build-up of the caterpillar's own body substance (Fig. 5.5, the author). The caterpillar uses part of the remaining energy for cellular respiration, and excretes the rest. The energy contained in the excrements remains temporarily in the ecosystem, but most of it is lost as heat after the organic material has been consumed by *detritus eaters* (detritus = crushed organic matter, italics added). The energy used for the caterpillar's respiration also leaves the ecosystem as heat. This is also the reason why one speaks of a flow of energy through ecosystems and not of an energy cycle. Only the energy that is stored as biomass by *herbivores* (herbivores, italics mine) (by building up endogenous substance or producing offspring) is available as food for secondary consumers. The efficiency with which animals or heterotrophic organisms in general use their food is calculated from the following Eq. 5.1.

$$\textbf{1. Production efficiency}: PE = \frac{Pn}{An} \times 100\% \qquad (5.1)$$

PE = net production efficiency
P_{n-} = Net secondary production
A_n = assimilated primary production

Production efficiency (PE) is the percentage of assimilated energy that is incorporated into new biomass. The rest is completely lost to the community in the form of respiratory heat. PE varies depending on the taxonomic group to which the organism belongs. Invertebrates generally have a high efficiency of 30–50%, losing relatively little energy as respiratory heat. Among vertebrates, *ectotherms* (in italics), whose body temperature varies with ambient temperature [...], have average PE values (around 10%), whereas *endotherms* (in italics), which expend considerably more energy to keep their body temperature constant, convert at best 1–5% of the assimilated energy into biomass. Microorganisms, including protozoa, are mostly characterized by a very high PE of 50% or more. (Begon et al. 2017, p. 416)

Urry et al. (ibid.) describe the production efficiency equation as follows and refer to Fig. 5.5:

Net secondary production is the energy stored in zoomass (building up the body's own substance). Assimilation, on the other hand, includes all the energy taken in and used for growth, reproduction and respiration. Accordingly, net production efficiency in heterotrophs is the fraction of energy stored in the assimilated food that is not used for respiration. For the caterpillar in Fig. [...] 5.5 (the author), the net production efficiency is 33 percent; 67 J of 100 J of

assimilated energy is used for respiration. (The energy lost in the form of undi-
gested material with the excreta is not included in the assimilation.) Birds and
mammals generally have lower net production efficiencies than other groups of
animals; they range from one to three percent; this is because mammals and
birds use a lot of energy for ectothermic ([…], alternately warm, the author) and
have net production efficiencies of about ten percent. Insects and microorganisms are even more efficient, with net production efficiencies averaging 40 percent or more, as Table […] 5.4 (the author) illustrates.

In addition to the *production efficiency of* organisms, the *consumption efficiency* and *assimilation efficiency* also determine the relative importance of
energy pathways.

The proportion of net primary production that flows along each of the possible
paths of energy depends on the transfer efficiency from one step to the next. In
order to predict a pattern of energy flow, we need only know three categories of

Fig. 5.5 Energy distribution of a herbivorous animal within the food chain. Less than
17% of a caterpillar's food is actually used for secondary production (building up its
own body substance). (Adapted from Urry et al. 2019, p. 1669. Photo Caterpillar:
Source: Rosa-Maria Rinkl, https://commons.wikimedia.org. CC4.0 (accessed Aug
29, 2019))

Table 5.4 Production efficiency of different animal groups, ordered by increasing efficiency

Group	Production efficiency P/A (%)
1. Insectivores	0.86
2. Birds	1.29
3. Small mammals	1.51
4. Other mammals	3.14
5. Fish and social insects	9.77
6. Invertebrates (invertebrates other than insects)	25.0
7. Non-social insects	40.7
Invertebrates without insects	
8. Herbivores	20.8
9. Carnivores	27.6
10. Detrivors	36.2
Non-social insects	
11. Herbivores	38.8
12. Carnivores	47.0
13. Detrivors	55.6

Begon et al. (1998, p. 508), after Humphreys (1979)

transfer efficiency [...] (as shown in Fig. 5.6, the author): consumption efficiency, assimilation efficiency and production efficiency. (ibid., p. 415)

In addition to formula (5.1) (see above), production efficiency, the formulas (5.2) for consumption efficiency and (5.3) for assimilation efficiency follow:

$$\textbf{2. Consumption efficiency}: KE = \frac{In}{Pn-1} \times 100\% \qquad (5.2)$$

KE = Consumption efficiency
I_{n-} = consumption from a compartment of the next higher trophic level.
P_{n-1} = Total production available per trophic level.

Consumption efficiency (KE) is the proportion of the total productivity available at one trophic level that is consumed by the next trophic level up. For primary consumers, KE corresponds to the fraction of joules (or of organic carbon, the source of potential energy) produced per unit area and time as NNP that finds its way into the digestive tract of herbivores. In the case of secondary consumers, it is the percentage of herbivore productivity that is eaten by *carnivores* (carnivores, italics mine). The remainder dies without being eaten and enters the decomposer system. Realistic averages for

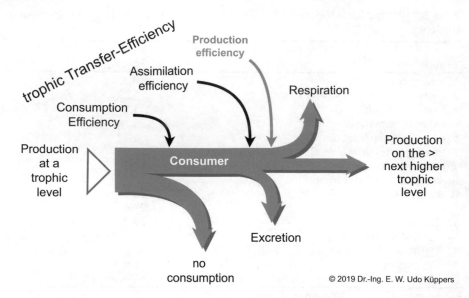

Fig. 5.6 The relationship of trophic transfer efficiency to its three components: consumption efficiency, assimilation efficiency and production efficiency. (Adapted from Begon et al. 1998, p. 416)

herbivore KE are about 5% in forests, 25% in grasslands, and 50% in phytoplankton-dominated communities. For carnivores, KE varies from 25% to nearly 100% (Begon et al. 2017, p. 416).

$$\textbf{3. Assimilation efficiency}: AE = \frac{An}{In} \times 100\% \tag{5.3}$$

AE = assimilation efficiency
A_{n-} = assimilated portion of food energy for growth or work.
I_n = food energy ingested by a trophic level consumer.

Assimilation efficiency (AE) is the proportion of food energy absorbed into the digestive tract by consumers of a trophic level that is assimilated by the intestinal wall and becomes available for growth or work. The remainder is excreted in the form of feces and enters the decomposer system. For microorganisms to state an AE is much more difficult because in their case food does not pass through an intestine and feces are not produced and excreted. Bacteria and fungi digest dead organic material externally and normally absorb almost all of the product: they are often attributed an AE of 100%. For terrestrial herbivores, *detritivores* (dead organic matter not yet humified, italics mine) and *microbivores* (tiny animals specialized in eating fungi or bacteria, italics mine), the AE is typically low, 20–50%. It is somewhat higher in aquatic herbivores and high in carnivores – around 80% and higher. The way plants shift their production to roots, wood, leaves, seeds and fruits also affects their utility for herbivores. Seeds and fruits can be assimilated with an efficiency of not less than 60–70%, leaves with about 50%, while the AE of wood sometimes does not exceed 15%. Begon et al. (2017, p. 416)

Total trophic transfer efficiency from one trophic level to the next is simply calculated as KE × AE × PE and represents the percentage of energy or organic material transferred from one trophic level to the next. The concept dates back to the pioneering work of Raymond Lindemann in 1942, after which it was assumed for many decades that trophic transfer efficiency was about 10%. Some ecologists even spoke of a "10% law". However, there is certainly no law of nature that ensures that exactly one tenth of the energy that goes into one trophic level is transferred to the next level up. As an example, consider a zebra feeding on grasses in a savanna ecosystem. As mentioned above, the KE for herbivores in grasslands is usually 25%. The AE for herbivores feeding on grass is relatively low, perhaps in the range of 20%, and the PE of an endothermic animal, such as a zebra, reaches about 3%. This gives a trophic transfer efficiency of KE × AE × PE = 0.15%, which falls short of 10% by two orders of magnitude. The transfer efficiency from zebras to lions is somewhat greater, but it is also far less than 10% due to the low PE of vertebrates. Even with high values for KE (80%) and AE (80%), the transfer efficiency from zebra to lion is only in the range of about 1%. Begon et al. (2017, p. 417)

> **ATTENTION!**
> - The figures in the efficiency equations of various animal groups are to be seen under the framework conditions of a complexly interconnected and dynamic further development. Analogies with the (mono)causal, not infrequently linear, efficiency equations underlying progress that occur in technology are therefore self-evident.

> **POSTULAT**
> - The principles of dynamic "holistic strategies" for efficiency determination of organisms in nature are perfectly adapted to their environment. Under similar holistic boundary conditions, the principles of nature trump the technically focused principles of causal strategies for efficiency determination of products and processes in technology and economy by orders of magnitude.

5.2.4 Basic Features of the *Ecosystemisation of* Nature and a Comparison with Technology

All the aforementioned production performances and working principles of nature are based on fundamental criteria that enable myriads of producers, consumers and destroyers, under the most stringent quality controls and relentless quality selection, with networked strategic approaches, to produce high-performance solutions with technical efficiencies that are second to none. Table 5.5 compares 20 of these fundamental criteria of nature with corresponding technical criteria. However, without going into further detail on the individual arguments on both sides of the tension field and the advantages and disadvantages of the two "eco-strategies*", some of which are self-explanatory, two general, completely different approaches can be derived:

> For the nature strategy: Decisive for holistic sustainable progress is *input management!* – What is decisive is what goes in at the beginning of a process → Resource efficiency.

> For the technology strategy: Decisive for economic short-term progress is *discharge management!* – What is decisive is what comes out at the end of a process → Resource exploitation.

Table 5.5 Comparative characteristics in the field of tension between nature and technology

Nature	Technology
Differentiated diversity	Powerful simplicity
Differentiation	Standardization
Dynamic stability	Destability
Embedding	Exploitation
Dialogue capability	Manipulability
Flexibility	Speciality
Holistic	Fragmentation
High order structure	Low order structure
Communication	Mastery
Relevance	Accuracy
Sensory system, self-regulation	Power and active control technology
Polyculture	Monoculture
Variety of options	Efficiency of selected options
Cybernetics	Mechanics
Networking	Isolation
Circuits	Segment processes
Compatibility	Need satisfaction
Structural wealth	Standard dimension
Interaction	Impact
Reciprocity	Rigid hierarchy

After Dürr (1999, p. 264), order changed d. d. A

5.3 Three Central Life Flows and One Universal Feature of Evolution

5.3.1 Only *'Open'* Systems Are Capable of Evolution: In the Truest Sense of the Word

All organisms are open systems of nature. The progress of their development therefore depends in a decisive way on their abilities to bring energetic, material and informational (communicative) processes into a life-advantageous context. Organisms are only capable of evolution as open systems. On this point, Ebeling and Feistel[3] write (see also Feistel and Ebeling 2011; Ebeling 1991; Ebeling and Feistel 1982):

> It is characteristic for evolution that it has to accept a high entropy production for the construction of new structures, whereas a minimum entropy production is aimed at for the preservation of established structures. Evolutionary processes can therefore only take place in the nonlinear range at a great distance from thermodynamic equilibrium. They are associated with a sudden increase in entropy production. For such processes a chaotic push is characteristic, which removes the system from the vicinity of equilibrium.

In Fig. 5.7, let us consider in general terms the evolution of organisms on our planet. From our central energy source, the sun, a stream of energy-rich radiation flows onto the earth. This supplies the organisms with the necessary primary energy, which build up low-entropy structures through various transformation processes by means of self-organization and thus generate order. The prerequisite for these self-organized processes is not only the supply of higher-value energy but also the discharge of lower-value energy – entropy – into the environment.

By using evolutionarily developed principles, the evolution of organisms progresses from generation to generation, whereby the flow of energy is irreversible, but the material and communicative processes take place via interconnected circular processes.

Figure 5.8 shows the focus of self-organization using the example of the *human* organism, again with the three central flows of life: energy, substances, information (communication).

---

[3] https://www.bertramkoehler.de/GR1.htm (accessed Jul 29, 2019), see also: https://www.researchgate.net/publication/309379699_Physik_der_Selbstorganisation_und_Evolution (accessed Jul 29, 2019).

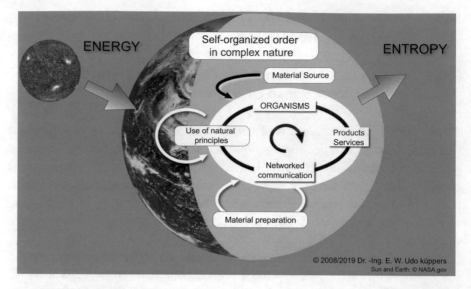

Fig. 5.7 Evolution and self-organization of organisms on Earth (principle representation)

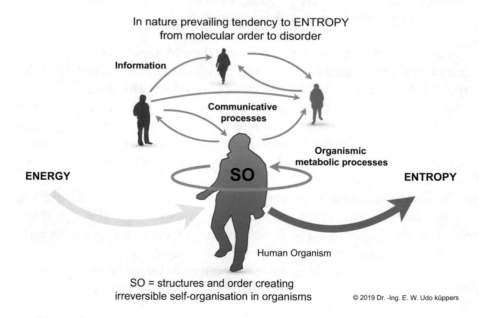

Fig. 5.8 Self-organisation using the example of the human organism

Ebeling (1991, pp. 36–43) describes the central importance of self-organization for evolution through 11 working principles, four of which are reproduced here:

1. **Principle of entropy export (pumping principle):**

Entropy is a basic quantity for all macroscopic systems; it is a measure of the value of the energy contained in the system and at the same time a measure of the disorder of the particles that build up the system. Self-organization is only possible if the system can export entropy (see Figs. 5.2 and 5.3, respectively). Since entropy is a measure of the value of energy, a system can export entropy exactly when high-value energy [...] is supplied to it and low-value energy [...] is withdrawn.

2. **Principle of energy transformation:**

Self-organizing systems are characterized by chains of energy transformations. The building of ordered structures is linked to high-value forms of energy [...]. A part of the supplied energy is always converted into a specific high-value form inside [...].

3. **Principle of supercritical distance:**

Self-organization is only possible when the distance of the system from equilibrium exceeds certain critical values. It occurs only at a distance from equilibrium, its occurrence is associated with discrete transitions [...].

4. **Principle of non-linearity and feedback:**

Self-organization requires nonlinear dynamics of the system, which is usually due to feedback effects. The scale of nonlinearities is particularly rich in chemical reactions and in hydrodynamic processes.
Brief note on *feedback,* as a necessary process in biological systems (see also Küppers 2019, p. 108 ff.).

In processes or control processes, whether in nature or in technology, feedbacks – RCs – are commonplace. The feedbacks that exist per se in biological systems or organisms, as well as in the technosphere, can by definition act both "positively" and "negatively" or as a combination of both.
  In a developing system organism, positive and negative feedbacks are in a balanced and dynamic relationship. Positive RK alone lead to a non-linear

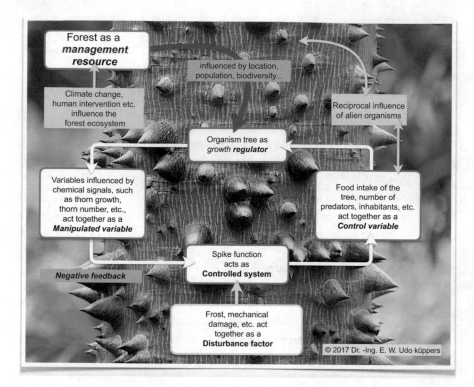

Fig. 5.9 Habitus of the kapok tree (Ceiba pentandra) with superimposed cybernetic function. These are details of a biological (biocybernetic) control loop, illustrated on a tree that helps itself against food-seeking animals with a special defence mechanism using thorns. Findings from research in control technology are brought together here with biological insights in the form of model concepts. The snapshot of the biological control loop should not, however, obscure the true material, energetic and communicative processes of the organism in nature, which cannot all be depicted in detail and which in reality take place in an unimaginably complex manner

increase towards the system limit or to a non-linear decrease to a state without effect and progress. Negative RK in combination with positive RK have the tendency to dynamically stabilize a complex system without breaking the system boundaries.

A recurrent process of feedback is found in biological networks, as indicated above. Figure 5.9 shows the defensiveness of a kapok tree against injury, represented by a *negative* system-stabilizing feedback loop. Figure 5.10 shows the characteristic but simplifying pattern of a predator-prey relationship, where several feedback loops interact dynamically in such a way that all organisms involved (fox, rabbit, plant) have a chance to develop further. This is where

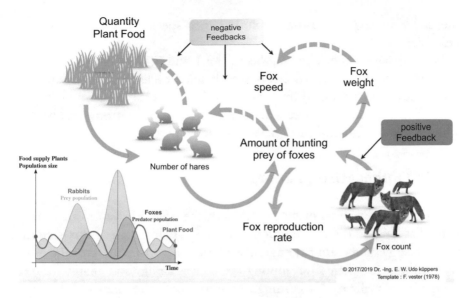

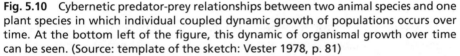

Fig. 5.10 Cybernetic predator-prey relationships between two animal species and one plant species in which individual coupled dynamic growth of populations occurs over time. At the bottom left of the figure, this dynamic of organismal growth over time can be seen. (Source: template of the sketch: Vester 1978, p. 81)

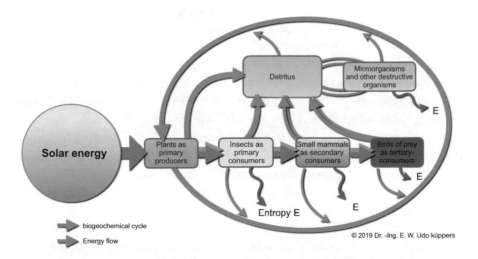

Fig. 5.11 Basic energy and material flows between the sun as an energy source and the organisms of nature. (Adapted from Urry et al. 2019, p. 1663)

the clever strategy of natural evolution in a species association shows its best side.

With a view to the following Sects. 5.3.1 and 5.3.2, Fig. 5.11 shows an overview of the basic structure of biogeochemical cycle processes and physical energy flows of nature and its organisms. The irreversible energy paths running from the energy source sun in the direction of entropy can be seen, whereas material process paths have circular properties.

5.3.2   Optimal Use of Energy

Open evolvable systems of nature are permanent energy transformation systems that temporarily create order. The energy-rich radiation fluxes emitted by the energy source sun penetrate every order- and structure-creating organism. All position themselves among themselves along food chains, whereby energy is optimally used according to the specific requirements of the organisms in their biotopes.

In biological control networks, where many independent decentralized organisms are linked to each other and to the environment, only energy-efficient short impulses in the form of chemical and physical signals are needed to react in a suitable way that sustains life.

Self-organization is the spontaneous emergence of relative order within a system, without external influences. Higher order results from lower order. Organisms create the *bioenergetic conditions for* this as open systems in interaction with the environment. In biological systems, for example, coherent new variations in form and structure occur in the course of development, which act on themselves – in the form of system-internal "control impulses". Here we also speak of emergent macroscopic forms and structures, which, to describe it again clearly, are *not an* addition of internal "building blocks" of the organism. Self-organization is also directly related to *self-preservation*. Biological systems are also called *autopoietic systems* (Varela et al. 1974, p. 187 ff.) because they possess these two properties, which lead to a fundamental biological principle of organization.

The driving force of all these processes is solar energy. In retrospect, Chap. 3 shows by Fig. 3.1 exemplarily this course of energy, which ultimately leads via the multitude of transformation processes to a global energy devaluation – increase of entropy.

As simple as it is to draw the conclusion that life would not exist without energy, the analysis of energy processing in networked organisms is complex. It must also be taken into this calculation that energy is also required to allow the energy conversion systems of nature to come into being in the first place, so that they can carry out their structure-creating work.

> Exactly at this criterion, that also energy is needed to create the prerequisite to convert energy, many technical energy-conversion-systems fail at their energy-consumption-calculations, which deliver completely unrealistic results about the real consumption of energy.

5.3.3　Loss-Free Stock and Material Processing

> Nothing is superfluous in nature!

This statement is the central guide for substance- and material-processing processes in nature.

The "entrepreneurial" goal of every organism is to strengthen its ability to survive in the permanently changing environment. To grow, the green leaves of trees *(producers)* absorb sunlight as low-entropy radiation (photosynthesis, respiration) and transport nutrients from the soil via the roots. For rabbits and snails as plant eaters (herbivores), the green leaves are tasty treats. Songbirds use the tree trunk to build a living cavity. Other animals such as spiders or snakes are carnivores. This also includes birds. They like, for example, the worms hiding under the bark of the tree, which in turn feed on smaller animals. Birds have developed "special tools" in the form of special beaks in order to be able to enjoy the worm morsels. Finally, there are the omnivores, which include crows, pigs, bears, and even humans. Taken together, they count as *consumers.* To protect themselves against excessive feeding damage, trees produce chemical defense substances. The bark as the outer part of the bark (dead bark layer) is at the same time a packaging protection against rain, hail or snow and a depot for toxins that can harm the organism. At the end of a life cycle, the low-entropy substance of organisms decays and is broken down by microorganisms *(destruents)* into molecular organic components and minerals, which in turn are supplied to producers as building materials. All in all, the biotic community exists and grows through high-level biocybernetic interdependencies. In a very simple form, the structure of substance and material processing of organisms outlined above in Fig. 5.2 reflects the interconnected dynamic process between producers, consumers, and destructors. Each organism processes the materials that are useful to it in its own way, with the networked influences of other organisms ensuring that they all operate in a state of steady state.

The particular challenge of substance and material processing for organisms is the limited resource capacity. Over time, this has led to the development of special techniques of the highest effectiveness and efficiency, as described, among other things, in Chap. 4. Figure 5.12 shows that innovative further developments in nature are fundamentally possible with a limited resource capacity. The substance and material natural pathway is compared with that of technology.

The intricate, self-organized and well-tuned cybernetic processing – and their energetic flows – on the micro- and macro-levels of the natural path leave no doubt that evolution has found sustainable paths of progress for its organisms:

> The natural path with its material-technical processing uses a minimum of raw materials in order to achieve a maximum (optimum) of biodiverse, functional, per se nature-compatible and material-composite results.

Analogous to this is our own technology path:

> The technical path with its material-technical processing uses a maximum of raw materials to achieve a variety of special, functional partly inferior ecological results.

Figure 5.12 also symbolizes – in view of the anthropocene impacts as well as the massive changes in our habitats due to climate change – the increasing drifting apart of both development paths, caused by ourselves.

Specifically

From a holistic point of view, we are not only destroying the course of natural developments in a multifunctional way, we are also creating new foundations for technical processes that further advance the destruction of nature. What a *short-sighted, misguided vicious circle of* human activity!

Here, too, it is clear:

> It is better to learn from nature's principles before nature increasingly shows and teaches us how it responds to our short-sighted misguided vicious circles.

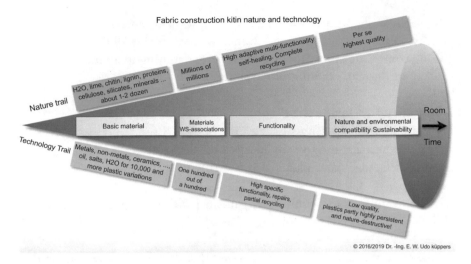

Fig. 5.12 Comparison of substance and material paths in nature and technology

5.3.4 Intelligent Communication of Plants, Animals and Humans

If we humans were to pay more attention to the cross-species communication networks between plants and animals with their warning or key signals (Sect. 4.12) and at the same time optimise our own ability to communicate according to promising rules, then social problems would presumably be dealt with in a more preventive rather than an aftercare manner through more cooperative rather than confrontational behaviour. Some of us have had the foresight to recognize this and have formulated rules that go back to the Greek philosophy of Aristotle. For the present, let us mention just three of many, Paul Wazlawick, Jürgen Habermas and Friedemann Schulz von Thun. We will discuss these rules of human communication with related sources in detail in the practical examples in the main chapter II. In this chapter, we will primarily consider the evolutionary predispositions for communicative decision-making processes in humans with their *"two brains" in the* case of communication between organisms!

The following three topics of plant communication are taken from Mancuso and Viola (2015, pp. 99–118) and are excerpted or given in spirit for brevity. Animal communication is briefly touched upon, human-animal communication is supplemented and very briefly addressed until we illuminate humans themselves and their communication.

To begin with, we use the relatively simple *"sender-receiver model"* developed by *Claude E. Shannon* and *Waren Weaver* in the 1940s as a basis for defining communication among plants, before arguing in a much more complex way for human communication later on.

Communication Inside the Plant

Plants use at least 20 senses to register their environment (see Sect. 4.12), whereas we humans use five. In addition to determining numerous parameters and collecting data, plants also use this data directly for their survival advantage.

The communication within an organism plant, or animal or human provides us with countless examples.

> If the roots of a plant [...] were to discover that there is no more water in the soil, or that a leaf is being attacked by plant pests, then the transmission of this information to the rest of the plant would be vital – because any delay in the transmission of information would endanger the life of the entire organism. So the transmission of this message would be a must. (Mancuso and Viola 2015, p. 84)

Unlike animals and humans, plants do not have a central nervous system through which electrical impulses are transmitted to the brain in the case of a stimulus, e.g. a cut on the calf, whereupon a cascade of internal body processes is set in motion to close the wound.

> [...] plant organisms (do) not even have the biological structures that are required for animals transmit the electrical signals and transmit information from the periphery to the central system: Plants do not possess nerves. But as we have seen [...] (above), the transmission of information is just as significant for plants as for animals, and sometimes just as urgent. Information from roots or leaves is essential to the whole organism, but it can ensure survival only if it is transmitted quickly. (Mancuso and Viola 2015, p. 85)

The vascular system of plants is capable of conducting electrical, hydraulic and chemical signals. Mancuso and Viola (2015, pp. 85–89):

> [...] (The plants) thus possess three independent and sometimes complementary systems,. which are suitable for short and long distances. In this way, they can reach parts of the plant that are only a few millimetres or even several metres away. Let's look at how these systems work. The first system is based on electrical signals. It is used very frequently and is not very different from that of animals

or humans, although it has been "personalised" for plants. For, as we have seen, plants have no nerves, and therefore do not possess the special tissue which in animals serves to transmit nerve impulses. At first glance, this seems like a real problem. How are plants supposed to send out electrical signals if they lack the special tissue to transmit them? The plants have come up with something. On short distances, they use openings in the cell wall, so-called plasmodesmata (from the Greek *plasma, meaning* structure, and desma, meaning band), and transmit the signals from one cell to the next via these plasma bridges. For longer distances, for example from the roots to the leaves, they rely on their "vascular system." [...] Plants, like animals, have a hydraulic system through which they transport various substances; it functions much like our vascular system, except that it has no central pump. Plants do not have a heart because their bodily functions are not located in special organs. But through a kind of circulatory system – arteries and veins, so to speak – they transport fluids from the bottom to the top and vice versa. The tissue is called "xylem" when it transports substances upward, and "phloem" when it transports substances downward. [...]

Incidentally, electrical signals also flow comfortably and relatively quickly through the complex vascular system. In contrast to the rather slow chemical messengers, they transmit important information, such as about water conditions in the soil, posthaste. [...] The three internal signalling systems (electrical, hydraulic and chemical, the author) thus work in a complementary manner.

Communication Between Plants

Plants prefer to communicate via chemical molecules and even develop their own "body language", similar to the way we express ourselves through gestures, facial expressions and body posture. Mancuso and Viola (2015, pp. 90–99):

> They (the plants, the author) also communicate by touching or a certain posture they take towards neighbours. This is especially true for their roots, but also for above-ground parts. For example, when plants race to grow to escape shade, they adopt different postures toward their neighbors. Another example of the "gestural" communication of plants is what the French botanist Francis Halle (born 1938) calls the "shyness of tree crowns". Some tree species conspicuously avoid touching the treetops of their closest neighbors. This is not true of all species, however. Many are far from shy and unabashedly make themselves at home in each other's crowns. However, some plant families such as the beech family *(Fagaceae), the* pine family *(Pinaceae)* or the myrtle family *(Mirlaceae)* are extremely reserved and have something against "encroaching" neighbours. Take a look upwards if you happen to be walking through a pine forest! Or have you ever noticed that the crowns of even closely spaced pine trees always keep a

certain distance – as if every touch were unpleasant for them? Why this happens and how the trees manage to do it is still unclear. [...]

Plants interact on several levels and, like animals, reveal different characters. So could we say that there are combative, aggressive, cooperative or shy plant species? Of course. But that is not all. Plants behave like animals in other ways too. Which is not surprising, because basically all living things pursue the same goals and therefore in all likelihood also choose similar means to achieve them. [...] We do not normally assume that plants maintain kinship or family relationships because we believe that this is the sole preserve of more highly evolved animals or humans. But plants can not only undoubtedly recognize relatives, they are also exceedingly friendly to them. [...] Those who can identify individuals with whom they share a particularly large number of genetic characteristics benefit in terms of evolution, ecology and behaviour. He can, for example, better organize his territory and does not have to waste resources uselessly in fights with family members. It can avoid mating with close relatives and, *above all*, benefits indirectly from the success of all others with similar genetic makeup. [...]

The fact that plants exhibit altruistic behaviour is a new, groundbreaking discovery that actually allows only two possible and equally revolutionary conclusions. Either plants are altruistic because they are much more highly developed than previously assumed, or altruism and cooperation are already found in simple life forms whose struggle for existence has so far allegedly known only one winner: the strongest. But no matter how you look at it: Either way, plant communication with the help of roots would have a concrete evolutionary goal, namely to distinguish between family and strangers, between friend and foe.

This also includes the fact that plants are able to enter into a life partnership with organisms of other species for mutual benefit, the so-called *"mutualistic symbiosis"*.

We often find, for example, mycorrhizae (from the Greek *mykes,* fungus, and *rhiza,* root): symbioses between the underground part of the forest mushrooms we love to collect and eat and the roots of various plants[...] The fungus supplies the roots with minerals such as phosphorus, which is always in short supply in the soil, and in return receives the high-energy sugar produced by photosynthesis. (ibid., p. 95)

So much for sedentary organisms that fight incessantly to defend their habitat against invaders, quite the opposite of mobile animals that, if necessary, seek new habitats when the declared enemy threatens to become overpowering.

Communication Between Plants and Animals

As immobile organisms, plants are dependent on the assistance of other organisms if they want to distribute their seeds and thus ensure their species survival. This is often done by air or water, but not infrequently by animals, which they do in particular in the case of defence (see Mancuso and Viola 2015, p. 99).

When a plant notices that an insect is settling on it and nibbling its leaves, it immediately starts to defend itself. Of course, it must first recognise the enemy in order to develop an appropriate defence strategy. Plants normally defend themselves with chemical weapons, that is, they produce special substances that make them unappetizing, indigestible or even poisonous to the plant pest. However, they do not waste unnecessary energy and produce the "deterrent" substances only in the threatened leaf and its immediate surroundings – in the hope that this will dissuade the insect from its intention. [...]

Plant decisions are usually based on the following calculation: How can a problem be solved with an effort that is as small as possible and as large as necessary? This calculation and the corresponding strategy are often crowned with success. In the case described above, the insect will taste a leaf or two and then seek its fortune elsewhere as quickly as possible: That means victory all along the line!

The plant can easily repair the damage by simply sprouting new leaves, and the loss will not hurt it much because, as we know, neither its functioning nor its survival is threatened by the absence of even considerable parts. [...] If the insect continues to nibble unimpressed, or if several hungry mouths settle on the plant, then it can also do otherwise: then it produces "deterrent" substances in all the leaves and alarms – by volatile chemical substances,

which it releases into the air – its neighbouring plants, so that they join in the defence. It can also summon reinforcements. [...]. (ibid., pp. 100–101)

The time and process of pollination is a central event for plants because they are particularly dependent on communication with animals. Mancuso and Viola (2015, p. 107 f.):

Insects are particularly useful for pollination – in the service of the so-called entomophilous plants (from *éntomo, Insekt,* and *phílos*). However, they are by no means the only ones to take on the delicate task of transport. There are also "zoophilous" plants, which are pollinated by other animals (Greek *zóa*), ornithophilous plants, where pollination is done by birds (Greek *ornítes*) such as hummingbirds or parrots, and even chiropterophilous species, where bats (Greek *cheiropterói*) play courier. Bats? That's surprising. But they are used as pollen bearers by many American desert cacti such as the famous *Joshua Tree*. Other zoophilous plants rely on reptiles (various *panda nut species* on geckos, for example), marsupials or even primates.

The richness and sophistication of plant communication, which also offers interesting approaches for our own communicative behaviour, is far from being fully explored. Just think of the multitude of plant senses that we can cleverly use to track down natural phenomena, if we have not already abandoned valuable plant organisms and their communication secrets to our destruction of nature.

Animal Communication

On the basics of animal communication, Hickman and co-authors write (Hickman et al. 2008, pp. 1182–1186):

> Only through communication can one animal influence the behaviour of another. Compared to human language with its enormous communicative potential, non-human communication in the animal kingdom is severely limited in its extent and expressive power. Animals can communicate through vocalizations. Smells. Touch and movement. In fact, any available sensory modality can be used for the purpose of communication. This gives richness and variety to communication between animals.
>
> Unlike human languages, which consist of words with more or less strictly defined meanings and which can be regrouped again and again to create a virtually unlimited range of new meanings and representations, the repertoire for communicating other animal species consists of a comparatively limited set of signals as a rule. Each signal conveys one, and only one, message. These messages are not subdivided or regrouped. to construct new or even novel messages. A single message, however, may convey to the receiver several units of information relevant to him.

Using three examples, Hickman and co-authors attempt to make clear the different types of animal communication (ibid):

Example 1 The chirping of the cricket

The chirping of a cricket announces to unfertilized females the presence of male crickets, their species affiliation (males of different species have different "songs"), sex (only males produce sounds), location (place of origin of the sounds), and social status (only a male cricket capable of defending the area around its burrow can chirp from this one fixed location). This information is of high importance for females and fulfils a biological function. However, there is no way for the male cricket to modify his vocalizations to convey additional information about available food, predators, or other aspects of the habitat that might increase his prospects for mating, the survival of the potential mating partner, and thus both fitness.

Example 2 The sex pheromones of moths

Attracting mating partners in silkmoths *(Bombyx mori)* is an extreme example of stereotyped communication from a single message that has evolved to serve a single biological purpose – reproduction. Virgin female silkmoths possess specialized glands that produce and secrete a sex attractant to which males are highly sensitive. Adult male silkmoths smell with their protruding, bushy antennae, which are covered with thousands of sensory hairs that serve as receptors [...]. Most of these receptors respond to the odorant bombykol ([...] hexadecadien-1-ol, a long-chain unsaturated alcohol) and to no other substance.

To attract males, the females sit quietly in one place and emit a tiny amount of bombykol, which is carried away by the wind. When a few molecules of this substance reach the antennae of a male silkmoth, it is thereby stimulated to fly off against the wind in search of the emitting female. The search is initially aimless. If the male comes more or less by chance within several hundred meters of a female. It perceives a concentration gradient of the attractant. Guided by this concentration gradient, the male finally flies purposefully towards the female and copulates with her immediately after they have found each other.

Example 3 The language of honeybees

One of the most highly developed and therefore most complicated non-human communication systems is the abstract symbolic language of bees. Honey bees communicate the locations of food sources (= nectar sources) to their conspecifics when these sources are too far away to be easily located by individual bees themselves. Bees communicate through a sequence of movements known as a "dance", which takes two main forms. The form with the highest information content is the *tail dance* [...] (italics mine). Bees perform these dances most frequently when an animal returns to the hive from a rich food source with nectar in its stomach or pollen grains in the pockets formed by hairs on its hind legs (pollen baskets). [...]

The tail gait points […] at the same angle to the food source as the angle of the current sun land relative to the food source.

The distance to the food source is also encoded by the dancing bee. If the food source is close to the hive (less than 50 m away), the signaling bee performs a simple dance called a *round dance* (italics added). The bee returning from foraging simply beats a circuit of one full turn. The round dance conveys the message that food can be found in the vicinity of the hive.

Another form of communication is that of display, as practiced, for example, by the impersonation behavior of blue-footed boobies of the Galapagos Islands, or the courtship behavior of the capercaillie, or countless other species, often using means that appear strange to sophisticated, which are left to the relevant literature for explanation (Dörfler 2019; Urry et al. 2019; Kappeler 2012).

Communication Between Humans and Animals
Anyone who dedicates himself to the subject of communication between humans and animals cannot avoid mentioning the great behavioural scientist Konrad Lorenz (1903–1989). In order to duly address the scientific achievements of the behavioral scientist and Nobel laureate, it would take far more than a few lines in the context of this book. Therefore, only a few of his works are highlighted, which deal with the language of animals and communication between humans and animals, such as:

- 1949: He talks to the cattle, the birds and the fish. New edition 1998, dtv, vol. 20.225
- 1950: How man came to be a dog. dtv vol. 329
- 1963: The so-called evil. Zur Naturgeschichte der Aggression. New edition from 1998, dtv
- 1965: On animal and human behavior. From the development of the theory of behavior. Collected essays from the years 1931–1963. Volumes I and II, Piper, Munich, Zurich and others.

It should not remain unmentioned that Lorenz theses on the behavior of animals (family life of gray geese, with numerous analogies to humans), for which he received the Nobel Prize for Medicine in 1973 and other of his research results and his political stance were not uncontroversial. However, he has never moved away from his basic positions critical of civilization and progress ("Verhausschweinung des Menschen" (Eckhard Fuhr on 21.12.2015 at welt.de)).

Interesting to mention is Lorenz' collaboration with the science theorist Erhard Oeser as well as the marine scientist and theoretical biologist Rupert Riedl – and last but not least with the Austrian-born philosopher Karl Popper, from which the *Evolutionary Epistemology* was developed. It is primarily concerned with how animals and humans perceive [...] with which an individual experiences the world.[4]

Biologists Hickman and co-authors (2007, p. 1186) describe "communication between humans and other animals" and "animal consciousness" from a contemporary perspective as follows:

One uncertainty in studies of animal communication arises from the difficulty of understanding which sensory channels an animal uses for this purpose. Signals can take the form of visual displays, smells, vocalizations, tactile vibrations, or electrical currents (in certain fish). Establishing interactive communication between humans and other animals is even more difficult, as the researcher must translate meanings into symbols that an animal can understand. Moreover, humans are poor social partners for the majority of all animals. However, such communications are not impossible: dogs can easily learn the meaning of many different hand signals and respond with the appropriate actions.

Animal consciousness is a general term for mental performances including perception, thinking, and memory. Many biologists consider some of the mental processes in the brains of other animals to be similar to those in humans. Recent studies of consciousness in nonhuman primates and in gray parrots and some other bird species have produced intriguing results.

An ego consciousness is one aspect of the broader concept of consciousness. Author Donald Griffin (2001) has argued in two books that many animals have self-consciousness and can think or reason. However, this feat of higher brain activity has been demonstrated in very few animal species. The ability of great apes, parrots, and some other animal species to develop language-like skills is significant in this regard, as they reveal cognitive performances that allow us to engage in communicative contact with them. The possibility that animals might possess a thinking ability and have (some) ego-consciousness has opened a new window in behavioral research and given new meaning to animal research more generally. However, studies on the thinking abilities and consciousness of animals other than humans are and will remain highly controversial for the time being.

Wachsmuth (2013) extends – contemporary with the increasing digitalization of our society – natural communication through artificial intelligence, in the

---

[4] https://oe1.orf.at/artikel/214500/Zu-Lebzeiten-eine-Ikone-heute-umstritten (accessed Aug 19, 2019).

understanding between humans, animals and humanoids. This book thematizes, among other things:

> Sign languages and the connection between signs and meaning as well as the effect of spontaneous body movements and posture for the transmission of messages. The question of the extent to which animals and humans can exchange meaning is exciting. Are apes able to use symbols and communicate intentionally, do they perhaps even have consciousness? And will a machine creature like Max one day be able to converse with us from the perspective of an "I" of its own? Machines with "lives of their own" – can and should they exist? Could they become empathetic partners of humans? Would that take something away from humans? Wachsmuth's book asks important questions and takes a broad view: it looks back into a long evolutionary history and ahead into an exciting future.

In the summary to the chapter on "The Behavior of Animals", Hickman et al. (2007, p. 1188) write:

> Communication. often considered the essence of social organization. is a means by which animals influence the behavior of other animals. To do this, they use sounds, smells, visual signals, touch or still other sensory modalities. Compared to the richness of human language. other animals communicate by means of very limited signal repertoires. One of the best known examples of animal communication is the dance language of honeybees. Birds communicate through calls and songs as well as through visual display. Through ritualization, simple movement patterns have evolved into conspicuous, usually species-specific, signals with defined meanings.

Human Communication

Humans in every respect as organisms occupy a dubiously special position in evolution in general and in communication in particular. In general, as self-reflective organisms, we are capable of questioning our own actions and, if necessary, drawing and practicing conclusions in one – advantageous – or another – disadvantageous – direction for us. Whether we always make appropriate use of this useful self-reflection as an evolutionary achievement remains to be seen. For this purpose, two significant examples of human communication can be highlighted on a highly topical occasion.

The first – negative – example shows how a clear lack of self-reflection and obsession with power has led to the accelerated destruction of our biodiversity, which is essential for survival, by a single person – to the detriment of humanity.

The second – positive – example shows the initiative of a young person, which in the meantime has grown into an earth-wide movement among schoolchildren, against precisely this destruction of nature.

Example 1: Massive and accelerated destruction of the Amazon rainforest as one of many major carbon dioxide (CO_2)stores and oxygen producers on Earth.

Jair Bolzonaro, president in Brazil since 2018 and thus "master" of Brazil's Amazon rainforest, demonstrates by virtue of his office how a single policy-maker can succeed in massively destroying the livelihoods of indigenous peoples and, moreover, of all people (Vaughan 2019).

Organized deforestation under the protection of the Brazilian government, cultivation of monocultures as animal feed for cattle, whose meat ends up transported over 10,000 km in European grocery stores, is the increasing chain of sinister links against all reason.[5] It is the in detail small but from a temporal point of view enormous area of three football fields per minute,[6] according to the World Wildlife Fund (WWF)[7] it is two (a football field has according to the rules of the German Football Association – DFB – the minimum area of 4050 m² and the maximum area of 10,800 m²). Under WWF assumption and a minimum area of 4050 m², 8100 m² of Amazon rainforest is destroyed per minute. According to WWF, the Amazon basin has an area of 6.7 million square kilometers, whereas the approximately 8000 m² of forest destruction per minute seems small. Even calculated over a year (365 days, 525,600 min), this results in an area of about 4×10^3 km², which is lost forever as a life-giver for a multitude of organisms; compared to the previously mentioned 6.7 million total area, this is acceptable – isn't it?

What makes this calculation, with which stakeholders of the logging industry presumably argue, look so unspeakably trivial, is the fact that the dynamic complex life networks of the Amazon rainforest are completely ignored. Extremely superficial – not to say extremely stupid – arguments are communicated here, which are far beyond a real value of the real food webs necessary for existence for the biodiverse wealth in the Amazon basin, if it is at all possible to value nature with its perfect principles and regularities in a purely calculative way.

Although the destruction of the Amazon rainforest has been going on for years, the communication of a single politician further aggravates the

---

[5] https://www.zeit.de/politik/ausland/2019-05/brasilien-umweltschutz-jair-bolsonaro-wald-zerstoerung (accessed Aug 15, 2019).

[6] https://www.zeit.de/2019/33/abholzung-amazonas-regenwald-brasilien-spekulation-jair-bolsonaro (accessed Aug 15, 2019).

[7] https://www.wwf.de/themen-projekte/projektregionen/amazonien/zustand-und-bedeutung/ (accessed Aug 15, 2019).

situation, which, to articulate it clearly once again, in terms of the climate change taking place in the Anthropocene, affects all of humanity and thus each and *every one of us.*

Example 2 : The pupils' initiative *"Fridays for Future"* against the lethargy of adults in positions of responsibility to finally act consistently and take action against the destruction of nature instead of burdening today's children and young people with the burdens and consequences of adult failure in this matter of survival.

Who is this group of – now Earth-wide – activists for sustainable livelihoods? On the associated website are clear and rightly communicated arguments (font adapted):[8]

WE ARE FRIDAYS FOR FUTURE.
The climate crisis is a real threat to human civilization – addressing the climate crisis is the main task of the twenty-first century. **We demand a policy that does justice to this task.**

Fridays for Future: These are all those who take to the streets for our climate. The climate strike movement is international, non-partisan, autonomous and decentralized. Join us and become part of our movement!

As in the negative example before, in this positive case of communication it is a single person, the 16-year-old Swedish climate activist Greta Thunberg, who took the initiative to actively oppose the policy of "denial", arguing:

Why should I be studying for a future that soon may not be, when no one is doing anything to save that future?
Why build on a future that will soon no longer exist?

Our answer to this question is the climate strike: we are striking for effective policies that do justice to the scale of the climate crisis. We have ten years to reach our goals. **Let's go!**

In the name of science: More than 27,000 scientists in German-speaking countries alone are behind us and support our demands. We demand nothing more from politics than the consideration of scientific facts.

These demands are clear and unambiguous. In contrast to the negative example of communication described above, the demands to treat nature and the environment better than has been done so far have no legally secure basis.

---

[8] https://fridaysforfuture.de (accessed Aug 15, 2019).

Only an earth-wide insight into the necessity of protecting our basis of life instead of destroying it is the impetus for a global, partly radical reversal in the thinking and actions of today's decision-makers in politics and business. The increasing destruction of biodiversity alone, including almost certainly many unknown species, which can and already do provide us humans with new medical agents and innovative nature-compatible technical developments, e.g. through systemic bionics (Küppers 2015), already contain enough arguments for a forward-looking new way of thinking and acting, the results of which even inveterate destroyers of nature and the environment will benefit from.

From an evolutionary communicative point of view, the conditions that have led to people deciding the way they do are of particular interest. Who is aware that humans have two brains, or rather two nervous systems, in their bodies?

Does that surprise you?

We already know from Sect. 5.1 that in our head there is a complex working brain or a neuronal nervous system, the *central nervous system – CNS –* which, with an unimaginable number of about 86 billion neurons and almost 10,000 connections per neuron, decisively determines our life – and thus our consciousness. From a medical-neurological point of view, processes of thinking and thus also of communicating, in which feelings also play a role, are primarily assigned to our central nervous system. It is all the more astonishing that we, as a planning, foresighted and self-reflecting and apparently rational organism, not infrequently feel a *gut feeling* when we talk to each other, from which we derive decisions, for example whether our conversation partner is likeable, less likeable or the exact opposite, whether we should make an important professional decision or not etc. etc. Rationally, we cannot justify this; however, we are left with a feeling of security or insecurity, of well-being or discomfort, which in quite a few cases guides us in how we continue to behave towards our communication partner. We often experience that excitement or irritation hits us in the stomach, that unpleasant things are heavy in our stomachs or that we are in love and feel "butterflies" in our stomachs. We recognize our stomach or our gastrointestinal tract as organs of our body in which something is going on.

In reality, the abdomen is an intelligence center.

It contains as many nerve cells as a pet's brain. We have about 200 million nerve cells in our gut. That's about as many as [...] a dog has in its cerebral cortex. [...]

Not only are there millions of neurons hidden in our gut, but also many 100 billion bacteria that secretly influence our personality.

The introduction quoted in advance and the last explanation are taken from the film by Céline Denjean: "Le ventre notre deuxième cerveau" © ARTE France – Inserm – Scientifilms 2013. The intensive medical-neurological scientific research on the functional characteristics of the enteric nervous system is intended to reflect the following excerpt of a series of scientific papers (Boesmans et al. 2019; Kulkarni et al. 2018; Gershon 2018; Mayer 2016; Heuckeroth and Schäfer 2016; Lake and Heuckeroth 2013; Holschneider and Puri 2008; Furness 2007; Costa et al. 2000).

Gershon's discovery of the second nervous system is crucial for the analysis of our evolutionary behavior of communication processes. Although we have intuitively argued with our so-called *gut feeling* in communicative processes for a long time, without bringing clear causal conclusions into the field for it, the hope lies in future analysis results of the enteric neuron network to be able to understand these feelings better.

Overview

Excursus in anticipation of the practical chapters under II:
   In this context, the research of the underestimated phenomenon in the human body, the linking of head and abdominal brain, is particularly relevant to the increasing robotics research or Humanoid research of interest (Küppers 2018), which is increasingly mastering programmed functional movement and action processes and is on its way to teaching humanoids to feel, as in the Karlsruhe Research Center for Computer Science, with the help of the six-legged walking robot Lauron,[9] or in autonomous driving (McDuff and Kapoor 2019), or in programmed emotions (Broekens 2018).

What was previously named the intelligence center is called the *"Enteric Nervous System – ENS"* and is directly connected to the central nervous system in our head via the large *vagus nerve, the vagus nerve.*

It was Michael Gershon, professor of pathology and cell biology at Columbia University in New York, who claimed that "[…] we carry a sensitive and intelligent organ in our abdomen." (02:15′–02:20′) (Gershon 1999, 2001).

---

[9] https://www.fzi.de/forschung/projekt-details/lauron/ (accessed Aug 18, 2019).

Our intestinal wall is covered with 200 million [...] nerve cells. They are distributed along the entire length of our alimentary canal and enable us to digest. [...] Our two brains have a lot in common. [...] The outsourcing of the second brain has a background similar to that of the PCs on our desks. By moving it to the periphery, to the gut, the head brain works more effectively. It does not need to be connected to the digestive brain via thick cables, nor does it need to be expanded by several hundred million neurons.

Michel Neunlist of Inserm, Institut des Maladies de l'Apapareil Digestif in Nantes, France is considered a recognized expert on the enteric nervous system.

> He has studied the evolution of our two brains, with surprising results. The term second brain is not quite correct, because for me it is the first original brain. In fact, the first primitive organisms consisted only of a digestive tract, from which the enteric nervous system then evolved. [...] Evolution [...] created our cranial brain to enhance our nutrition. At the same time, eyes and ears evolved and facilitated foraging. Without this division of labor, we would still be spending all our energy today digesting, digesting, and digesting again.

It was only through our subsequently developed head brain that, over time, we too were able to communicate. Both brains are connected – as already mentioned – via the vagus nerve, using the same neurotransmitters. The neurotransmission is the language and the neurotransmitters are the words the neurons use to communicate. *Serotonin* is one of those words. In the head, it acts on the *hypothalamus* (a section of the diencephalon) and thus controls our emotions. In the stomach, serotonin is produced more than 95%, determines the rhythm of intestinal activity and regulates our immune system.

Evolution is just extremely ingenious. It has therefore also perfectly coordinated the development of our two brains in terms of time and function. This evolutionary precision took place over long periods of time and, as is well known, without a target, far beyond human life spans. But we as present living humans at the beginning of an anthropocene age have it in our hands to draw our lessons from this and from further principles of nature, especially for our children and grandchildren!

Looking at our work activities in the time of a neoliberal capitalism in democracies (see, among others, the instructive book by Rainer Mausfeld (2019), *Angst und Macht*), in which still the physical from the mental work experiences a separation, it should be clear that the direct connection of both brains, which use the same "words" for a neuronal communication, identify us and other organisms as *holistic* living beings. It would therefore continue to be not only a wrong decision, but an unparalleled disaster we are heading for, if we use the digitalization of our society to further force the separation between human physical and mental work and at the same time create replacements by robots or humanoids, which may one day, through algorithmically programmed electromechanical work, combined with emotional algorithms, degrade humans, as a means (point) – instead of perceiving them as the center. This danger of a communication disaster from the present point of view is still fictitious, but for how long?

5.3.5 Self-Organization Through Skilful "Negative" Feedback

Nature is, of course, also a veritable treasure trove for processes of self-organization, ranging from the smallest bacterial colonies to plant and animal and human organisms. (Blech 2010).

It is worth noting that bacterial colonies in our gastrointestinal tract also contribute to our health or viability. Westerhaus (2012) describes it like this:

> In the human intestine alone, there are more than 1000 different species of bacteria. The total weight of the colonizers is estimated at 1.5 kilos, per gram there are up to a trillion germs.

Extrapolated to the calculated total weight of all microorganisms – the technical term for this is *"intestinal microbiome"* – in the human intestine, the result is the unimaginable number of 1.5×10^{14} germs! One might be tempted to say that humans are their own ecosystem. We form a kind of living community with our microbiome, where negative feedback contributes to the stability of our condition. So-called intestinal maloccupation or food intolerance sets off a cascade of *positive* and *negative* feedback processes, in which negative feedback ultimately contributes to the decisive restoration of a stable state of health.

Figures 5.9 and 5.10 show self-organization processes involving negative feedbacks (shown as dashed lines in Fig. 5.10), of three organisms that cyclically maintain their species growth under *normal* environmental conditions.

How a self-regulating process works through positive and skilful negative feedback is something we have all certainly experienced in ourselves. Here is an example (Küppers 2019, p. 51):

Let's look at our breathing or respiratory rate. When an untrained runner is running, the breathing rate will increase very quickly with increasing running speed *(positive reinforcing feedback)*, because oxygen is increasingly needed, up to a limit that forces the runner to reduce his speed due to lack of capacity *(negative, opposite weakening feedback)*, whereupon the breathing rate is reduced again and normal values settle in. Trained runners also regulate their oxygen supply through the variable breathing rate. However, they do this much more effectively and in a more adapted manner, which allows them to cover longer distances without reaching their absolute performance limit.

But now let us turn to the term *"negative"* feedback, where we include the term *"positive"* feedback for the sake of completeness.

Negative Feedback: Balanced Feedback

A control loop or control system is controlled by negative feedback. Feedback occurs when the output signal of an information processing system is fed back to the input, creating a closed loop. (ibid, p. 48)

Negative feedback occurs when the feedback circulation loop has a weakening effect on the so-called reference variable of a circulation function, whereby the circulation cycles, which are usually repeated several times, lead neither to a standstill nor to an overload of the system, thus possessing a stabilising function, the state of which is known in nature as the *"steady state"* stage or *"steady flow"*.

Positive Feedback: Reinforced Feedback

In contrast to the negative feedback designed for balance and stability, there is also a so-called positive feedback. As a rule, circulation cycles of positive feedback are run through several times; in a control loop, this can lead on the one hand to a standstill of the control function, and on the other hand to a one-sided build-up beyond the physical limit of the system – *overload* – and thus ultimately to the destruction of the system itself.

A healthy human organism regulates its body temperature to 37 °Celsius. In a state of rest, without great physical exertion, this is just as successful as under physical stress, for example during a running sprint, which briefly leads to an increase in temperature above the core temperature of 37 °C. In Fig. 5.13 this process is marked by the blue feedback loop, as a "negative" feedback stabilizing the temperature.

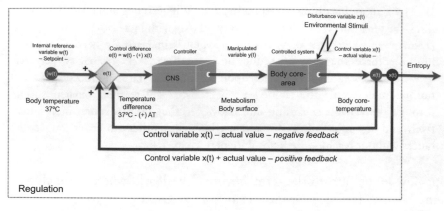

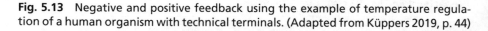

Fig. 5.13 Negative and positive feedback using the example of temperature regulation of a human organism with technical terminals. (Adapted from Küppers 2019, p. 44)

In contrast, in a sick human organism, triggered by an external disturbance variable, in the form of a bacteria-infected insect bite or worse as sepsis (a blood poisoning), a "positive" feedback loop (red feedback loop) can set in, which overrides the body's own temperature regulation for stabilization. This is where the limit of the body's own self-organized regulation for temperature stabilization can be exceeded. Only external medical treatment can help to restore the body's own regulation processes!

That regulation or control processes have a cybernetic background is shown by the definition of cybernetics by Norbert Wiener, who was the first to recognize the value of a "negative" feedback loop and described it on the cover text of the second, revised and supplemented edition of his book on *cybernetics* (1963):

Definition of cybernetics according to Wiener (1963):

Theory of communication and control processes in machines and living organisms.

It goes on to say:

This means that cybernetics is used to summarize the diverse efforts to unite intelligence, psychological, sociological, biological and, more recently [1960s, the author] medical research projects.

Feedback has been the salt in the soup of evolutionary organismic developments for billions of years. Norbert Wiener is credited with discovering the importance of "negative" feedback for the stability of a system, however structured it may be, although at the beginning he was not thinking of the immeasurable number of biological applications, but – due to the times – of military ones.

Both feedback processes among organisms are particularly evident in analyses of food chains or food webs or material cycles, the observation of habitats and their growth limits. It is the *"Principles of Ecology"* (Odum, original 1989, German editions 1991, p. 1999) that describe such processes. This becomes clear in another natural example, the development of a *climax society*. Odum (1991, p. 196) describes how biocoenoses and ecosystems harmonize:

> Communities and ecosystems go through a development from a juvenile to a mature stage that can be compared to the growth and maturation of an individual organism; however, patterns and control mechanisms are quite different, [...]. Most of us are aware that there is ecological succession (a temporal progression of plant and animal communities at a site), because it occurs continuously everywhere in nature. What is less well known is that these changes follow certain patterns which, in the absence of major disturbances, are predictable. When an area becomes available for the development of a living community (for example, when a crop field is abandoned and returned to nature), different plants and animals gradually colonize that area in a series of temporary transitional stages called seral stages. Over time, more permanent communities develop until a mature or climax stage is reached that is in equilibrium (steady state) with or determined by the regional climate and the local substrate, water and topographic conditions.

The developmental stages from a grassland via overgrown shrubs to a pine forest and finally to a deciduous forest (mixed oak or beech forest) as a climax forest are interspersed with positive and negative feedbacks. So-called pioneer plants set an enormous pace to fill the unvegetated soil areas with higher organisms again by positive feedback. However, this only happens over a limited period of time (negative feedback), after which new seral stages take over the growth progress until the next higher seral stage or the final stage of a climax society.

Another example of natural negative and positive feedbacks is shown in the following scenario, see also Fig. 5.14:[10]

---

[10] https://de.wikipedia.org/wiki/Nahrungskette (accessed Aug 21, 2019).

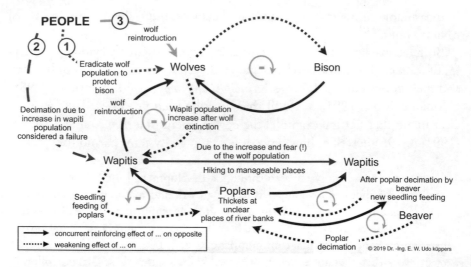

Fig. 5.14 Positive and negative feedbacks in the food web of the Yellowstone National Park biotope, USA

Overview

In the American Yellowstone National Park, wolves were persecuted in the first decades of the twentieth century and exterminated by 1926 in order to secure the population of bison. As a result, an extreme density of wapiti deer was observed in the area, which in some regions were densely grazed like cattle. After attempts to limit the density of wapitis by shooting them failed, the wolf was reintroduced in 1995 and placed under protection. Now the following effects occurred:

- Along the riverbanks, thickets of poplars grew up in some places instead of grass. Closer examination showed that these were obscure places. These were now obviously avoided by the wapitis, which had previously suppressed the poplars by their feeding on the seedlings. The decisive factor here was obviously not so much the density limitation of the wapiti population by the feeding pressure of the predator wolf (as expected according to the textbook), but simply the fear of the wapitis of the wolves, i.e. a change in behaviour. Such indirect effects are highly significant in numerous ecosystems, but are often neglected by the fixation of ecosystem research on production and energy turnover (Peckarsky et al. 2008, footnotes replaced by author names, the author.).
- Attracted by the poplar thickets as a food resource, the beaver, which had become extinct in the national park, began to migrate into the area again and soon reached high densities. Thus, the presence of the predator wolf was shown to be crucial for the occurrence of the beaver via indirect effects (quasi: around two corners) (Ripple and Beschta 2004, footnotes replaced by author names, the author).
- No one could have predicted this.
- Beaver and wapiti combined may eventually suppress softwood forests again. (Baker et al. 2005, footnotes replaced by author names, the author).

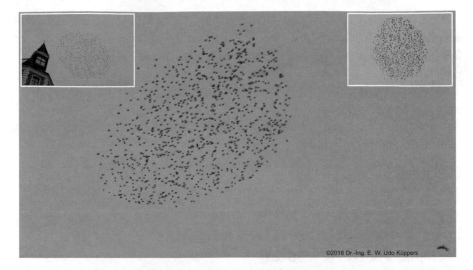

Fig. 5.15 Starling flock over summery Berlin in August 2018. little information is needed for each individual starling to better protect itself against potential predators (bottom right in Fig.) when flying in flocks rather than as individuals

Figure 5.15 shows the three – all failed – attempts by humans to interfere with or regulate a functioning food web in Yellowstone National Park, as described previously.

Populations linked by negative feedback would have settled over time without deliberate human intervention through the premature extirpation of wolves, similar to the predator-prey relationship depicted in Fig. 5.10.

Lupus est homo homini, non homo, quom qualis sit non novit. For man is a wolf to man, not a man. This is true at least as long as one does not know himself.[11]

What the Roman comedy poet Titus Maccius Plautus wrote about 254–184 B.C. for human behavior towards conspecifics, applies in the present example even more to the intervention of man in a nature that is not understood! Here man is – in a figurative sense – a wolf to the organisms of nature, without knowing exactly their functionalities and values.

Other examples of self-organization including negative feedback in nature are flocks of fish or birds, the collective behavior of ants in search of food, and so on.

Figure 5.15 shows – somewhat coarse-grained – images of a flock of starlings presumably being observed from a distance by a bird of prey (lower right image).

---

[11] https://de.wikipedia.org/wiki/Homo_homini_lupus (accessed Aug 21, 2019).

So-called swarm intelligence (Kroll 2013, p. 375 ff.; Horn and Gisi 2009; Vehlken 2009, pp. 125–162) recognizes three basic rules that must be followed by each individual in the swarm to ensure safety in the swarm:

1. Move towards the center of your fellow species that you see around you
   – *Cohesion, Cohesion*
2. Move away from your fellow species as soon as someone gets too close to you
   – *Separation, separation*
3. Move in roughly the same direction as your neighboring species…
   – *Alignment, adjustment*

5.4 Eight Basic Biocybernetic Rules According to Frederic Vester

Frederic Vester's understanding of the complex workings of nature, with guidance for a better understanding by thinking and acting humans of their own environment, he communicated in an inimitable manner. From his numerous publications (including Vester 1974, 1975, 1976, 1978, 1985, 1989, 1991, 1999 and more) Vester recognized "Our world as an interconnected system" and the (bio)cybernetic interrelationships in nature, from which he derived eight *basic biocybernetic rules (including Vester, 1983, pp. 66–86)* – not least for human use of sustainable error-tolerant action. We will now take a closer look at these rules – slightly modified by the author.

Section 5.3.5 has already anticipated the importance of negative feedback in conjunction with positive feedback. Therefore, we will concentrate here on the graphical representation of feedback processes, underlying effect functions between the arguments of the system elements and practical examples.

> Basic biocybernetic rule 1: Negative feedback dominates positive feedback.

5.4.1 Feedback and Its Functions

Growth-oriented organisms, both as individual organisms and in species assemblages, are linked to a large number of feedback processes that have both reinforcing and balancing effects. We would like to take a closer look at some of the special features underlying these processes, both in nature and in technology. Due to the enormous complexity of nature's networks, only very simple functional relationships of feedbacks will be explained in the following, which do not occur in nature in such an isolated way, but which contribute to the understanding of the process.

> The simplified presentations on feedbacks primarily serve the understanding about the course of the effect relationships between the arguments of system elements. In nature more than in technology, these types of feedbacks are involved in complex dynamic food networks or technical-economic processes, which in the breadth of their application and presentation would go beyond the scope of this book.

Positive Feedback and Its Functions

Figure 5.16 shows a series of positive feedbacks with associated functional progression between the arguments or system elements, while Fig. 5.17 presents examples from nature. All of the feedback examples shown exhibit a "positive" tendency in the result, although one of the examples also exhibits a positive,

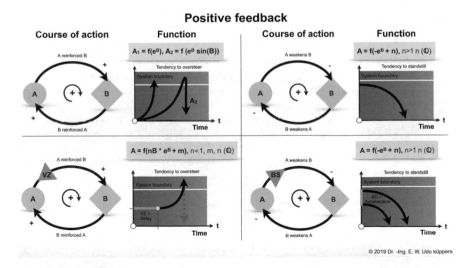

© 2019 Dr. -Ing. E. W. Udo küppers

Fig. 5.16 Positive feedbacks and their exemplary functions

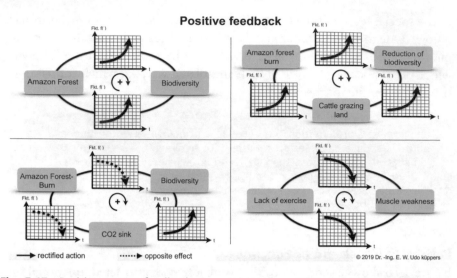

Fig. 5.17 Positive system feedbacks in combination with individual negative feedbacks in organisms and their exemplary functions

reinforcing tendency due to two – generally even-numbered – negative feedbacks in the compound. Thus, the entire feedback loop has a positive effect.

Negative Feedback and Its Functions

Figure 5.18 shows a series of negative feedbacks with associated functional progression between the arguments or system elements, while Fig. 5.19 presents examples from nature. Odd numbers of negative feedbacks in combination with positive feedbacks, as shown by the three examples in Fig. 5.19, always result in a negative, system-stabilizing overall feedback, which, for example, approaches biotope boundaries (system boundaries) asymptotically or in a growth-regulated manner without massively exceeding them.

Where humans are integrally involved in feedback processes, caution must be exercised and corresponding uncertainties in the functional sequence must be taken into account.

In the complex dynamic biogeological networks of nature between plants, animals and geological conditions, this caution is not necessary, as long as the flow processes take place in a natural way – without human intervention and without massive impact damage from storms, floods, etc.

Paradoxical situations of feedback arise and are regrettably on the rise when it is recognizable and comprehensible that man is the cause of natural problems and at the same time is supposed to act as their problem solver.

Finally, two examples of ecological interconnections (Figs. 5.20 and 5.21) provide a deeper insight into the complexity of feedbacks that can only be

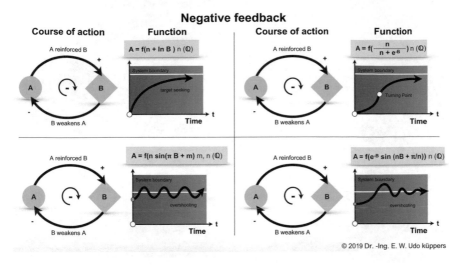

Fig. 5.18 Negative feedbacks and their exemplary functions

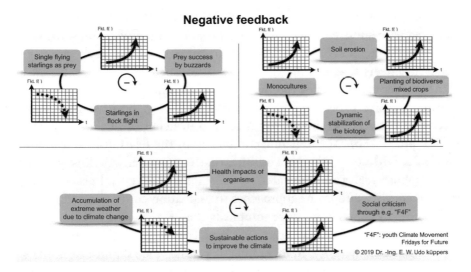

Fig. 5.19 Negative system feedbacks in organisms in combination with individual positive feedbacks and their exemplary functions

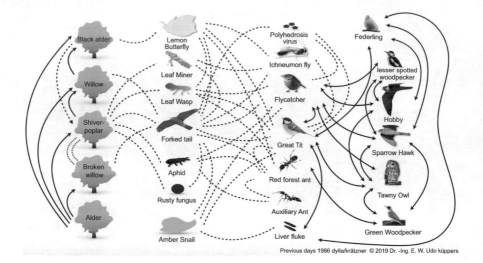

Previous days 1986 dylla/krätzner © 2019 Dr. -Ing. E. W. Udo küppers

Fig. 5.20 Example of the organismic ecosystem. (Adapted from Dylla and Krätzner 1985, p. 85)

captured with sufficient precision – even if not to the full extent – using systems analysis methods. Figure 5.14, on the other hand, shows an interdependency of effects that is still relatively easy to analyse, but is by no means complete; nature still holds too many secrets for that.

The extensive cybernetic network structure, with positive and negative feedbacks between organisms, are described in detail in Dylla and Krätzner (1985, pp. 77–86). The dynamic complexity of the relationship structure in Fig. 5.20, with its qualitative and quantitative biocybernetic linkages, show that a robust analysis of the interrelationships can only lead via the path of systems analysis, and linear approaches to evaluation come to nothing.

Figure 5.21 shows even more complexly than Fig. 5.20 the cybernetic network structure between organisms in a large ecosystem, with relatively low to medium species diversity. Reindeer play a central role as land-use animals in the tundra and taiga food system regions. Vogel and Angermann (1979, p. 229) write in this regard:

> [...] biocenotics (*) examines (in a biotope, the author) the interrelationships of organisms. Thus, food chains can be identified that link entire series of species (e.g., green plant-herbivores – small predators – large predators – necrophages (scavengers, n.d.)). the food chains are linked by unspecialized forms with versatile food, the totality of all interwoven food chains forms the food system (A). (to be seen in Fig. 5.21, n.d.)

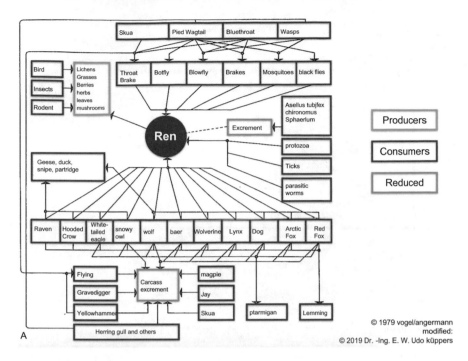

Fig. 5.21 Example of an organismic ecosystem in tundra (cold steppe) and taiga (boreal coniferous forest and northernmost forest formation on Earth) with special reference to the reindeer. (According to Vogel and Angermann 1979, p. 228)

Biocenosis describes a community of organisms of different species in a limited habitat *(biotope)*. Both together – biocenosis and biotope – form an ecosystem. Biocenosis and biotope together form the ecosystem. In this ecosystem, a variety of interdependencies or interrelations arise, which also interact with the inanimate (abiotic) nature. Ultimately, this results in an ecological steady state.

As mentioned above, biocenotics – or *synecology* – investigates these interactions.

5.4.2 Growth-Independent System Function

Basic biocybernetic rule 2: The system function must be independent of growth.

Despite the 170 billion tonnes of organic matter or biomass produced annually (Sect. 5.2.3), growth processes in nature are coordinated by

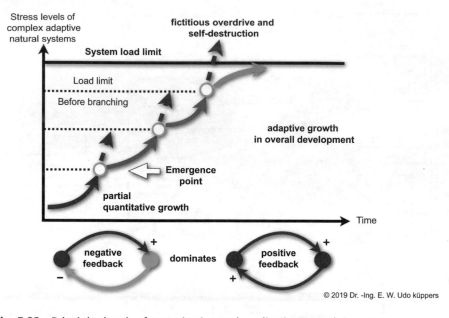

Fig. 5.22 Principle sketch of quantitative and qualitative growth in nature

regulated, flexible and cycle-oriented processes (see also biocybernetic rule 6) in such a way that they never lead to the destruction of a biotope under their own steam.

Figure 5.22 shows the basic growth process of a natural development over several emergence stages, as is known from the ecosystem development of a forest as a climax society (Sect. 5.2.1).

A growth example (Odum 1991, p. 197) shows: Over a period of more than a hundred years, natural succession evolved from fallow farmland, to grassland, to shrubs, to pine forest, to deciduous forest, the final stage of development, with successive increases in species abundance and pair density. Each successional stage of development has typical species with their own growth rates, which, once a new higher stage of development with increasing species diversity is reached, transitions into a new growth cycle. This adaptive growth process is one of several quality features that ensure the survivability of natural systems.

Another controlled adaptive growth example is known from (zoological) metamorphoses in the insect kingdom. Surely each of us has seen caterpillars of butterflies on trees in spring or summer, which do nothing but eat food and grow to a certain size. The point at which they stop growing and become butterflies by pupation is evolutionary. If the caterpillars were to continue growing regardless of this growth limit, which of course does not happen, the organism would not be able to survive. This already vaguely suggests a comparison with the growth philosophy of humans in the technosphere, which we will discuss in Chap. 6.

A third example of adaptive growth through continuous (zoological) metamorphosis is the development of frogs, via aquatic tadpoles to terrestrial organisms (Portmann 2000).

A fourth and final example from botany illustrates the ability of roots, shoot axes, and leaves to transform by adapting to particular habitat conditions (botanical metamorphosis) (Bell 1991, pp. 194, 224, 300).

We can sum up at this point: Growth in nature takes place in a very controlled manner. There are no material reservoirs created in the course of growth that are stored and later processed as needed in growth processes, nor are there landfills full of materials that no longer create value. Ultimately, adaptive cybernetics, and in particular the "negative" feedback in the food webs, make a decisive contribution to this.

5.4.3 System: Function-Oriented Instead of Product-Oriented

Basic biocybernetic rule 3: The system must be function-oriented instead of product-oriented.

Survivable systems [...] function-oriented instead of product-oriented. This allows for far greater flexibility and adaptation to change. [...].

The mitochondria, for example, tiny power plants in our body's cells, have the task of controlling the metabolism of substances and energy. They can use one and the same cycle to burn carbohydrates to carbon dioxide as well as to branch off to the production of amino acids. Even bacteria produce different enzymes, and hence products, depending on the "millieu" and environmental conditions in order to cope with the respective situation. A principle that is typical for the smallest as well as for the largest biological "work processes" (Vester 1991, p. 72).

Since the discovery of the "second brain" with its clever gut decisions by Gershon and Erde (1981) in the 1980s, the question has been pursued whether the tens of billions of bacteria in the gastrointestinal tract not only possess multifunctional properties in their milieu, but the gut microbes are also capable of influencing our social behavior and thus our emotions, which are generated in the limbic system of our head brain and expressed via the cerebral cortex through fear or joy. (Mayer 2016, p. 156 ff.).

> Bacteria are functionally oriented in many ways in our bodies – or should we say more precisely: we play the roles in the cosmos of our bacteria that are assigned to us by the functional divergence of the bacteria!

Similar functional orientation or functional diversity to that found in bacteria exists everywhere in nature, for example in protein building blocks, the proteins, as universal building or operating materials in organisms, or the diversity of membrane functions, e.g. as transport function, signal transmission and communication, extraction and conversion of energy, and so on. (for membrane functions, see Boujard et al. 2014, p. 162) (Fig. 5.23).

The dynamic functional orientation of organisms in nature is naturally perfectly adapted to the habitats. In addition, organisms know how to deal highly efficiently with the energy provided, with the substances available and through skilful information processing or communication. At all levels of

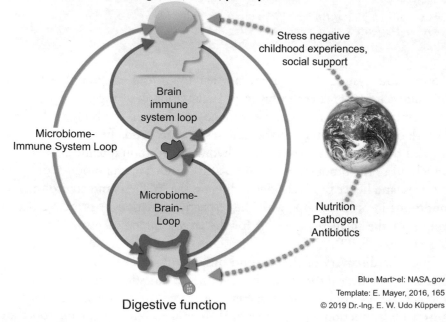

Fig. 5.23 Coupling and feedback between gut-microbiome-brain axis and environment. (According to Mayer 2016, p. 165, slightly modified by the author)

development, functions are linked to structures. If the functions change due to external influences, this is also associated with changes in the structure.

Evolution has therefore managed to develop so many variations from the basic function of locomotion in the animal kingdom, to create perfectly adapted functions and structures of locomotion in deserts, rain forests, steppes, polar regions, oceans, in the air, wherever living organisms are to be found, in other words, on the whole globe.

The same basic principle of the function locomotion is recognizable in the subgroups of flying, swimming, running, climbing etc.

Let's look at a fictional example:

Assuming "companies" of nature wishing to "market" the function of flight, there is an abundant choice of innovative "product developments" available, such as those of ultra-lightweight construction of missiles and wings, the flight capability of gliding, soaring, hovering, energy-efficient dressing flight on long distances, injury minimization in collisions without total failure, highly economical "fuel consumption", etc. Even with a specialization on only a few species, an enormous development potential with variant-rich innovative flight "products" would be available. Costly entrepreneurial competition for the "best" flight products would be dispensable, because in nature all flight functions are energy-optimally adapted. Their innovative "natural product range" could be varied or extended without further ado with few changes or low material or energy "cost burdens", due to the basic functional orientation of the "natural enterprise", adapted to the specific environmental conditions of the use of the "products". Only a minimum of impulse energy is needed to make the "product range" even more lucrative, of course all within the framework of an adapted and cyclical growth. Every "company" would be able to develop sustainably and innovatively.

Because of the way in which, in our fictitious example, "nature companies" make "flight products" for different habitats, with function-rich flight techniques – from insects, hummingbirds, songbirds, forest birds, seabirds, birds of prey, etc., and so on. – on the general target path of the function of flying, variably adapted and innovatively "developing, producing and selling", the thought could also mature that we wake up in our increasingly crisis- and catastrophe-ridden technosphere, and not only take note of the ingenious achievements of nature, but also use its principles for our sustainable continued existence. We will come back to this in Chap. 6.

5.4.4   Use of Existing Forces

Biocybernetic basic rule 4: Use existing forces energetically cleverly instead of using new forces energy-wasting.

The only energy source that is available to us organisms is – as already mentioned several times – our sun, which has been active for about 4.5 billion years. By virtue of its radiant energy and its still long-lasting life span, which is estimated at about 6 billion years, we do not need to worry about the loss of this energy source.

At the beginning of its development – whether this happened 4 or 3.5 billion years ago is unimportant – evolution had only this natural source of energy at its disposal, which the first living beings, in the form of micro- and later macro-organisms, knew how to use for further processing through skilful energetic conversion processes.

The principle of the *"energetic transformation chain"* was born and has remained until today. The higher organisms climbed in evolution, the more versatile and skilful were their energetic utilisation processes, which they applied – adapted to the living environment. As so-called "open" systems, organisms at all stages of development process energy for their own body structure and bodily functions. They create "order in the organismic system" for a limited lifetime at the expense of the increasing disorder around them and us. Self-organization processes in organisms and within groups of organisms promote this order and cohesion. Countless natural examples of this exist as irrefutable evidence of life evolution directed by energy cascades.

As a rule, organisms, e.g. South African springboks as prey of cheetahs, only need a short energy impulse as an escape reflex, in order to then use their full energy reserves for the escape. They are probably very alert, but are not permanently under energetic "full load" when cheetahs are in their environment. Just like the hunting cheetahs, they know how to allocate their energy reserves optimally.

This small nature-example is to show nothing else than the adapted and skilful handling of the available energy, that is stored in the bodies of the two animals in the muscle-cells and is called up when needed.

Forces, in whatever form they are effective in nature, whether as solar-thermal, aero-, hydrodynamic, mechanical or electrical energy (e.g. of the electric eel) are used optimally for the purpose. How else could a bearded vulture with low jumping energy from a rock with minimal wing beats as initial impulse power act, with a wingspan of almost three meters, without a single power-consuming, energy-draining wing beat of its board-like broad wings, fly circling through the air for several hours, if it does not skillfully use the aerodynamic energy of updrafts? All other organisms demonstrate this talent as well, except one species: man.

The use of external or counter forces, to pass them on with low impulse energy and to use them sensibly to save one's own strength is a principle that was perfected in martial arts by the Japanese Judo and Jiu-Jitsu master Jigoro Kano in the transition from the nineteenth to the twentieth century in Japan.

"Winning by yielding" or *"minimum effort with maximum effect"* was and is the guiding principle of this sport. Possibilities, how we can divert our energy-wasting, thus exorbitant costs and consequential problems of energetic processes in our technospheric environment also according to this nature and martial arts principle, we also discuss – as often communicated in the previous place – in the coming practical Chap. 6.

5.4.5 Multiple Use: Functional Diversity

> Basic biocybernetic rule 5: Products, functions and organisational structures should be subject to multiple use.

Multiple use in nature is a principle of advantage that extends from the microscopic smallest organisms, such as bacteria or building blocks of life, such as genes and proteins, to the macroscopic range of "giants of life", such as 100 m high sequoia trees.

> (The) *multiple use (is) a* basic principle without which nature would have gone bankrupt already on the molecular basis and thus probably already some billion years ago. Already in the smallest organisms one can observe how polypeptides and other macromolecules are built up in such a way that they can be used for the most diverse functions: for the construction and degradation of hormones and enzymes, for the control of gene activity or the pain process. Even single molecules are already used by the living cell for multiple roles: functional diversity. (Vester 1985, p. 83)

Evidence for functional divergence in microscopic nature was provided by an international working group around the Giessen anatomist and cell biologist A. Meinhardt (Filip et al. 2009), who found multiple use or *"functional divergence"* on a ribosome (macromolecule in a cell). In a short institute information of the University of Giessen on 27.01.2009 re: "Schwarzarbeitendes"

Protein "wirkt entzündungshemmend" the following is reported (https://www.uni-giessen.de/ueber-uns/pressestelle/pm/pm11-09,   accessed   on 23.9.2020):

> The human genome consists of surprisingly few genes. One way to increase the functional diversity of the proteins encoded by these genes is to "moonlight" proteins. This refers to the fact that some proteins can carry out secondary activities in addition to their main task. In particular, proteins that build the ribosome – the protein factories of the cell – are known to do this. The research team from Germany (Gießen and Aachen), Sweden (Stockholm) and the USA (Yale) has now identified a "side job" for the ribosomal protein S. 19 (RPS19) in mammalian cells (namely the anti-inflammatory effect in case of injuries, the author).

Under Chap. 15: "Long-distance migrants and the multiple fates of emigrated neural crest cells" in Müller and Hassel (2018, p. 442), both authors write:

> Not a few of the [...] signal substances we meet again, if one follows the path-finding of axons of nerve cells [...], of sprouting blood capillaries [...] or the tubules of outgrowing lungs or kidney tubules [...]. There are many overlaps, which prove the high conservation of proven signaling systems, but also the *economic multiple use* (italics added) of a once laid track of attractants, here for blood vessels and neurons.

Junker[12] (2010, p. 1), in his contribution to "Evolutionary Developmental Biology", points out some "[...] recent astonishing discoveries in developmental biology and genetics [...]". Under the headings of *multiple use,* he writes:

> Similar proteins up to entire cascades of proteins connected in series in metabolism and signal transmission are used several times, in some cases even many times. A famous example are the lens crystallins, the main component of the lenses of lens eyes. In the cell metabolism of other organs of the same animal, these also fulfil tasks such as heat shock proteins or in the metabolism of sugars or alcohol.

Another example leads us into multiple use or functional divergence in soils or arable soils, wherein microorganisms perform multiple tasks, e.g. nitrogen and carbon mineralization as well as much more (Ottow 2011, pp. 81–121). Ottow writes about this in the preface to his book "Microecology of Soils":

---

[12] https://www.wort-und-wissen.de/artikel/a05/a05.pdf (accessed September 02, 2019).

Soils are dynamic products of complex interactions between soil-dwelling organisms (plants, microorganisms and animals), the parent rock and the climate (weather patterns). Consequently, soils are always site-specific and characterized by characteristic biocenoses. [...]. The diversity and abundance of microorganisms (*bacteria, archaea,* protozoa), fungi, myxomycetes, oomycetes, algae, diatoms, rotators, tardigrades, oligochaetes, enchytraeids, nematodes, spiders, insects, molluscs, mammals and viruses in the upper crust is enormous. Soils are, as a result, habitats *par excellence.* [...]. Our earth is, down to great depths, an intensively animated sphere of morphological life diversity and unimagined wealth of metabolic processes. Most microorganisms are still neither known nor cultivable. Their genetic and functional diversity surpasses any imagination.

Nature is not so quick to *show* its *cards.* This is true on both a large and a small scale.

The leap to higher dimensions leads us into the macroscopic realm of nature. A species-rich forest is used in many ways by the networked plant and animal biodiversity for their special survival strategies. Individual tree species of the forest derive multiple benefits through perfectly adapted biotic communities (symbioses), see also *Basic Rule of Biocybernetics 7.*

Nature does not know stupid monocultures with even more stupid one-sided growth strategies. Mixed forests or wild mixed orchards (orchard meadows), or those imitated by humans, show in many ways how soil can be sustainably used in multiple ways from a holistic point of view, and thus how its soil quality can be preserved over a long period of time. Even in the microcosm, as just described, multiple use is nothing unknown, e.g. in the case of genes and proteins.

It is a significant characteristic of survivable systems to provide their products and processes with a survival advantage through multiple benefits. All this happens on the basis of dynamically stable networks, behind which is hidden not a constraint but an exceedingly expedient strategy. Vester writes in this regard (1991, p. 76):

The principle (of multiple use, the author) corresponds to the energy and effort-saving mode of operation of nature, where the pollination of flowers is coupled with the nourishment of insects, where the earthworm not only serves the birds as food, but at the same time aerates the soil, where the leaves regulate the humidity between plant and air, but also provide photosynthesis. We find the same rational principle in us in the coupling of sexuality with social tasks such as partner cohesion and aggression inhibition, where the release of sex hormones not only serves reproduction, but at the same time strengthens the immune system and disease defenses, protecting us from infections and cancerous diseases.

5.4.6   Circular Recycling

> Basic rule of biocybernetics 6: Use of closed-loop processes instead of linear processes for material recycling.

Cycle-oriented process sequences also correspond – how could it be otherwise – to an energy-efficient, effort- and consequence-minimizing mode of operation of nature. Let us recall Fig. 5.2, with the manifold linking of different organismic work processes, represented in a concentrated way by the cycle symbols *P-K-R*, which stand for *produce-consume-reduce. In* contrast, technical processes are still largely subject to the dictates of linear work processes of *P-N-E*, i.e. *produce-use-dispose*, despite many approaches to a so-called "circular economy". This "linear economy" or "waste or discard economy" is anything but sustainable in the sense of holistic provision for the future, but more on this in Chap. 6.

Nature possesses evolved structures for cross-species energy conversion, material-processing, resource-efficient and application-optimized networks of action, as well as information networks – decentralized "high-performance networks", if you will – that every organism, every species, every genus, every family, every order, every class, and every phylum, with countless intermediate categories, knows how to use perfectly for itself.

Once again: Nature knows no degree of waste or wastefulness. Both terms originate from the idiosyncratic human linguistic usage and were or are therefore unsuitable for a survival-promoting biocybernetic economy of nature. The reverse is true if we "*obey*" nature, which means nothing other than making advantageous use of its ingenious principles. The solutions to this are open and freely available; one only has to look closely.

5.4.7   Symbiosis or Mutual Benefit

> Basic biocybernetic rule 7: Use symbioses for mutual benefit.

Cross-species cooperation or interspecific interaction, for mutual benefit, represents *symbiosis* in nature. Such cooperative behaviors are distributed among diverse relationships in the animal and plant kingdoms, as we subsequently demonstrate with concrete examples.

Cooperation and competition are two evolutionary behaviors that are closely related, at least in sociobiology, the science of the biological adaptations of animal and human social behavior, as Voland (2013, p. 2) elaborates.

> The change from one behaviour to another is then an expression of a conditional strategy. It contains a rule for adopting behaviors appropriate to the situation, for example, according to the motto: "When competition increases, become aggressive; when competition decreases, become cooperative!" We are dealing here with strategic flexibility. If conditions change, behavior changes – but not in the sense of contingent randomness, but according to biologically evolved rules. (ibid., p. 12)

This type of situational cooperation within a species is another, equally evolutionarily crucial, as that of symbiosis, which always occurs across species. Campbell et al. (2006, p. 1410) put it this way:

> Interspecific interactions that are positive for both species involved are called symbiosis or mutualism. Such relationships sometimes require co-evolutionary adaptations in the partners, as changes in one species are likely to affect the survival and reproductive capabilities of the other. [...] (Examples include) symbiotic nitrogen fixation by bacteria in the root nodules of legumes, cellulose degradation by microorganisms in the gut of termites and ruminants, photosynthesis by unicellular algae in the tissues of coral polyps, and the exchange of nutrients in mycorrhizal plant roots.

Another example of a plant-animal symbiosis, between acacia trees and ants, is described by Campbell et al. (ibid.):

> Certain Central and South American acacia species have hollow spines that are inhabited by stinging ants of the genus *Pseudomyrmex*. The ants feed on the sugar syrup secreted by the nectaries on the leaf stalks and on the excretions of the protein-rich beltlet corpuscles (which look orange, d. A.) at the tips of the leaflets. In return for "shelter and food," the ants attack anything – in any way – that touches their host plant. They sting other insects, remove fungal spores and other organic material from leaf surfaces, and nibble the shoots of other plants that grow into the foliage crown of Acacia trees.

In his small and readable booklet, Bernhard Borgeest (1997) has compiled 24 symbioses involving 24 trees from 24 countries. As stationary creatures, trees cannot simply run away when danger threatens. As was shown in Sect. 4.12, they have a rich set of sensory protection instruments, including symbioses.

Reading Borgeest's symbiosis stories, it also becomes clear what extraordinary relationships people have with trees in their country, at least in times when people perceived plant – and likewise animal – organisms as partners even more intensively than is the case today at the beginning of the third millennium. Borgeest (1997, pp. 7–8):

> It seems that trees also take root in the souls of people. Anyone who has ever lain in their shade as a child and listened to the stories told by the wind and leaves understands how the largest of all plants can become the epitome of home and security. Some trees seem to convey this familiarity more than others. For many nations seem to grow together with certain species in a special way. What the birch is to the Russian, the maple is to the Canadian. The red maple leaf is emblazoned on the country's flag like a logo that is supposed to ensure the *corporate identity of* a state. And the essence and aesthetics of Japan are reflected in the blossoms of the cherry tree, whose leaves die in beauty because they sink to the ground with no fading.
>
> A tree and its land – symbioses where natural history and cultural history meet.

That the trees described by Borgeest, in addition to their symbioses, can also provide useful insights into current environmental problems, some of which are already being used, is illustrated by five examples:

- Ginkgos are planted as street trees in large Chinese cities because they are surprisingly resistant to exhaust fumes (ibid., p. 46).
- *Suitable for 999 purposes:* The coconut palm. What the ancient Fijians practiced can serve as a model for the engineers of today: Total recycling through suitable material compositions in new technology (ibid., p. 57).
- An environmentally friendly floor covering and insulating material can be obtained from cork oaks (ibid., p. 61).
- The wood of redwood trees is both hard (core) and soft (bark). It contains natural insecticides and fungicides through the substances Sequirin A and B. In addition, it is flame-retardant due to the lack of resin and therefore a thoroughly interesting material for environmentally compatible house insulation (ibid., p. 66).
- Swiss stone pines, which grow at high elevations to near timberline, tolerate sulfur dioxide (SO_2) and ozone (O_3) better and longer than many other tree species (ibid., p. 93).

Biocyberneticist Frederic Vester formulates his view of symbiosis as follows (Vester 1999, p. 70):

> From biocybernetic findings it follows [...] that a system develops most advantageously in symbiosis with its environment. This means that it forms suitable substructures, itself becoming part of a superordinate structure and interacting with it.

Even if functioning symbioses are necessarily limited to small spaces, and decentralized structures and diversity with broad functionality seem necessary, a central statement for evolutionary nature management (Vester 1999, p. 139):

> The ecological and economical sense of the symbiosis is that it leads to a considerable saving of raw material, energy and transport of all elements involved and thus also relieves the environment.

> **A Brief Look Ahead**
>
> Even though technospheric progress has made life and work easier and more comfortable for us through a multitude of practical results since the industrial revolution in the middle of the eighteenth century, and has even generated partial prosperity, it also imposes enormous burdens on us, e.g. through the consequences of excessive exploitation of raw materials, destruction of biotopes and widespread establishment of monocultures:
>   Looking ahead to the increasing material, energy and information problems in the Anthropocene era that our societies, and thus we ourselves, will face, the process of symbiosis will play a significant role in the future viability of life support on our planet.

Therefore, we take the time to look at this important survival criterion of evolution in detail, from different professional perspectives, in order to learn from it for the management of our problems in the present and the future.

It was the young PhD student in biology, Lynn Margulis[13] (1938–2011), who significantly advanced a revolutionary new insight into the process of life (the German botanist Andreas F. W. Schimper[14] postulated in 1883 the symbiotic origin of chloroplasts, which are organelles of the cells of green algae and land plants that perform photosynthesis). An organelle is the structurally

---

[13] https://de.wikipedia.org/wiki/Lynn_Margulis (accessed 04 Sep 2019).
[14] https://de.wikipedia.org/wiki/Andreas_Franz_Wilhelm_Schimper (accessed 04 Sep 2019).

definable region of a cell. In 1905, the Russian biologist Konstantin S. Merezhkovsky[15] devised the precursor hypothesis to the *"endosymbiont theory"* made famous by Margulis.[16]

Margulis describes her work on *endosymbiont theory* as follows (Margulis and Sagan 1997, pp. 96–97):

> Organisms enter into many forms of symbiosis. One of the most impressive is the intimate connection known as endosymbiosis. In this, one living thing – microbe or larger – lives not only close to (or even constantly on top of) but within another living thing. In endosymbiosis, organic living things nest together. Endosymbiosis is like a long-lasting sexual relationship, except that the partners belong to two different species. Some endosymbioses have actually become permanent.
>
> Bacteria, usually the masters of symbiosis, are the best endosymbionts for at least four reasons: First, because they have been forming stable bonds with each other for billions of years, they can establish very durable associations. Second, their tiny bodies permanently distribute and acquire genes. Therefore, they can respond rapidly to genetic changes. Third, bacteria experience individuality only in a limited way; vigilant antibodies are absent – consequently, an "infection" can become the basis for a lifelong association, a mutual evolution, and is not fought as in an animal with an immune system. Fourth, the large chemical repertoire of bacteria leads to metabolic complementarity. This is less often observed in associations between already highly individualized members of plant and animal species. Of course, over time, some plants and animals may become as close as some bacteria.

In this description of bacterial benefits, let us also remember our own cosmos of bacteria in the microbiome of the gastrointestinal tract, described in Sect. 5.3.5.

> Symbiosis creates new individuals. Without bacteria in the gut, humans could not produce B or K vitamins. Cows and termites would be helpless without the floating fermenters in their digestive systems – ciliates and bacteria that decompose grass and wood. Algae living inside transparent whirligig worms feed their hosts so well that their mouthparts atrophy. The green worms that have had their mouths stuffed, so to speak, sunbathe rather than struggle to find food. The endosymbiotic algae even convert the worm's uric acid into food.

---

[15] https://de.wikipedia.org/wiki/Konstantin_Sergejewitsch_Mereschkowski (accessed 04 Sep 2019).
[16] https://de.wikipedia.org/wiki/Endosymbiontentheorie (accessed 04 Sep 2019).

There are thousands of other partnerships. For example, all of the approximately 20,000 lichens originated as symbiotic associations of fungi with algae or cyanobacteria.

However, the most important symbioses led to the eukaryotic cell (cell with a true nucleus, the "domain" of all living things, d. A.). Most protoctist cells and all plant, animal and fungal cells contain mitochondria (they are true masters of *functional divergence*, as they are involved in multiple processes in the organism, d. A.). Respiration with oxygen, which keeps members of the four youngest kingdoms of organisms alive, takes place in these particular organelles. (Like organs in a body, organelles are special functional structures of a eukaryotic cell.) Mitochondria look like bacteria. In the larger host cell, they even grow and divide on their own. As is now thought, they are descended from bacteria – but after more than billions of years of living together, they can no longer survive on their own outside the cell boundaries. (ibid.)

Ecosystem scientist E. P. Odum (1991, p. 190) writes of symbiosis:

We now come to those relationships that might be called "positive interactions." Charles Darwin's emphasis on the "survival of the fittest" has focused attention primarily on competition, predator-prey relationships, and other negative interactions. However, as Darwin himself pointed out, cooperation for mutual advantage is also widespread in nature, and it is also of great importance for natural selection.

This – actually self-evident – clarification, if one looks retrospectively at the works of Schimper, Mereschkowski and Margulis, that competition and cooperation are equally important for the progress of life, is postulated in the social environment, as Darwin's contemporary, the philosopher and sociologist Herbert Spencer (1820–1903) already *ingloriously* proved, one-sidedly in favour of competition – among humans (!). Regrettably, this public-spirited portrayal continues – especially in the economic environment – right up to the present day.

Cooperative behavior is one – if not the – prerequisite for avoiding conflict. It is far less stressful and risky than confrontational behaviour.

If, for example, two people face each other whose mutual communication sends out positive, i.e. emphatic, signals, it seems inconceivable – under evolutionary conditions of these positive interactions – that conflict situations will abruptly arise. Exceptions confirm the rule, initiated by the irrationality of human communication behaviour. Examples of pseudo-cooperation are also not uncommon in flora and fauna, but they will not be discussed further here.

What forms of cooperation do we now know from nature?

> Positive interactions between two or more species can take three different forms, possibly in an evolutionary series. *Commensalism* (+0) is a simple form of positive interaction in which one species benefits while the other has neither harm nor benefit.
>
> If two species benefit each other but are not dependent on each other (i.e. can survive separately without further ado), one can speak of *cooperation* (++). If the relationship finally becomes so close that it is vital for both species, we have *symbiosis* (++).
>
> Commensalism is particularly common between small mobile and larger sedentary organisms. On seashores, such relationships can be observed particularly well. Virtually every worm burrow, clam, and sponge harbors various "uninvited guests" (crustaceans, annelids, small fish, and so on) that need the protection or leftover food of their hosts but neither harm nor benefit their partners. Those who like oysters and open them in large quantities will occasionally come upon a small, delicate crab living in the mantle cavity of the oyster. Known as "shell guards," these crabs are usually commensals, or "fellow eaters," but sometimes they go too far and eat their host's tissues (Christensen and McDermott 1958). It is a small step from commensalism to parasitism on the one hand and mutual aid on the other. Commensals are no less host-specific than parasites, although some are found associated with only a single host species.
>
> Symbiosis, in a sense the highest form of cooperation, is widespread and extremely important. Many pairs or groups of species live together as obligate partners for mutual benefit (none can live alone). In an indirect way, symbiotic (mutualistic) relationships are also beneficial to the overall ecosystem. For example, the […] symbiosis between plants and nitrogen-fixing microorganisms not only benefits both partners, but also plays a key role in the life-sustaining nitrogen cycle. In most cases, two species enter into a symbiotic relationship that are taxonomically distinct (and not even "distant relatives"), but both can provide vital "goods and services" that the respective partner requires. Two examples, ruminants and their panenflora, and termites and their intestinal flagellate, […] (are) related to the detrital food chain […]. In both cases, the microorganisms convert cellulose to fats and carbohydrates that the animals can utilize, and the hosts provide the microorganisms with habitat and protection from competitors and predators. (Odum 1991, pp. 190–191)

The fascination that apparently emanates from symbiotic (mutualistic) behaviour was already described in a compact and structured way 70 years ago by Herbert Brandt (1949). Table 5.6 shows the various interactions of symbioses in nature, without being able to explicitly go into the numerous examples.

In the extensive chapter on symbiosis and mutualism, Begon et al. (1998, pp. 329–356; and 2017, pp. 291–303, respectively) described numerous

Table 5.6 Classification and occurrence of symbioses*1

Symbioses	Organisms
Plants with plants	Bacteria and bacteria
	Bacteria and algae
	Algae and fungi
	Bacteria and higher plants
	Fungi and higher plants
Plants with animals	Bacteria and animals
	Algae and animals
	Mushrooms and animals
	Higher plants and animals
Animals with animals	Protozoa and multicellular organisms
	Lower animals and lower animals
	Lower animals and vertebrates
	Vertebrates and vertebrate animals

After H. Brandt (1949)

examples that underscore the breadth of evolutionary application. The current 2017 edition also addresses the costs and benefits of mutualistic relationships. Comparable to Table 5.6, Begon et al. (1998) describes various symbioses in broad application, compiled in Table 5.7. Begon et al. (1998, p. 329) write introductory to the subsequently listed symbioses in Table 5.7:

> The Earth's biomass is largely made up of mutualists, and the spectrum of mutualistic (alternating) relationships is very broad.

Under this perception, our overexploitation of nature and thus the destruction of entire biotopes with their biodiverse riches takes on an even higher significance on a fictitious scale of destruction of life in nature.

Whether it is the aforementioned examples of symbioses reaching into the depths of hot hydrothermal vents, complemented by the examples of deer and geese, ravens and wolves, tropical walnut and golden hare, yucca moth and flower, etc., etc., described by Brandstätter and Reichholf (2017).Symbioses in nature offer a rich treasure that, through skilful transformations, including with the support of *"systemic bionics"*, can remedy our ecological social and economic problems in the technosphere, or at least steer them in a more sustainable direction.

Even if, by definition, symbioses *always* involve at least two different species, and within a species, e.g. between humans, one can rather speak of cooperation, humans, as the trigger of many destructions of nature and at the same time as their potential problem solver, also always remain connected with *external* nature, the animal and plant world, in some symbiotic way. *Internally,*

Table 5.7 Classification and occurrence of symbioses

Symbioses	Organisms
Mutualisms based on behavioural patterns	Ants and plants
	Cleaner fish and their customers
	Honey Indicator And Honey Badger
	Shrimps and gobies
	Anemonefish and sea anemones
Mutualisms in agriculture	Homo sapiens has a mutualistic relationship with useful plants and animals
	Keeping of caterpillars by ants
	Beetles and ants as mushroom growers
Mutualisms in seed and pollen dispersal	Mutualism in seed dispersal the dispersal of hook-bearing or sticky seeds by animals is *not* mutualism-unlike dispersal by fruit eaters
	Pollination mutualisms – insects and flowers
Mutualisms with inhabitants of the digestive tract	The rumen – flora and fauna
	The termite gut – cellulose degradation by protozoa
Symbiosis and mutualism Within animal tissue or cells	Insect mycetocytes (insect body cells) and bacteria
Mutualisms between higher plants and fungi	Mutualisms between plants and fungi are *true* mutualisms, as in:
	Leaves of healthy winter wheat and at least 200 different species of fungi
	Mutualism of plant roots and fungi: mycorrhiza
Photosynthetically and chemosynthetically active symbionts of invertebrates	Freshwater polyps can feed both autotrophically (self-feeding) and heterotrophically (feeding on other organisms)
	Animals near hydrothermal vents Live in symbiosis with chemosynthetically active bacteria
Mutualism between fungi and algae: Lichens	Lichens, a mutualistic relationship of fungi and algae or cyanobacteria seem to have evolved several times in evolution
Nitrogen fixation by mutualists	Mutualism between nodule bacteria and legumes (leguminous plants)
	Nitrogen fixation Mutualisms with non-legumes
	Nitrogen-fixing mutualists in ecological succession
The evolution of subcellular structures from symbioses	The spectrum of associations in the plant and animal kingdoms is extremely diverse

After Begon et al. (1998, pp. 329–356)

we are naturally symbiotically active through the bacterial cosmos in our enteric nervous system. *Nevertheless!*

- *Through the human "symbiotic" impulse to develop strategies that reduce bee mortality or restore higher species biodiversity, we indirectly benefit from the resurgent harvest of plant foods.*
- *Through the human "symbiotic" impulse to develop strategies that reduce mindless food monocultures as a result of industrial vicious circles or replace them completely with clever use of mixed cultures that can also sustainably feed people without consequential costs and problems, we will reap economic and at the same time ecological benefits.*
- *Through the human "symbiotic" impulse to develop strategies that "systematically" utilize the natural potential of medicinal remedies, we may be able to mitigate or avoid disease more effectively than before, thereby restoring our health and resuming activities.*
- *Through the human "symbiotic" impulse to develop strategies that stop the poaching and indiscriminate shooting of big game in Africa, we are indirectly rewarded through conservation and the possibility of seeing these species in the wild rather than exclusively in zoos for a long time to come.*
- *Through the human "symbiotic" impulse to develop strategies that use animal and plant survival strategies to our advantage, we respect their existence and life and gain innovative techniques that probably would never have been achieved without this indirect, indirect symbiotic relationship.*

These few of umpteen examples of symbiotic indirect relationships between humans and plants or animals show, beyond the formal definition of symbiosis, what is possible and doable when we combine our foresighted intelligence with the ingenious principles of nature to keep it alive.

From this perspective, we can certainly benefit a great deal from the symbioses in the animal and plant kingdoms. However, it is crucial to recognise these sustainable advantages of a win-win situation, which is not always easy to initiate because it also has a long-term effect, and to have the will – above and beyond all mechanical, partly mindless bureaucratic control processes – to implement them practically.

5.4.8   Design or Shaping

Basic biocybernetic rule 8: Orientation of product, process and organizational design to biological design.

- **FEATURE** Irregularity within regularity – a basic principle of biological design: more familiar and soothing than geometric uniformity.

Organisms perfect for themselves this principle of nature highlighted by Vester (1999, p. 140). Trees all grow via a root system, the trunk, the branching branches with leaves, which all look similar on a tree but are never completely the same. The growth principle of trees is similar but always different in nuances. Even in interconnected biogeonic environments, the meandering design of free-flowing river courses is similar but not congruent. Mass living quarters of weaver birds in southwestern Africa are all the same at first glance, yet always individually different in design upon closer inspection. Transparency, lightness and resilience are universal characteristics of biological design experiments. In addition, the skilful material composites used for this purpose are 100% recycled at the end of their life or use.

Even the natural infrastructure, the construction of transport routes, e.g. in villages of indigenous peoples in Africa, who feel particularly connected to nature, is intuitively done on an irregular network of paths of minimal (regular) length (Schaur 1991). Again, Vester (1999, p. 140):

> Biological design takes into account endogenous and exogenous rhythms, uses resonance and functional fits, harmonizes system dynamics and enables organic integration of new elements […].

Do you feel a difference in the way you walk through a mixed forest with a biodiverse richness full of sounds on the one hand and through a forest laid out as a grid-like monoculture of fast-growing woods for technical processing on the other? If not, then you have already moved far away from the natural rhythms and resonances of nature – of which you are a part.

Nets in nature are another example of perfect biological design using minimal materials, high stiffness and irregular regularity, which have spread especially in the realm of insects (Stern and Kullmann 1996; Kullmann et al. 1975). Other authors have explored design and form in nature – and technology – over many years, some of which are mentioned as representative of in-depth study (Nachtigall and Pohl 2013; Vogel 2000; Tsui 1999; Wunderlich and Gloede 1977; Patzelt 1972).

One could conclude, with a comparative view from all this: The organism is the *center of* biological design and form, whether as an expression of its own functional performance or as a shaper of its environment and not – as so often with the human organism: *means – point.* Paradoxically, the human organism is itself part of nature, but has moved further away from it than any other organism.

With the eight described Vester's basic rules of biocybernetics this chapter ends, in which an attempt was made to give you, the dear reader, a further, partly pictorial insight into the power spectrum of nature and to show mechanisms that lead to it.

The following Chap. 6 now moves into the field of bionic technospheric application, where insightful examples will be given on how to learn from nature's ingenious principles for progress, *without* inevitable self-inflicted consequential problems that we still have to "deal with" at present and in the near future.

References

Baecker, D. (1994) Postheroisches Management. Merves, Berlin

Baker, B. W. et al.: (2005) Interaction of beaver and elk herbivory reduces standing crop of willow. In: Ecological Applications. Band 15, Nr. 1, 2005, S. 110–118.

Brandstätter, J.; Reichholf, J. H. (2017) Symbiosen. Das erstaunliche Miteinander in der Natur. Matthes & Seitz, Berlin

Begon, M. E. et al. (1998) Ökologie. Spektrum Akad. Verlag, Heidelberg, Berlin

Begon, M. E. et al. (2017) Ökologie. 3. Aufl.; Spektrum Akad. Verlag, Heidelberg, Berlin

Bell, A. D. (1991) Plant Form. An Illustrated Guide to Flowering Plant Morphology. Oxford University Press, Oxford, New York, Tokyo

Berkhoff, K., B. Karstens & J. Newig (2004) Komplexität und komplexe adaptive Systeme – Ansätze des Santa Fe Instituts. In: Bundesministerium für Bildung und Forschung, Steuerung und Transformation, BMBF, Berlin: 101–107.

Blech, J. (2010) Leben auf dem Menschen – Die Geschichte unserer Besiedler. Rowohlt-Verlag, Reinbek b. Hamburg

Boesmans, W. et al. (2019) Structurally defined signaling in neuro-glia units in the enteric nervous system. GLIA, 2019, 67:1167–1178.

Borgeest, B. (1997) Ein Baum und sein Land. 24 Symbiosen. Rowohlt, Reinbek b. Hamburg

Boujard, D. et al. (2014) Zell- und Molekularbiologie im Überblick. Springer Spektrum, Berlin, Heidelberg

Brandt, H. (1949) Symbiosen. Kosmos-Bändchen Nr. 181, Franckh'sche Verlagshandlung, W. Keller & Co., Stuttgart

Broekens, J. (2018) A Temporal Difference Reinforcement Learning Theory of Emotion: unifying emotion, cognition and adaptive behavior. Originally announced July 2018, pre-print, https://arxiv.org/search/cs?searchtype=author&query=Broekens%2C+J

Campbell, N. A. et al. (2006) Biologie. 6. Aufl., Pearson, München

Christensen, A. M.; McDermott, J. (1958) Life History and Biology of the Oyster Crab, Pinnotheres ostreum. In: Biol. Bull. 144, 146–179

Costa M.; Brookes, S. J. H.; Henning, G. W. (2000) Anatomy and physiology of the enteric nervous. System. Gut 2000; (Suppl IV)47: iv15–iv19

Dörfler, E. P. (2019) Nestwärme. Was wir von den Vögeln lernen können. Hanser, München

Dürr, H.-P. (1999) Naturrecht und Menschenrecht. Mensch, Technik und Natur. In: Sauer-Sachtleben (Hrsg.) Kooperation mit der Evolution, S. 256–265, Diederichs, München

Dylla, K., Krätzner, G. (1985) Das ökologische Gleichgewicht.in der Lebensgemeinschaft Wald. Quelle und Meyer, Heidelberg

Ebeling, W. (1991) Chaos – Ordnung Information, 2. Aufl. Harry Deutsch, Thun, Frankfurt a. M.

Ebeling, W.; Feistel, R. (1982) Physik der Selbstorganisation und Evolution. Akademie Verlag, Berlin

Egner, H.; Ratter, B. M. W.; Dikau, R. (2008) Umwelt als System – System als Umwelt? Systemtheorien auf dem Prüfstand. Oecom, München

Feistel, R.; Ebeling, W. (2011) Physics of Self-Organization and Evolution. Wiley-VCH, Weinheim

Filip, A.-M. et al. (2009) Ribosomal protein S19 interacts with macrophage migration inhibitory factor and attenuates its pro-inflammatory function. J. Biol. Chem. doi:https://doi.org/10.1074/jbc.M808620200.

Furness, J. B. (2007) Enteric nervous system. Scholarpedia, 2(10):4064.

Gershon, M. D. (2018) Development of the Enteric Nervous System: A Genetic Guide to the Perplexed. Gastroenterology, February 2018, Volume 154, Issue 3, Pages 478–480

Gershon, M. (2001) Der kluge Bauch. Goldmann, München, Original 1998 The Second Brain, HarperCollins, New York

Gershon. M. D. (1999) The second brain. A Groundbreaking New Understanding of Nervous Disorders of the Stomach and Intestine. HarperCollins Pub., New York

Gershon, M. D.; Erde, F. M. (1981) The Nervous System of the Gut. Gastroenterology 1981:80:1571–94

Griffin, D. A. (2001) Animal Minds. Beyond Cognition to Consciousness Rev and Exp. University of Chicago Press, Chicago, USA

Harmon, L. D. (1973) The Recognition of Faces. Scientific American, Vol 229, No 5, p. 70–83

Hauschildt, J. (1990) Zielbildung und Effizienz von Entscheidungen. In: Fischer, R.; Boos, M. (Hrsg.) (1990) Vom Umgang mit Komplexität in Organisationen: Konzepte – Fallbeispiele – Strategien, S. 131–147, Universitätsverlag Konstanz

Herculano-Houzel, S. (2009) The Human Brain in Numbers: A Linearly Scaped-uo Primate Brain. Front Hum. Neurosci. 3:31, Nov. 9

Heuckeroth, R. O.; Schäfer, K.-H. (2016) Gene-environment interactions and the enteric nervous system: Neural plasticity and Hirschsprung disease prevention. Dev Biol. 2016 September 15; 417(2): 188–197.

Hickman, C. P. et al. (2007) Zoologie. Pearson, München.

Hickman, C. P. et al. (2008) Zoologie. 13. akt. Auflage, Pearson, München

Holschneider, A. M.; Puri, P. (Ed.) (2008) Hirschsprung's Disease and Allied Disorders. 3. Ed. Springer, Berlin, Heidelberg

Horn, E.; Gisi, L. M. (2009) Schwärme – Kollektive ohne Zentrum. Eine Wissensgeschichte zwischen Leben und Information. Transcript, Bielefeld

Humphreys, W. F. (1979) Production and respiration in animal population. In: Journal of Animal Ecology, 48, 39–66

Junker, R. (2010) Evolutionäre Entwicklungsbiologie. Internet-Beitrag (s. a. a. O., Kapitel 5.4.5)

Kappeler, P. M. (2012) Verhaltensbiologie, 3. Aufl., Springer, Heidelberg

Kroll, A. (2013) Computational Intelligence. Eine Einführung in Probleme, Methoden und technische Anwendungen. Oldenbourg, München

Kullmann, E. et al. (1975) Netze in Natur und Technik. IL8. Institut für leichte Flächentragwerke, Universität Stuttgart.

Kulkarni, S. et al. (2018) Advances in Enteric Neurobiology: The "Brain" in the Gut in Health and Disease. The Journal of Neuroscience, October 31, 2018 38(44):9346–9354

Küppers, E. W. U. (2019) Eine transdisziplinäre Einführung in die Welt der Kybernetik. 258 S., Springer Vieweg, Wiesbaden

Küppers, E. W. U. (2018) Die humanoide Herausforderung. 464 S., Springer Vieweg, Wiesbaden

Küppers, E. W. U. (2015) Systemische Bionik. Springer Vieweg, Wiesbaden

Lake, J. I.; Heuckeroth, R. O. (2013) Enteric nervous system development: migration, differentiation, and disease. Am J Physiol Gastrointest Liver Physiol. 2013 Jul 1; 305(1): G1–G24.

Leser, H. (1997) Landschaftsökologie. Ansatz, Modelle, Methodik, Anwendung. UTB. Stuttgart.

Mancuso, S.; Viola, A. (2015) Die Intelligenz der Pflanzen. Kunstmann, München

Margulis, L.; Sagan, D. (1997) Leben. Vom Ursprung zur Vielfalt. Spektrum Akademischer Verlag, Heidelberg, Berlin, Oxford

Mausfeld, R. (2019) Angst und Macht. Westend, Frankfurt a. M.

Mayer, E. (2016) Das zweite Gehirn. Wie der Darm unsere Stimmung, unsere Entscheidungen und unser Wohlbefinden beeinflusst. Riva, München

McDuff, D.; Kapoor, A. (2019) Visceral Machines: Risk-Aversion In Reinforcement Learning With Intrinsic Physiological Rewards. Published as a conference paper at The International Conference on Learning Representations ICLR 2019

Müller, W.; Hassel, M. (2018) Entwicklungs- und Reproduktionsbiologie des Menschen und bedeutende Modellorganismen. 6. Aufl., Springer Spektrum, Berlin

Nachtigall, W.; Pohl, G. (2013) Bau-Bionik. Springer-Vieweg, Berlin, Heidelberg

Odum, E. P. (1989) Prinzipien der Ökologie. Spektrum der Wissenschaft, Heidelberg

Odum, E. P. (1991) Prinzipien der Ökologie. Spektrum der Wissenschaft, Heidelberg

Odum, E. P. (1999) Prinzipien der Ökologie. 3. Aufl., Thieme, Stuttgart

Ottow, J. C. G. (2011) Mikrobiologie von Böden. Springer, Berlin, Heidelberg

Patzelt, O. (1972) Wachsen und Bauen. Konstruktionen in Natur und Technik. VEB Verlag für Bauwesen, Berlin

Peckarsky, B. L. et al.: (2008) Revisiting the classics: Considering nonconsumptive effects in textbook examples of predator – prey interactions. In: Ecology. Band 89, Nr. 9, 2008, S. 2416–2425.

Portmann, A. (2000) Metamorphose der Tiere. In: Adolf Portmann: Biologie und Geist, Burgdorf, Göttingen

Rasper, M. (2000) Lust und Qual der Erkenntnis, natur&kosmos, Juni, 46 ff.

Ratter, B. M. W.; Treiling, T. (2008) Komplexität – oder was bedeuten die Pfeile zwischen den Kästchen? In: Egner, H.; Ratter, B. M. W.; Dikau, R. (2008) Umwelt als System – System als Umwelt? S. 23–38

Ripple, W. J.; Beschta, R. L. (2004) Wolves and the Ecology of Fear: Can Predation Risk Structure. Ecosystems? In: BioScience. Band 54, Nr. 8, 2004, S. 755–766

Schaur, E. (1991) Ungeplante Siedlungen. IL39. Institut für leichte Flächentragwerke, Universität Stuttgart.

Stern, H.; Kullmann, E. (1996) Leben am seidenen Faden. Frankh-Kosmos, Stuttgart

Tsui, E. (1999) Evolutionary Architecture. John Wiley & Sons, New York

Urry, L. et al. (2019) Campbell Biologie. Pearson, München

Varela, F. G.; Maturana, H. R.; Uribe, R. (1974) Autopoiesis: The organization of living systems, its characterization and a model. In: Biosystems, vol. 5, Issue 4, May 1974, Pages 187–196

Vehlken, S. (2009) Fish & Chips. Schwärme – Simulation – Selbstoptimierung. In: Horn, E. et al. Schwärme Kollektive ohne Zentrum. Eine Wissensgeschichte zwischen Leben und Information. Transcript, Bielefeld

Vester, F. (1999) Die Kunst vernetzt zu denken. Ideen und Werkzeuge für einen neuen Umgang mit Komplexität. DVA, Stuttgart

Vester, F. (1991) Ballungsgebiete in der Krise. vom Verstehen und Planen menschlicher Lebensräume. dtv Sachbuch 11332, aktual. Neuauflage aus 1983, dtv, München

Vester, F. (1989) Leitmotiv Vernetztes Denken. Für einen besseren Umgang mit der Welt. Heyne, München

Vester, F. (1985) Neuland des Denkens. Vom technokratischen zum kybernetischen Zeitalter. Dtv Sachbuch 10220. dtv, München

Vester, F. (1983) Balungsgebiete in der Krise. dtv, München

Vester, F. (1978) Unsere Welt – Ein vernetztes System. Klett-Cotta, Stuttgart

Vester, F. (1976) Phänomen Stress. DVA, Stuttgart

Vester, F. (1975) Denken Lernen Vergessen. DVA, Stuttgart

Vester, F. (1974) Das kybernetische Zeitalter. Neue Dimensionen des Denkens. S. Fischer, Frankfurt a. M.

Vaughan, A. (2019) Land grab in the Amazon. New Scientist, vol 243, Issue 3240, 27. July, Page

Vogel, S. (2000) von Grashalmen und Hochhäusern. Wiley-VCH, Weinheim

Vogel, G.; Angermann, H. (1979) Taschenbuch der Biologie. Band 1, VEB G. Fischer, Jena

Voland, E. (2013) Soziobiologie. Die Evolution von Kooperation und Konkurrenz. Springer, Berlin, Heidelberg

Wachsmuth, I. (2013) Menschen, Tiere und Max. Natürliche Kommunikation und künstliche Intelligenz. Springer Spektrum Akad. Verlag, Heidelberg

Westerhaus, C. (2012) Allerbeste Freunde. Der Mensch und seine Mikroben. wissenschaft im Brennpunkt, Feature, Deutschlandradio, 21.10.2012

Wiener, N. (1963) *Kybernetik. Regelung und Nachrichtenübertragung im Lebewesen und in der Maschine.* Zweite, revidierte und ergänzte Auflage. Econ-Verlag, Düsseldorf 1963 (287 S.), Original: Wiener, N. (1948) *Cybernetics or Control and Communication in the Animal and the Machine.* Übersetzt von E. H. Serr, E. Henze, Erstausgabe: MIT-Press, USA

World Bank Group (2018) What a Waste. International Bank for Reconstruction and Development/The World Bank, 1818 H Street NW, Washington, DC

Wunderlich, K.; Gloede, W. (1977) Natur als Konstrukteur. Edition Leipzig

Part II

Beyond the Exhaustive Wealth of Technospheric Maximum Solutions

In Part I: "The Inexhaustible Wealth of Evolutionary Adaptive Solutions" you were able to inform yourself, dear reader, in a variety of ways about how nature, with its "evolutionary toolbox" full of highly efficient, perfectly coordinated "tools", helps to strengthen the survival of organisms. Interestingly, the most highly evolved species – ourselves – is also one of them. Stupidly, we do a lot – too much – to successively destroy our life-supporting basis, the biodiversity-rich nature around us. Much of this is done by inner drive via short-sighted, misguided thinking and actions whose destructive effects have reached unimaginable proportions. Vicious circle routines continue to determine the course of dominant economic processes and goals. This often happens without any consideration and at the expense of the ecological, over billions of years grown, organismic ingenuity to survive, in which inevitably also the social, the cooperative of all organisms – including us – is affected.

While it became clear in Part I that only dynamic and coordinated, organismic, error-tolerant and growth-regulated management leads to the excellent natural achievements that we can marvel at today, we are using the opposite development strategy in the technosphere: individual, error-prone, growth-excessive management full of disposal problems and accumulated consequential costs. Viewed holistically, even the creative and progressive successes in some areas in our society do little to change this statement.

We have also come to realize that the ingenious natural solutions that inform the content of this book, and whose principles may well be, and already are, of service to us in our own advancement, are not conjured from a "nature's copying machine." They demand our intelligence and creativity, our ability to look with foresight beyond self-drawn boundaries of scientific and other disciplines, and to recognize that even excellent detailed solutions of which humans are capable can never unfold sustainably – in the original sense of the word – without a holistic view of the connected environment. They exhaust themselves in a relatively short time and are replaced by other excellent detailed solutions with all the – often burdensome – consequences that are associated with them.

It is the humanly devised strategy of consistent economic maximisation with maximum exploitation of exhaustible resources. Techniques dedicated to the use of renewable energy are – despite marginal progress in detail – from a holistic, i.e. economic, ecological *and* social point of view, still only the playthings of economics and politics (Küppers 2013, p. 190–209).

This is all the more true for the "Education Republic of Germany" that has been proclaimed by politicians in Germany for decades, with the remarkable current result of a glaring and increasing shortage of trained teachers in primary schools (Klemm and Zorn 2018), to name just one of many educational deficiencies in our country. Eric Assadourian (2018, p. 3–20) is even more explicit in the World Watch Institute's study, EarthEd: "Rethinking Education on a Changing Planet," which gets to the heart of the extent of misguided education on an Earth in transition.

The "digital frenzy", which seems to be spreading over the German education sector with great delay in time and expertise, is so far only the famous "drop in the ocean". Nature with its principles can provide systemic, i.e. holistic and progressive answers to these and many other self-inflicted problem areas of our society, if it is wanted.

10.1   References

Assadourian, E. (2018) EarthEd: Rethinking Education on a Changing Planet. Chapter 1, The World Watch Institute, W. W. Norton & Comp., New York, London.

Klemm, K.; Zorn, D. (2018) Lehrkräfte dringend gesucht. Bedarf und Angebot für die Primarstufe. Bertelsmann Stiftung, Gütersloh.

Küppers, E. W. U. (2013) Denken in Wirkungsnetzen. Tectum, Marburg.

6

Ways Out of the Trap of Short-Sighted Technospheric Design Routines

Abstract Short-sighted, misguided development strategies are the vicious circles of a system, whether it occurs as part of nature, the economy, politics or society. In quite a few cases, they develop into repeated habit or routine loops with a high destructive potential, from which it is difficult to escape.

The extremely complex dynamic development processes of nature have learned over billions of years to keep their partially occurring exponential growth devil's circles – e.g. in pioneer plants – in check by system-stabilizing "negative feedbacks" or "angel's circles" and skilful demarcation of boundaries. Viewed holistically, growth takes place sustainably and in conformity with the system.

The situation is completely different in the technosphere with growth strategies. Their individual goals can be assigned to the keywords maximization of growth and profit, exploitation of resources, accumulation of consequences or short-term value creation, including not inconsiderable burdensome consequences for societies.

Ways out of the vicious circle of habit loops and "cemented" routines are only possible if these are broken and replaced adaptively and dynamically by new routines with a sustainable effect. To this end, a large number of exemplary process sequences and functional principles exist in nature, the concrete applications of which in the technosphere are presented as examples in Chap. 7.

© The Author(s), under exclusive license to Springer Fachmedien Wiesbaden GmbH, part of Springer Nature 2022
E. W. U. Küppers, *Ingenious Principles Of Nature*,
https://doi.org/10.1007/978-3-658-38099-1_6

Economic deficits may dominate our headlines,
but ecological deficits will shape our future.
Original: Economic deficits may dominate our headlines,
But ecological deficits will dominate our future.
Lester Brown et al. (1986, p. 23) State of the World 1986,
W. W. Norton & Comp., New York, London

6.1   The Attractive Power of Habit

In the vernacular, it is not uncommon to hear the smug expression: *"Routine is the beginning of calcification"*. As is so often the case, this is a superficial generalization of activities, usually in connection with repetitive or monotonous work, where, for example, manual work is repeated over a long period of time and may lead to physical and mental stress. Ultimately, however, it is always the individual case of human stress that counts. Here, in the course of digitalization and the use of collaborative robots, human monotonous work is increasingly being reduced or replaced, which relieves people physically and mentally and entrusts them – ideally – with new, less monotonous work (Küppers 2018, p. 326 ff.).

The realization that the "technical"[1] services of nature possess an extraordinary potential of incomparable effectiveness and efficiency, as was attempted to show in advance in Part I of the book, has been known for a long time. It is just that this realization – despite manifold practical evidence, for example through bionic research and development – has not yet penetrated technospheric practice to the same extent as it should. This is because many of the technospheric strategies for product, process and organisational development are *linear* continuations of basic decisions once made and controlled in order to maximise economic growth. The flip side or positive, thus destructive feedback of this mindless economic strategy is enormous excessive exploitation of limited resources of our earth through simultaneous accumulation of environmental and social problems, with socially relevant additional costs! Figure 6.1 outlines this process, which – from a purely economic point of view – can be classified as *"attractive force of habit"* over and over again the same routine process, as long as the harmful effects (resource exploitation, environmental destruction, increasing consequential burdens in societies,

---

[1] When the term "technical" performances of "nature" is chosen, "technical" stands collectively for all variations of performances, including the interconnected organizational processes of nature, and are not limited, for example, to mechanical techniques.

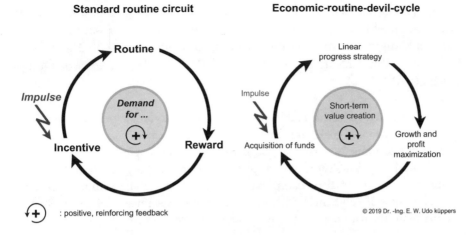

Fig. 6.1   Standard routine cycle and vicious cycle: power of economic habit

etc.) as ends justify the means and the generations of citizens too often accept it without complaint, because a part of them also profit proportionally from this kind of growth.

The one-sided *"attractive power of habit"* of economic progress just described has its limits where compliance with higher-level goals, such as preserving biodiversity, preserving fresh air and clean water, preserving our naturally fluctuating climate, etc., becomes essential for survival. The dominant economic objective of "growth above all else" has long since become obsolete for our survival and must be replaced by adaptive sustainable growth.

The indisputable disadvantages of mindless economic growth are as obvious as the adapted, fault-tolerant and sustainable growth of nature.

For billions of years, evolutionary nature has been highly successful with this *adapted growth strategy*, both qualitatively and quantitatively. Its ingenious principles are a perfect substitute for *technospheric "dead-end growth"*.

At this point it should be clarified once again that it is *"principles of nature"* that are transferable technospherically and by no means trivially copied natural solutions!

> Technospheric *"dead-end growth"* can never solve our overarching environmental, biodiversity and climate problems in the long run – it is more likely to make them worse.

> Principles of *"biosphere-adapted growth"* can contribute significantly to sustainably mitigating, and if necessary avoiding, our overarching environmental, biodiversity and climate problems.

The power of habit, from a current perspective, looking into an Anthropocene future of self-inflicted environmental disasters, natural destruction, and misguided myopic growth strategies can be classified as a "dark power of habit" according to Becker (2019, p. 187; see also Duhigg 2012, 2014).

> The **dark power** of habits: Habits often continue to dominate people's behavior even when people's goals and intentions are against them [...]. Strong habits often just continue unperturbed, no matter what the people concerned decide to do [...] – typical everyday examples of this are smoking, going to bed too late, being unpunctual or eating badly. Friedrich Nietzsche summed up the stubbornness of habits nicely with the well-known statement "Many are stubborn about the path once taken, few about the goal."

Habits are behaviours with characteristic properties (ibid., p. 187; cf. Ronis et al. 1989). This applies to all people, in our context especially to social decision-makers in politics and business, who have the power, instead of stubbornly sticking to misguided routines, to build new routines with sustainable value- rather than cost-oriented goals.

Becker (2019, p. 185) writes in this regard:

> People do not make completely new decisions over and over again, but tend to repeat decisions even when framework conditions change. This makes past behaviour one of the best predictors of future behaviour [...]. The power of habit is therefore beyond question – and this applies to all areas of life. People form habits in their daily lives very quickly. These relate to work behaviour (the political leadership of countries and business leadership, n.d.) as well as diet, shopping, the use of technology or sport. About half of people's daily behaviour is shaped by such habits [...].

Changing a once well-rehearsed behavior or routine, even if it makes sense – as postulated in this book – will be difficult to break, but it can be done step by step. and have a lasting beneficial effect. This requires much less energy than embarking on completely new paths with partly unknown risks that have no reference point in the past.

This is the typical behavioural reason why many of our decision-makers in politics and business tend to trust in the old, the tried and tested, even if it is demonstrably wrong. Even gradual progress on a generally misguided path does not change this much. The currently in focus automotive industry, which despite undeniable climate pollution by CO_2-emission in vehicles, constructs them ever *larger, heavier and more polluting* (type SUV = Sport Utility Vehicle), which is advertised e.g. by Peugeot as "dream car for everyone with guaranteed driving fun", can only be judged as unrealistic and far beyond the preservation of a life-sustaining nature. It is a defining example of how neo-Darwinism with its extreme goals – even bigger, even heavier, even more polluting, etc. – is spreading its harmful effects into the present. Other inglorious examples can be found in agriculture (use of highly toxic substances to maximize yields), animal husbandry (factory farming in unreal, sometimes deadly environments), energy supply (flawed strategies and self-blocking for sustainable regenerative change processes), and many other areas of work and life.

The *"positive force of habit"*, which – as in the economic sector – is characterised by almost self-running growth phases of success and has for a long time led politics and industry even through interim periods of poor success, are increasingly forced by the above-mentioned overriding goals of our habitat to take new paths of development. It is clear that this will not succeed with the old tools.

> Continuing habits or routines saves energy. Changing habits or routines requires energy.

According to studies conducted by psychologist Gerd Gigerenzer and co-authors (2012, 20) on the behavior of managers in upcoming decisions, it was found that the majority (50% in the first study, 76% in the second) rely on their gut feeling, but none would admit it publicly. It was also found that the higher the hierarchical level, the more managers relied on their gut feeling.

These results are also indicative of the *"power of habit"* practiced by managers, because it requires only a minimum of energy, whereas conscious, sometimes elaborate "controlled decision-making processes" require significantly more energy resources. Presumably, today in 2020, only eight years later, a current comparable study on the routine behaviour of managers would lead to similar results, which is more realistic than can be ruled out in view of the increasing risks in relevant economic sectors.

In view of our increasing destruction of *the* earth as our basis for living and working, there is no way around a fundamentally new development strategy of sustainability and strengthening of survivability, in which a *"positive power of habit"* is only established when it is possible to leave the entrenched technical *vicious circle routines*[2] and to build and maintain new routines that are dynamically adapted to the goals.

6.2   Transition to New System-Adapted Dynamic Habit Loops in Our Technosphere

We shape and design our products, processes and organizations, apart from their expediency, on the basis of nature! – Not nature shapes its evolutionary results on the basis of technology!

This statement may sound trivial at first glance. However, it is quickly put into perspective if we take into account the consequences of our actions in dealing with nature and the destruction of its enormous, as yet largely undiscovered range of services.

It is the global vicious circle of a technical – partly dubious – progress and irretrievable destruction of nature that routinely takes place. Breaking this cycle and re-leaning on the services of nature in a cooperative way means: replacing the old harmful routines with new, sustainable beneficial routines.

Strengthening nature's ability to survive is the overarching goal of technospheric activities. Otherwise, in the final analysis, we will also destroy ourselves.

Leaning back on the achievements of nature, however, does not mean lapsing into nature-enthusiasm, as critics like to put it. The path of the *"new routines"* leads rather via a holistic analysis of the indisputably overwhelming, qualitatively excellent and quantitatively perfectly regulated performance spectrum of nature, the principles of which far outshine much of what we devise and handle.

---

[2] Exclusively vicious cycle routines, after repeated circular passes, break technospheric system boundaries and ultimately lead to the weakening, stalling, or destruction of these systems.

Forming and shaping according to a *basic biological design,* as Vester (1991, p. 82) put it, is a benchmark for new routines. Furthermore, Vester speaks of an *"irregularity in regularity"* (Vester 1999, p. 140), the opposite of what many, reduced to technically mindless – based on basic economic expediency – forms products express.

This is *not a* generalization of human design routines, because productivity and functionality in technology often require standardization.

However, where technically designed environments have an effect on people's physical and psychological behaviour, where people learn, live and work, communicate with each other, spend their leisure time and much more, this is true in quite a few cases:

> Man is means – period.

Although politicians and economists like to boast about positioning the *human being at the centre of* their strategies and practical activities to be undertaken, the bitter realisation often remains that exactly the opposite occurs with a time lag, as expressed in the preceding mnemonic. This personality, this way of dealing with people in today's society should be particularly emphasized because it is not new, but has been cultivated for decades as an instrument of capitalist, neoliberal growth-fixated processes. (cf. Neuberger 1990). Whether the euphoria of some, through the use of collaborating or autonomous robots in industries, administrations or private spheres, will again see humans as the centre of events, is highly doubtful.

> At this point of the book it should be clear that nature with its ingenious principles, as effective as they can contribute to sustainable progress in the crisis-ridden technosphere, are ultimately at the mercy of political and economic decisions, thus decisions of humans. This reveals a dilemma!

Figure 6.2 demonstrates the juxtaposition of the vicious circle shown in Fig. 6.1 in the form of an economic habit loop with a sustainable alternative.

This kind of visualized "power of habit" will accompany many examples of biological-technological transfer in Chap. 7, in which ingenious models of nature present their *"positive power of adaptive dynamic habit"* for technospheric problem solving.

On the left in Fig. 6.2, the familiar, economically shaped routine cycle from Fig. 6.1 can be seen. Without regulating limits to growth – superimposed

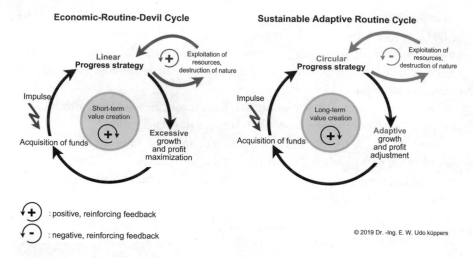

Economic-Routine-Devil Cycle   **Sustainable Adaptive Routine Cycle**

Linear Progress strategy — Exploitation of resources, destruction of nature — Impulse — Short-term value creation (+) — Acquisition of funds — Excessive growth and profit maximization

Circular Progress strategy — Exploitation of resources, destruction of nature — Impulse — Long-term value creation (+) — Acquisition of funds — Adaptive growth and profit adjustment

(+) : positive, reinforcing feedback

(−) : negative, reinforcing feedback

© 2019 Dr. -Ing. E. W. Udo küppers

Fig. 6.2 Routine loops with a way out of the economically destructive vicious circle trap

by a second destructive cycle process, which is connected with the linear progress strategy – an increasing success for the economic beneficiaries takes place through *excessive* growth with profit maximization, but a significantly greater disaster for the interconnected nature as our basis of existence! Short-term exploitation of natural resources leads relatively quickly to their limits, especially since they are usually limited resources.

On the right in Fig. 6.2, this economically driven vicious circle is likewise contrasted with a reinforcing routine cycle of economic beneficiaries – with one decisive difference: through the change in routine, through a circular rather than linear strategy of progress, two "birds are killed with one stone" at the same time.

Firstly, the routine change also works for increasing, but now *adaptive,* growth and profit; secondly, through the coupled negative feedback loop, the routine change works against resource exploitation and nature destruction, thus for resource-conserving and nature-preserving solutions.

Economic success is maintained over the long term by this adjusted growth strategy than before the routine change. Nature will thank the people for it.

Evolutionary nature is full of positively reinforcing circulation routines. However, there are always negative, balancing forces at work with these in the complex network, which limit positive excessive growth drivers in order not to completely destroy a system. Sustainable growth results only from an adapted mixture of both feedback loops. This is particularly true for the technosphere.

References

Assadourian, E. (2018) EarthEd: Rethinking Education on a Changing Planet. Chapter 1, The World Watch Institute, W. W. Norton & Comp., New York, London

Becker, F. (2019) Mitarbeiter wirksam motivieren. Springer, Berlin

Brown, L. et al. (1986) State of the World 1986, The World Watch Institute, W. W. Norton & Comp., New York, London

Duhigg, C. (2014) Die Macht der Gewohnheit. Piper, München

Duhigg, C. (2012) The Power of Habit. Random House, London

Gigerenzer, G.; Gaissmaier, W. (2012) Intuition und Führung –Wie gute Entscheidungen entstehen. Bertelsmann-Stiftung, Gütersloh

Klemm, K.; Zorn, D. (2018) Lehrkräfte dringend gesucht. Bedarf und Angebot für die Primarstufe. Bertelsmann Stiftung, Gütersloh

Küppers, E. W. U. (2018) Die humanoide Herausforderung. Leben und Existenz in einer anthropozänen Zukunft. Springer Vieweg, Wiesbaden

Küppers, E. W. U. (2013) Denken in Wirkungsnetzen. Tectum, Marburg

Neuberger, O. (1990): Der Mensch ist Mittelpunkt. Der Mensch ist Mittel. Punkt. Personalführung, 1, 3–10.

Ronis, D. L., Yates, J. E., & Kirscht, J. P. (1989). Attitudes, decisions, and habits as determinants of repeated behavior. In A.R. Pratkanis, S. J. Breckler & A. G. Greenwald (Hrsg.), Attitude structure and function (S. 213–239). Hillsdale: Erlbaum.

Vester, F. (1999) Die Kunst vernetzt zu denken. Ideen und Werkzeuge für einen neuen Umgang mit Komplexität. DVA, Stuttgart

Vester, F. (1991) Ballungsgebiete in der Krise. vom Verstehen und Planen menschlicher Lebensräume. dtv Sachbuch 11332, aktual. Neuauflage aus 1983, dtv, München

7

Biosphere-Technosphere Transformations: Thirty Workable Resolutions

Abstract In this practical chapter, thirty ingenious natural solutions are high-lighted as starting points for technospheric applications. They are largely based on the ingenious examples from evolutionary nature described in the previous chapters. It is indisputable that nature also uses routine processes during its evolutionary stages, but always with caution, never absolutely and permanently, but changes the routine where necessary, e.g. taking into account system limits (e.g. when food is scarce) and in the sense of sustainable further development to strengthen survivability. In the technosphere, this automa-tism rarely if ever applies. If a habitual routine is initiated, then it is not infre-quently run through several times, even if the expected reward is no longer so lavish or even becomes a burden and thus persists beyond the time of a neces-sary routine change. The tendency of a vicious circle spiral becomes apparent with all the harmful consequences that can result. This applies in the private sphere (addiction to sugary sweets) as well as in the professional/socio-economic, political/social sphere (addiction to recognition, addiction to affir-mation, addiction to power, etc.).

While the interconnectedness of plants and animals in nature has relied on adaptive routines for billions of years, in the dominantly economically con-trolled technosphere it is the human being who activates the triggering stimu-lus in a routine cycle by means of an impulse in order to enforce a desire, which as a rule is also not infrequently crowned with "economic" success. If success fails to materialize or becomes a burden – as mentioned above – substitute measures are usually taken (layoffs, short-time work, outsourcing, etc.) in order not to diminish the expected success (turnover, profit). Many attempts are made to increase the economic measure, only one thing is omitted: the necessary change of routine! Apart from gradual changes in standard routines in economic and political-social processes, it seems that a rethinking of the "decision-makers" in economy and politics will only take place when the environmental pressure (climate change) and the social pressure (worldwide large-scale demonstrations of young people etc.) against the destruction of the environment and the future perspectives of our children and grandchildren form an effective force which can no longer be ignored by the "decision-makers" in our society. But even if insight should be or is shown by these fellow citizens, in many cases one has *"done the math without the host"*. The completely disastrous developing energy turnaround since 2011 in Germany; the inadequate results around the asylum problems in Germany since 2015, the manipulations and lies of the automotive industry with regard to the emissions scandal, 2016 and subsequent years, the completely inadequate international aid after hurricanes destroying lives and living space and much more are clear evidence of *promises* of social decision makers *without content!*

In doing so, countless practical examples show us, in all candor, the *"Golden Rule for changing habits"* (Duhigg 2012, p. 63, 2014, p. 93):

> You can't eliminate a bad habit, you can only change it.

The catastrophic state of our planet, to which we ourselves contribute to a large extent, is the global "reward" of our short-sighted strategies and activities, which are now being allowed to *pay off by* the trigger impulse for dominant economic goals and the adherence to misguided routines.

The global impulse and trigger for the following thirty specified bionic circulation processes is a package of ingenious natural principles. It unfolds in thirty specific trigger stimuli that can and should lead to new forward-looking routine processes, ultimately resulting in sustainable progress-strengthening development as a reward. In contrast to the short-sighted and problem-ridden technospheric routines, which are largely known by their type of "reward" and are not discussed separately here, it should be emphasized that the biologically influenced, bionic anticipatory routines retain their variability, react to given environmental changes in time – *change habit* – or adapt their routine sequence.

Each of the following practical examples is introduced with a corresponding illustration that highlights the specific routine cycle for bionic application, the "craving for …".

Thanks to the high degree of interconnectedness in nature, their "technical" services cannot be considered in isolation and – with one and the same principle – often affect different organisms (see, among other things, the self-cleaning effect of leaf surfaces in lotus plants, lady's mantle, cabbage plants, etc.). The "desire for …" is therefore described in general *terms* as a "desire for *diversity of application*" in the technosphere and used where it fulfils the given purpose in order to achieve a specific goal.

The initial impulse for the triggering impulse to achieve a sustainable solution for a task is *always* the biodiverse wealth of nature. The postulate applies here:

> Through billions of years of evolution, organisms have evolved with techniques, procedures, organisational structures and principles under the most stringent quality criteria, which are most suitable for survival in their present biotopes. The transfer of long-term proven ingenious natural principles into technospheric application can therefore only be of lasting advantage – in view of the increasing human influences due to anthropocene effects that are detrimental to life.

Finally, one circumstance must be pointed out that is essential in the transfer process between ingenious biospheric model principle and technospheric transfer – in addition to the often neglected but fundamental effect networking of a targeted technology solution:

Tip

Life develops under "room temperature".

Ingenious principles, powerful constructions and highly effective processes and procedures are the hallmarks of this life. Icy polar regions at < −50 °C and hot deserts at > +50 °C are conquered and developed thanks to perfect survival strategies of the organisms. Even in the vicinity of deep-sea vents, so-called hydrothermal vents or white and black smokers, from which hot gases of about 300 °C emerge from the Earth's interior and mix with water that is 2° cold, life is present.

In contrast, technospheric developments sometimes require high three- to four-digit temperature ranges with exorbitant energy input! It should therefore come as no surprise that bionic solutions, following the biological example of ingenious flying, swimming, climbing, running, jumping, levering, drilling, striking, material use, energy conversion, organisational properties and much more, take *time to* reach the level of performance – more realistically: to approach it step by step – that holistic natural solutions specify as a measure of quality. Moreover, it must be taken into account that nature, with its permanently improved "toolbox of development", pursues a generally different strategy of progress than we, with our creativity, develop and practice progress in the technosphere.

Many ingenious techniques of nature are still hidden from us or have already been destroyed by us, without us being able to guess what kind of help the former biodiverse wealth would still have held in store for us (see among others the increasing disappearance of the South American regular forest as a global oxygen producer, CO_2 sink and biodiversity hotspot,[1] or the Australian Great Barrier Reef[2] – also a biodiversity hotspot, with an extent the size of Italy – which, through postulated human-induced anthropogenic impacts, is warming seawater, causing increasing coral bleaching cycles, thereby losing its status as a biodiversity hotspot, et al. etc.). Only the indigenous nature-loving peoples, whose existence is in danger with the disappearance of their habitats, appreciate the natural wealth.

> Working against nature instead of with nature for sustainable technospheric developments is no longer an option in view of the disastrous reality on our planet!

The following examples show ways with options for technospheric developments from natural principles, which are effective and efficient, tested for a long time by the strictest quality criteria and have a high system relevance.

> Due to its evolutionarily developed species richness and its technical individual and holistic achievements – largely unachieved by humans – nature is our model for the technospheric future par excellence – even if not in its entirety.

---

[1] Biodiversity hotspot refers to regions of the Earth where there is a high density of endemic organisms that are under sustained threat; thereby posing a high potential threat to the Earth's climate and the continuation of interconnected life.

[2] http://www.gbrmpa.gov.au/the-reef/reef-health, GBRMPA = Great Barrier Reef Marine Park Authority (accessed 2019-09-29).

As human beings, we cannot – as Hieronymus Carl Friedrich Freiherr von Münchhausen is anecdotally said to have done in the past – pull ourselves out of the anthropocene swamp by our own hair. All that remains is for us to use our own minds in such a way that we change our mental routines, which are a prerequisite for urgently needed action in the sense of forestry sustainability and willingness to cooperate, to a much greater extent in the future than has been done so far, in a problem-preventive manner, before we continue to remain prisoners of our causally misguided thinking.

However, it is indisputable that these nature-inspired practical approaches are not necessarily conducive to the future path of nature-analogous development in times of highly dynamic changes in societies (digitalization, "artificial intelligence", energy transition, earth-wide anthropocene impacts – climate, littering of the oceans, politico-economically controlled burning of forests as CO_2 sinks, etc.) and not least due to the current increasing tendency of politics to focus on partly massive confrontation and war instead of cooperation and community and, last but not least, due to the currently increasing tendency of politics to focus on massive confrontation and war instead of cooperation and common ground, are not necessarily conducive to strengthening the future path of the nature-analogous development strategy.

It is astonishing when people, or better: social decision-makers, see their salvation for the future in short-sighted and misguided thinking and acting, although the associated consequential problems or catastrophes – to the detriment of all living beings, including humans – are increasing day by day.

Whoever listens to nature learns to think in *contexts*. After all, we owe our ability to live and survive to this day to the evolutionary strategy of nature, which of all things is highly endangered by our species, with its ability to think ahead.

The following chapters of the thirty technical applications each introduce the circular sequence at the beginning through a specific routine circuit, which ultimately, via *systemic bionics* as a transfer station, drives or is intended to drive goals that obey and strengthen the precepts of Carlowitzian sustainability, resilience, anticipatory problem prevention and adapted efficiency.

All the following subchapters of this chapter, beginning with Sect. 7.1, are introduced with a so-called "routine cycle illustration". In it, a principle of nature can be seen on the left and a routine cycle on the right, whose focus is on practical – systems bionic – applications that can be derived from the aforementioned principle of nature.

The "routine cycle mapping" at the beginning of each example is only the *basis*, the systemic *impulse* for sustainable thinking and action. In the case of strategic and operational processes to be carried out, however, a number of specific networked influencing variables become apparent which reflect the reality of the example to be dealt with more closely than causal (monocausal) approaches are ever able to do in a complex environment.

7.1   Darwin's Theory of Evolution: Extracted Evolutionary Strategies for the Optimization of Simple and Complex Multi-parameter Systems

In this first example, Darwin's theory of evolution (see Fig. 7.1, right) provides the impetus for how we can develop innovative product and process solutions for multiparameter systems in the routine cycle of optimization tasks in the technosphere.

What for evolution are so-called *evolutionary factors,* such as genetic mutation, crossing over, selection, gene drift, etc., are for *Evolutionary Algorithms –* EA – strategy variable, object variable, step size control, random factor, parent and offspring individuals and populations, and so on.

Table 7.1 shows a comparison of natural optimization factors and their technical counterparts according to Weicker (2007).

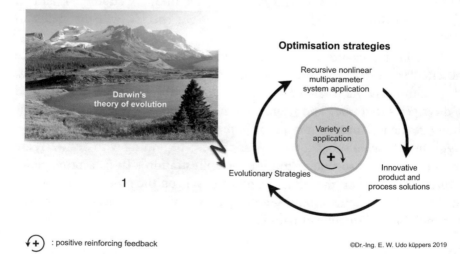

Fig. 7.1   Routine cycle of optimization strategies

Table 7.1 Comparison of natural and technical optimization factors

Evolutionary terms	Pendent with Evolutionary Algorithms
Population	Sum of several individuals, often as a fixed quantity
Individual	Candidate solution for an optimization problem
Genotype	Analogously for the total amount stored in the individual and
	Evolution manipulable information
Phenotype	Individual from the point of view of the optimization problem
Mutation	Small change to the genotype
Recombination	Combines two or more individuals together
Crossing-over	Synonym for recombination or as a designation for special recombination operators
Selection	In evolutionary algorithms, two types of selection can be distinguished
Fitness	Special calculation rule as an aid to selection, is also referred to as the goodness or goodness criterion of an individual
Genetic code	Mostly direct mapping (decoding function) from genotype to phenotype
Diploidity = double set of chromosomes in the cell nucleus	Special concept that is used, for example, for time-dependent problems
Gendrift	Mostly negative effect, which is to be avoided in evolutionary algorithms and requires countermeasures
Gene flow	Is used as migration in parallel evolutionary algorithms
Niching	Special technique sometimes used to prevent gene drift
Coevolution	Special technology used sporadically for different reasons
Lamarckian evolution*	Technique used in memetic algorithms, for example
	Memetic algorithms are based on a population-based approach to heuristic search with optimization problems

After Weicker (2007), Table 1.2, 15, slightly modified by the author

* Recently, a new field, *epigenetics* (Spork 2017; Kegel 2009/2018 et al.), has become established. There seems to be robust evidence that acquired traits from grandparents or parents from the environment, for example experiences, special lifestyles such as smoking, diet, etc., have an influence on the gene structure of the offspring. In his theory of evolution, Lamarck, in contrast to Darwin, had spoken of evolution of acquired characteristics, citing the example of the giraffe, whose neck grew longer and longer from generation to generation in order to reach the high-hanging fruit of trees. With this approach, however, Lamarck could not prevail against the fundamental theory of evolution of Darwin – strengthened by many practical evidences. He would certainly not be surprised if he could follow the trend of epigenetic inheritance, almost 200 years after his work.

Two broad areas emerge in the application of evolutionary algorithms:

1. Model simulations with EA
2. Practice Experiments with EA.

Under sufficiently accurate, mathematically ascertainable modelling, which also possess a large number of influencing parameters (object variable, external variable), EA show themselves to be superior to other known optimization methods such as gradient strategy, Hill Climbing, Monte Carlo, Newton-Raphson, Hooke-Jeeves, Simulated Annealing, etc., especially in a non-linear optimization space. Today's computer speeds allow millions of operations or evolutionary parameter variations in seconds.

The situation is quite different for optimization experiments of a practical nature. Where no modeling of the optimization task is possible, where simulations only yield incomplete solutions, where practical experience in combination with the experimental optimization cycles also plays an essential role. Last but not least, the experimental costs (personnel, machines, optimization devices, RHB (raw materials and supplies), energy, material, etc.) often decide on the duration and thus on the optimization progress or termination of the experiments. A continuous process as with a simulation, whose cycles can go into the thousands and more, is therefore excluded.

Tip

Experimental evolutionary optimization:
   In the optimization experiment with a limited budget, algorithm is paired with experience.

It is also shown here that very good progress successes can be achieved, even with low generation rates in the cycle of optimization. Furthermore, the experimental experience gives the feeling to stop at the right moment at a successful optimization cycle by a well-dosed termination criterion, in order to avoid a disproportionate additional effort with disproportionately high follow-up costs, which arise the closer the experiment approaches a (local), a fortiori an absolute optimum and the progress near an optimum is only minimal.

From a systems bionics point of view, the measurement of structures and processes of natural models is of fundamental importance for experimental optimal solutions in the technosphere. Here, taking specific boundary conditions into account, the starting point of the optimization can be set so skilfully in the initial phase of an experiment that the reference state to be improved according to the state of the art of the object or process to be optimized is already far surpassed at this stage, i.e. qualitatively improved!

The following three aerodynamic or fluidic experiments of the author with evolutionary strategies from the early days of evolutionary strategy (1970s/1980s) and from the near past (2006/2007, 2009/2010) prove this. Other EA examples, such as the *Traveling Salesman Problem, which* is to travel through hundreds and more places by the shortest path, have been impressively solved using Genetic Algorithms (see literature below). You, dear interested reader, will find a rich fund of sources for these as well as for many other applications of GA, across the disciplines and industrial areas, even without explicitly referring to them.

Under the umbrella term of *Evolutionary Algorithms* subsumed in the historical course of bionic optimization strategies, since the 1960s, *Evolutionary Strategies*[3] (Rechenberg 1973, 1994; Schwefel 1975, 1995.) *Genetic Algorithms* (Holland 1969, 1973, 1992; Goldberg et al. 1992, p. 198; Goldberg and Lingle 1985; Goldberg and Richardson 1987, and others) and *Evolutionary Programming* (Fogel et al. 1965; Fogel and Chellapilla 1998; Fogel 1988, 1999, and others).

Similar to nature with its networked adaptive control processes, which are still largely misunderstood, technical EAs work in optimization cycles on a much more trivial but nevertheless promising level than nature, which is rich in diverse experience, practices.

Figure 7.2 shows on the left the formal cyclic sequence of an evolutionary optimization algorithm. On the right in Fig. 7.2 a similar optimization cycle of the practical evolutionary strategic optimization of a fluidic problem from air conditioning and ventilation technology (Küppers 2007a, b, c) can be seen.

---

[3] Through his work as a PhD student in the 1970s and 1980s at the TU Berlin in the Department of Bionics and Evolutionary Strategy, the author had the great fortune to be part of a scientific team that was able to deal intensively with the science of bionics, which was still considered exotic at that time. As a result of bionic research, early bionic solutions for energy saving effects were developed using adaptive wing-winglet structures modeled after large birds of prey such as eagles, buzzards, vultures, etc. (Küppers 1983). These soar in thermal updrafts often for hours without any flapping of their board-like wings and thus without any waste of energy. The simple wingtip designs of today's aircraft are still miles away from the masters of flight in their flying skills. Practical results from research on fish fins as propulsion organs, neural network simulations, surface friction reduction by fish slime and other means, advances in evolutionary strategy and much more testified to a broad spectrum of research with a claim to pioneering engineering achievements in bionics in Germany.

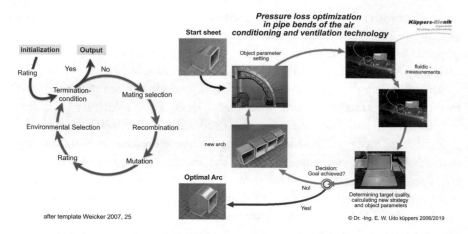

Fig. 7.2 Formal (Weicker 2007, p. 25) and practical cycle (Küppers 2007a, b, c) of an evolutionary algorithm or an evolutionary strategy with a practical example for the fluidic drag minimization of pipe or flow channel fittings, here 90° pipe bend

7.1.1 Evolutionary Strategic Optimization of an Aerodynamic Problem

The dream of man to fly without technical aids, as birds demonstrate to us in inimitable perfection every day, remains a dream. However, the creativity and intelligence of mankind has led to the fact that we have mastered a wide range of applications of the art of human flight, from gliding to rocket flight into space, albeit with far more complex techniques and energy input than the biological models.

The use of energy to propel various aircraft, with the necessary lift and associated drag generated by the airfoil geometry, is a major cost factor.

Aircraft wings generate a vertical upward lift force due to the flow of air around the usually upward-curved wings, which is counteracted vertically by the weight force of the flying body – including the weights of passengers and baggage.

The horizontal propulsive force of passenger and transport aircraft is usually generated by propeller or jet propulsion units supplied with fuel, and is in turn counteracted by the drag force of the air.

For efficient flight performance, the ratio of drag to lift – W: A – or their so-called dimensionless coefficients c_W: c_A, which corresponds to the tangent of the aircraft glide angle γ (gamma), is a governing physical quantity.

Without going further into the physics of flight at this point, we will focus on the drag component of the aircraft. From the notes before, we can conclude that the greater the drag force of an aircraft, the stronger the propulsive force must be in order to move forward.

The total resistance of an aircraft is made up of various individual resistances: Pressure resistance, frictional resistance and induced or edge resistance. The induced or edge drag is especially noticeable at the ends of the wings in the direction of the wing tips, because especially there a pressure equalization of the air from the upper side of the wing with its negative pressure and the lower side of the wing with its positive pressure takes place. The wider the wings are, the stronger this pressure equalization is. Especially in large passenger and transport aircraft, long-lasting circular air vortices are created at the wingtips, the circulation strength of which develops enormous force, so that following small aircraft have to keep their distance for minutes in order not to be in danger of crashing due to the vortices of "jumbos" flying ahead.

In certain phases of flight, this induced drag can amount to up to 50% of the total drag, which is why various techniques have been tried for decades to reduce this and other air resistance effective on aircraft.

How do our biological models, especially those with broad board-like wing surfaces, such as condors, eagles, etc., manage to defy the air resistance around their flying bodies, or to keep them in check? Again, the reader is referred to a wealth of literature sources. However, brief reference may be made to two techniques of induced drag reduction in birds:

1. Reduction of induced drag reduction W_i by wing stretching.
   The longer and narrower the birds' lift-generating arm and hand wings are, as seabirds practice – albatrosses, gannets and skuas are among them – the lower their induced drag in flight.
2. Reduction of the induced drag reduction W_i by variation of the wingtip geometry. This type of drag reduction is reserved for birds that have very wide wing surfaces. These include, as mentioned before, vultures, eagles, hawks, in general: all birds of prey.

As an early experimental example of bionic evolutionary strategic optimization, with the aim of reducing the induced boundary drag along the lines of birds of prey on an airfoil model in the wind tunnel, experiments were carried out with a 15-parameter optimization system consisting of five winglets and 3 angular coordinates each (angle of attack, angle of stagger and angle of spread) (Küppers 1983).

Figure 7.3 shows in advance two excellent master fliers of nature, whereby the Andean condor with its optimized wing geometry glides for hours in thermal updrafts without any power-consuming wing beat.

Fig. 7.3 shows on the left an Andean condor in lift-giving thermal gliding flight, with handfits spread out to reduce edge drag, and on the right a stork, also with broad flight wings and the technique of handfit spreading. Andean condor: https://pixabay.com/de/images/search/kondor/ (pixabay license, free stock footage, retrieved 10/01/2020). Stork: https://pixabay.com/de/photos/storch-fliegen-flügel-vögel-2432978/. (Pixabay license, free image, retrieved on 10.01.2020)

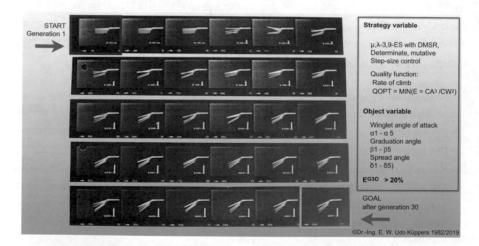

Fig. 7.4 First experimental evolutionary strategic winglet optimization with data on strategy and object parameters. (Küppers 1983, pp. 83–91)

Figure 7.4 shows the experimental generation course of the first evolution-strategic optimization on a model wing in the wind tunnel, with a Reynolds number around $5 * 10^4$. In fluid mechanics, it is the number that determines the ratio of inertial forces to viscous forces and is used in tube flows here wind tunnel flow -. Re is equal to an average flow velocity v_m times the pipe diameter d divided by the kinematic toughness v ($\mathbf{Re = (v_m \times d)/v}$). The critical

Reynolds number Re$_{krit}$ marks[4] the transition from laminar to turbulent flow. In this respect, the wind tunnel experiments take place in a turbulent flow environment.

The early bionic wind tunnel experiments clearly show a reduction of the energy-consuming edge vortices and thus impressively confirm the theory of a vortex-reduced or energy-efficient flight by means of winglets at the wing tips.

The visualization of the airflow at the beginning and at the goal of the ES optimization is shown in Fig. 7.5. It is complemented by the visualization of the edge vortex of a blunt wing tip without winglets.

Clearly visible is the spreading circulation vortex in the direction of the flow, which in the large format of "jumbo edge vortices" spreading out to form stately wake vortices, leaves small aircraft no chance to take off directly behind them.

This aerodynamic problem persists to this day, with experiments by aircraft manufacturers with small winglets (one to a maximum of two) at the wing

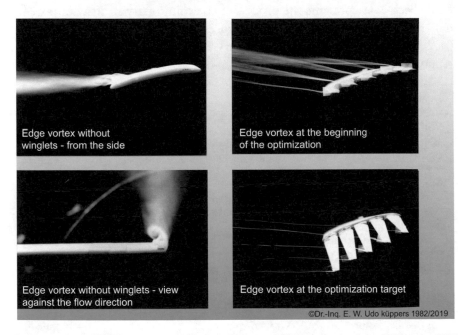

Edge vortex without winglets - from the side

Edge vortex at the beginning of the optimization

Edge vortex without winglets - view against the flow direction

Edge vortex at the optimization target

©Dr.-Ing. E. W. Udo küppers 1982/2019

Fig. 7.5 Flow visualization at wingtips with and without winglets. (Küppers 1983, pp. 94–96 and private photographs)

---

[4] For readers interested in fluid mechanics, this re-transition from laminar, uniform turbulence-free flow to turbulent flow can be easily illustrated by a water tap experiment. Here, the laminar flow of water at low water velocity can be seen, which becomes more turbulent after higher water speed suddenly turns into a turbulent water flow.

tips, which I'm sure each of you has seen before, not leading to a decisive breakthrough in significant vortex energy reduction and thus energy saving effects. Only relatively small energetic advantages could be achieved, but these at the expense of an increased interference resistance between wing and winglet.

The great breakthrough through novel wing geometries, in times of increasing energy sensitivity is still waiting. Because in the environment of earth-wide climatic changes due to fossil energy sources, kerosene as aviation fuel is also of fossil origin and promotes CO_2 -environmental pollution.

7.1.2   Evolutionary-Strategy Supported Optimization Coupled with the Meander Effect® of a Fluidic Air-Conditioning-Ventilation-Pipe System

While Sect. 7.1.1 presents the process of an evolutionary strategy in pure format, the evolutionary optimization process of a flow problem in this Sect. 7.1.2 links processes of the evolutionary strategy with those of natural flow structures, the so-called meanders of free-flowing waters (Fig. 7.6).

The Meander Design Principle of Flow Systems

Both the *energy efficiency principle* and the *meander design principle* are outstanding principles of nature for the reason that they act on nature as a whole, on the feedback network of animate and inanimate nature!

Meandering turns are a component of both animate and inanimate nature. Meandering turns, or optimal flow branching, effectively support both transport-energy efficiency in flow systems of organisms (blood cycles, gas cycles, food cycles) and the adaptive turns of flowing waters in the wild, which are interconnected with a variety of organismal activities. River courses as parts of circular processes strive to achieve a minimum of entropy production (minimum energy loss). The particular stream course of meandering[5] contributes to this.

Natural flow meanders are sections of nature's drainage networks through which energy flows and are characterized as "dissipative" steady-state nonequilibrium systems (flow equilibria).

In the wild, it can often be seen that a characteristic overshooting of the river's course occurs during so-called flow diversions. This natural feature is – bionically implemented – called the Meander Effect®.

---

[5] Figure 7.6 shows some examples of natural flowing meanders. Causallogically, one might ask: Why do flowing waters-even and especially free-flowing waters in flat landscapes where no obstacles of mountains or rocks would force them to curve-meander at all? Wouldn't it be more logical and energy efficient for the river to flow straight for as long as it could? This typical and humanly understandable question is based on the misconception that nature follows causal or monocausal rules. However, processes in nature are characterized by highly complex and thus dynamic states. This is also the reason for meanders, which make their energetically advantageous contribution to development in natural environments.

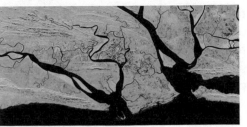

Wattpriele, North Sea, Germany   Kaladan Delta Myanmar, Burma

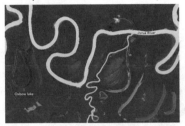

Amazon, Brazil   Vltava, Bohemian Forest, Czech Republic

Fig. 7.6 Free-flowing meanders in nature, https://pixabay.com/de/photos/mäander-moldau-böhmerwald-fluss-1316553/. (pixabay license, free stock image, retrieved on 10.01.2020), https://www.dlr.de/eoc/Portaldata/60/Resources/images/4_serv_geovis/4_serv_geovis_ausst/kaladan.jpg. (Source: DLR, CC-BY 3.0, retrieved on 10.01.2020), https://earthobservatory.nasa.gov/images/84833/meandering-in-the-amazon. (NASA Earth Observatory images by Jesse Allen, using Landsat data from the U.S. Geological Survey, retrieved Jan. 10, 2020), North Sea priels. (Own photo)

It is also a kind of evolutionary optimization to recognize and apply the geometric meandering flow courses as a natural model for technical, lossy flow processes, e.g. in pipe or channel systems. Particularly in the deflection areas of technical flow systems, for which so-called fittings are used (standardized pipe bends of various cross-sections with 30°, 60°, 90°, 120°, 180° circular arc deflection radius or branches are characteristic), physically inevitable high flow losses appear, which are composed of friction, detachment and deflection losses.

> Every technical standardised fitting of a pipe flow system, based on a circular arc deflection, already has a high proportion of potential energy losses due to this construction.

If natural solutions of meander shapes in flow systems were consistently used bionically, this would bring users significant energy cost saving effects. Even with initially more expensive manufacturing costs of the special pipe shapes, the potential and costs would outweigh the reduced energy losses (e.g. also due to the lower power consumption of drive units of flow systems) in the long term.

Economically reliable figures on the total potential of energetic losses in pipelines, whether with gas, liquid, solids or as multiphase flow as transport material, are difficult to find due to the differentiated applications; however, they are probably in the order of 7–8 digit Euro amounts per year. The example "Compressed Air Systems in the European Union"[6] shows in an older study from 2001[7] that in 80 out of 100 companies up to 100% more compressed air is produced than consumed, compressed air being the most expensive form of energy. Reliable figures on energy losses according to the current state of the art, by a new edition of the study, are not yet available.

The greatest energy efficiency potential lies in saving energy losses.

The evolutionary "meander optimization" of standardized circular arc pipe fittings (cf. Fig. 7.7) occurred according to the following and well-known circular sequence:

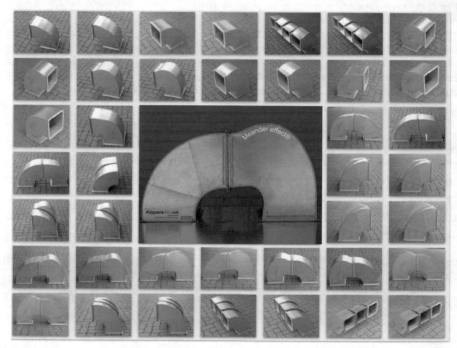

©Dr.-Ing. E. W. Udo Küppers 2006/2019

Fig. 7.7 Evolutionary pipe bend mutations with a start-finish comparison

---

[6] http://air.avexa.se/air/down/eu_compressed_air.pdf (accessed 2020-01-14).

[7] https://www.fluid.de/tipps-und-tricks/suche-nach-energieeffizienz-in-der-druckluft-lohnt-107.html (accessed 2020-01-14).

3.9-ES-3 Meander optimization of a standardized 90° pipe bend with square cross-section (200 × 200 mm²) of air-conditioning and ventilation technology

1. Definition of the start parameters (ES strategy and object variables)
2. Calculate first ES generation
3. Measurement of three contours from a selection of natural flow meanders with 90° flow deflection
4. Comparison of geometric shapes from ESG1 and natural meander
5. Experimental determination of pressure losses of all ES and meander bend geometries
   (Δp = p1−p2, pressure difference measurement from p1 flow section of the pipe bend and p2 downstream section of the pipe bend)
6. The three best values of the measurements (smallest Δp in the pipe bend) are determined as initial objects of the subsequent generation.
7. Repeat from step 3 to step 6 until the subjective termination criterion is reached.

Figure 7.8 shows a comparison of the four measured 90° mouldings of different cross-sections as the start and target geometry on the left; the trend in energy efficiency improvement as a function of the moulding cross-sections can be seen on the right.

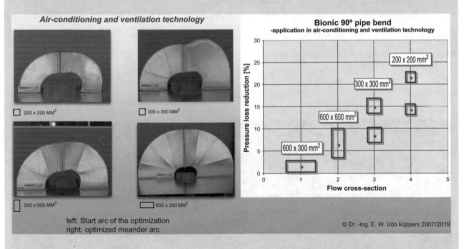

left: Start arc of the optimization
right: optimized meander arc

© Dr. -Ing. E. W. Udo küppers 2007/2019

Fig. 7.8 Pipe bends start-target comparison with trend in energy efficiency

7.1.3 Evolutionary Strategic Optimisation of a Pipe Transport System in Agriculture

Similar to the coupled evolutionary-strategic and meander-geometric approach in Sect. 7.1.2, an optimization of the flow course in the raw system of an agricultural seed drill took place in 2009/2010. The flow course of seed grains of different species, such as rye, cereals, rape, grass, peas etc. can be seen in Fig. 7.9.

The particular problem of the flow section again focused on a standard 90° pipe bend section in the middle of the seed transport line.

The aim of the optimisation was to transport the chaotic state of seed grains in a hopper through the two-phase flow path (air + seed) in such a way that, at the end, 36 individual seed grains are *evenly distributed* across the width into the soil by the machine's "singler" – irrespective of the machine's inclination and the unevenness of the soil. It was a challenge to achieve a uniform distribution structure from a chaotic initial state, for which the 90° deflection arc plays a central role.

The associated quality function Q of the optimization was: **Q = MIN (VK)**.

The dimensionless distribution coefficient VK, which interprets the uniform distribution of the seed grains over the width of the 36 parallel seed singles behind the machine, was to be minimised. The initial values for VK were specified by the company, as shown in Table 7.2 below.

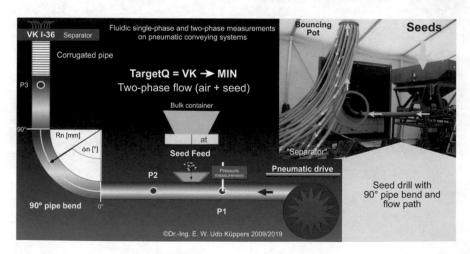

Fig. 7.9 Left: Sketch of the flow pattern in the seeder with quality function, right flow pattern in the seeder

Table 7.2 UK standard values with qualitative assessment

Assessment of the cross distribution	UK for cereals, peas, grass	VK for rapeseed
Very good	<2.0	<2.9
Good	2.0–3.2	2.9–4.7
Satisfying	3.3–4.5	4.8–6.6
Sufficient	4.6–6.3	6.7–9.4
Not sufficient	>6.3	>9.4

The results of all 90° Meander Effect® tube bend variations, with the aim of optimal i.e. minimum VC values for wheat and rape, point in the direction of two local minima:

1. **Meander-Effect® sheet no. 3**, with

    VK wheat from: 0.895 (rating very good)
    VK rape from: 4.583 (rating good)

2. **Meander-Effect® sheet no. 9**, with

    VK wheat from: 0.962 (1.024) (rating very good)
    VK rape of: 2.862 (2.935) (rating good to very good).

Figure 7.10 shows the VC qualities of all meander pipe bends with circular arcs as reference values, while Fig. 7.11 compares the two best meander pipe bend geometries with standardised circular arcs. Finally, Fig. 7.12 shows the meander pipe bend installed in the seeder. This clearly shows the practical consideration regarding the cost-efficiency comparison of this innovation. A meandering course of the two-phase flow is indeed guaranteed at the critical point of the 90° deflection and significantly reduces the VC compared to the VC of circular bends! However, an approximate meander solution was preferred to a 1:1 transfer of the best arc shape optimized by evolutionary strategy – precisely because of cost-benefit considerations and increased manufacturing costs. However, the optimum VK qualities of the Mäander-Effekt® sheets 3 and 9 were not achieved in this way. The practical benefit of the approximate meander solution was ultimately the decisive factor, rather than the evolutionary meander solution.

Optimal product or process solutions in the modeling or practical evolutionary optimization experiment ultimately always compete in the application with the cost-benefit ratio of the innovation.

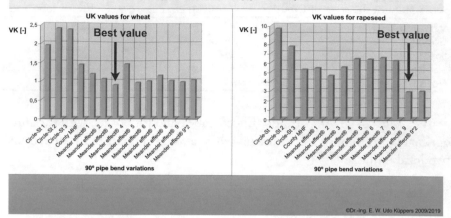

Fig. 7.10 VC qualities of all meander arcs for the seed varieties wheat and oilseed rape

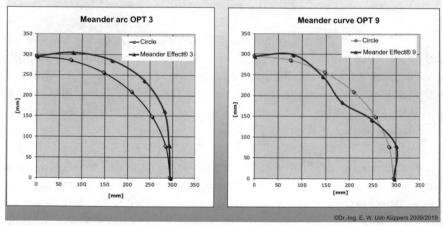

Fig. 7.11 The two best meander shapes of the experiments

Evolutionary strategies or evolutionary algorithms have conquered a wide field of applications for decades. In addition to the large field of flow systems, evolutionary optimizations in architecture, product and process engineering, the optimal design of networks, traffic engineering, storage technology, packaging technology, and numerous other fields of application show their extraordinary ability when it comes to wringing an optimal state out of a multi-parameter system in an appropriate time and quality. Model simulation and practice often go hand in hand.

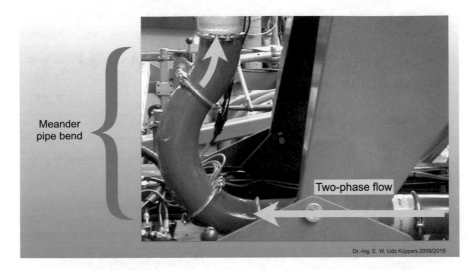

Fig. 7.12 Built-in meander bend in the two-phase flow section of the seed drill

If an attempt is made to optimize a worthwhile principle or design feature from the rich fund of nature for the technosphere, we are always dealing with a dynamic complex optimization problem. For a transformed technical, economic or social solution of natural models, the consideration of interconnected relationships around the optimization goal in both biospheres and technospheres plays a significant role.

Isolated causal strategies for solving complex – even less complex – optimization problems are therefore seldom long-lasting and certainly not sustainable in the sense of Hanns v. Carlowitz. Section 7.2 offers a systemic approach that involves the transfer of adaptive, naturally optimal solutions, taking into account interconnected relationships.

But also for purely technospheric or societal solutions in complex environments, Sect. 7.2 provides a good basis for a principled approach to problem solving of various kinds.

7.2   Networked Nature: Networked Technology – *"Systemic Bionics"* as a Forward-Looking Transfer Method for Sustainable Practice

In this chapter I would like to introduce you, dear reader, after the chapter introduction by Fig. 7.13, to the extension of the discipline *Classical Bionics*, in which the real environment of biological models and bionic transfer results is placed in the centre, according to the postulate:

> The decisive factor for a sustainable effect of bionic solutions is not so much the considered subject/object itself, but rather their interconnected relationships with each other!

Systemic bionics (Küppers 2015) was born from this. Excerpts from it are processed in this chapter. In the preface it says (ibid., pp. VII–VIII):

"The peaceful cluelessness of society in the face of a total influence …
   of certain forces has been unparalleled in recent years […]."
   With these words, among others, Viviane Forrester (1997) wrote an impressive plea against the cynical influence and power of the global economy, against the paternalism of citizens and entire societies, but also against the lethargy of

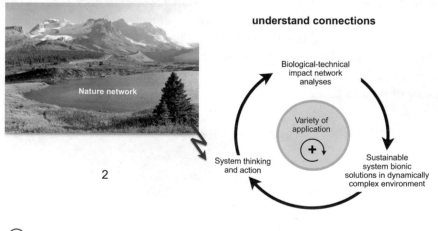

understand connections

Nature network

Biological-technical impact network analyses

Variety of application

System thinking and action

Sustainable system bionic solutions in dynamically complex environment

(+) : positive reinforcing feedback

© Dr. -Ing. E. W. Udo küppers 2019

Fig. 7.13   Routine circuit 2: Understanding connections

political decision-makers. Is Forrester's indictment still up to date? And above all: What does it have to do with Systemic Bionics?

This *peaceful cluelessness* and *paternalism of* the citizens has no less importance than the question of survival in view of the worldwide and not diminishing destruction of our nature.

Nature is, after all, our only basis for life and, incidentally, our only real reservoir of values for the raison d'être of bionics. Countless, qualitatively unsurpassed "survival techniques" are a gift of nature to us and a technical challenge at the same time. If we study the clever and skilful approach of the natural development strategy – with a view to the *interconnected interrelationships* – which has been in use for billions of years, and apply it beneficially and sustainably for us, we hold an unlimited resource of progress in our hands. To prepare it in a fault-tolerant way for practical projects in social spaces through systemic bionics presupposes a *holistic approach,* the overall social significance of which is not yet really perceived. It is therefore hardly surprising that a significant number of *certain forces* in the social environment still focus their acquired and trained (mono)causal thinking and actions on a limited living and working environment that seems useful to them, rather than on the holistic, clearly more complex reality that can only be grasped with a systemic view.

With the practicable, strongly complexity-reduced and thus manageable, causal, – often misguided, i.e. – instructions for action, social vicious circles are produced, which are not infrequently the trigger for consequential problems and catastrophes.

This is contrasted with holistic networked action guidance. The actual value of this so-called systemic solution strategy is likely to be seen less in short-term ephemeral successes and more in the strengthening of forward-looking long-term effects with significantly lower burdens, with the help of traditional Zen wisdom:

If you're in a hurry, take a detour.

Key concepts that accompany work with the *Systemic Bionics* method and that are fundamental to further development in the biosphere, but also have their raison d'être in the technosphere, where they are still all too often ignored, include (described in detail in Küppers 2015, pp. 9–22).

- Recognize system interrelationships and evaluate their functionalities holistically.
- Understand the cybernetics (regulation, information processing, etc.) of systems in nature and technology.
- Use processes of self-organization, self-repair or self-production effectively and efficiently.
- Perceive real complex conditions in the biosphere and technosphere and use them as a challenge for new sustainable developments, instead of replacing them – where so often – out of ignorance or conscious suppression,

with short-sighted, misguided linear-causal development paths with high consequential problems.

- Aiming for value-retaining, fault-tolerant solutions instead of chasing after short-term, seemingly cost-efficient solutions with time-delayed repair cascades of high energy and material expenditure as well as high follow-up costs, and much more.

Above all, there is the transferable central concept of *forestry sustainability*, (in the sense of Hanns Carl von Carlowitz, see above). Dieter Füsslein (2013, p. 251) writes in this regard:

> For Carlowitz, nature is the foundation on which the economy and society are built (today: strong sustainability), and for this he uses convincing ethical arguments.
>
> Carlowitz does not add up economic, ecological and social resources, and certainly he does not place three pillars next to each other, but he describes the reactions and correlations, their temporally, locally and functionally linked interconnections.
>
> In this way he brought together the natural law bases, the social insights and the demands on politics of his time to form a guiding principle.

Systemic bionics internalizes this holism in a strict sense. Even today, more than 300 years after Carlowitz, the guiding principle just described is still valid, in particular new, up-to-date demands on politics!

A Short Excursion into Sustainability and Politics

However, if you take a closer look at and analyse the practical results of political activities in Germany over the last 3–4 decades, you will notice – despite certain efforts by individual politicians – a frightening lack of interest in the forward-looking thoughts and transferable practices of Hanns Carl von Carlowitz.

- Whether it is the "educational republic of Germany" trumpeted broadly by politicians some 30 years ago, whose current results of PISA studies (2018,[8] 2019)[9] make disillusioning, mediocre result statements,

---

[8] https://www.bmbf.de/de/pisa-2018-deutschland-stabil-ueber-oecd-durchschnitt-10349.html (accessed 2020-01-13).

[9] http://www.oecd.org/berlin/themen/pisa-studie/ (accessed 2020-01-13).

- whether it was the abrupt political turn away from nuclear energy in 2011, after the nuclear power plant accident in Fukushima, Japan, about whose orderly and societal exit scenario and the consistent use of renewable energies is still being hotly debated 9 years later (2020),
- whether it concerns the social "gap" between poverty and wealth, which is still wide, long-term unemployment managed at a high level for years without today's politicians not only discussing but consistently implementing forward-looking solutions (5th BMBF Poverty-Richness Report 2017),[10]
- whether it is the surprising wave of migration of asylum-seeking people in need of help without possessions that has taken place in Germany since 2015, and to which policy-makers are still unable to come up with a sustainably effective solution five years later (2020) (Küppers and Küppers 2015).

Friedrich Engels (1820–1985) is credited with the – after the previous four descriptions on sustainability in politics – far-sighted statement:[11]
An ounce of practice is better than a ton of theory
Carlowitz's central and strong concept of sustainability also competes with two other sustainability concepts of rather short-sighted, misguided, and nappy-soft descriptive strength:

1. Sustainability dilemma

We are prisoners of our own mental models.
The *sustainability dilemma* is that today's sustainability is associated with completely different mental models and practical goals! The power of habit or the entrenched course of our thought routines largely controls our actions even when new insights, new perspectives or goals demand new routines.

Carlowitz mastered this change from entrenched routine (clearing the forest as a supplier of wood without preventing its environmental impact) to new innovative routine (clearing the forest as a supplier of wood in a forestry-sustainable way).

The new mental model led to strong sustainability in the forest, both to ensure electoral profits and to reduce the shortage problem of wood and more. Figure 7.14 shows the circular progression of the power of habit or the power of mental routine before and after a change of routine.

---

[10] http://www.bmas.de/SharedDocs/Downloads/DE/PDF-Pressemitteilungen/2017/5-arb-kurzfassung.pdf?__blob=publicationFile&v=2 (accessed 2020-01-13).
[11] https://www.zitate.eu/autor/friedrich-engels-zitate/117007 (accessed 2020-01-13).

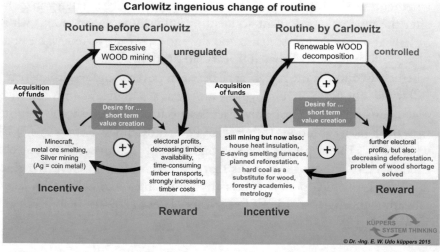

Fig. 7.14 Carlowitz's ingenious change of routine in eighteenth-century forestry

It can be seen in Fig. 7.14 on the left how the impulse to acquire money makes its way circularly via incentive, habit (routine) to reward and back to incentive until wood was scarce and procurement costs were high.

After Carlowitz, through his ingenious thinking, changed the cycle routine from excessive to sustainably renewable timber management and supplemented it with further incentives, as Fig. 7.14 on the right is intended to convey, not only the electoral salary was achieved, but also other amenities. In addition, the problem of deforestation and the increasing shortage of wood was solved.

Personal experience of how difficult it is to change once established routine cycles as mental models, because consequential problems have an increasingly negative effect on health, for example, is something I'm sure each of us has gone through. Cravings for something sweet often make it hard not to walk into a nearby candy store. Convenience despite weight gain is often the driving force to avoid regular exercise or regular exercise in general. Of such routine devil's circles that pretend sustainability in the wrong sense, many exist.

The close connection between systemic bionics and the idea of forestry sustainability is already understandable because this discipline draws its special techniques and services from nature, just as Carlowitz sees nature as the foundation on which the economy and society are built.

2. Sustainability paradox

The *sustainability paradox* is an expression of a product- or process-oriented quality improvement for environmental relief, which, however, is cancelled out by the quantity of the products – and even causes the environmental impact to increase again! This is the well-known and notorious growth driver *"REBOUND effect"*. Currently it can be seen in many products, such as passenger cars, whose energy consumption decreases from generation to generation, but is cancelled out again by increasing car weight, digital "upgrading" and so on. This is indeed productive "sustainability" in car manufacturing, but with extremely weak sustainability, far beyond the Carlowitzian idea.

The environmental or social responsibility often brought into play by companies largely fizzles out and remains until recently more of a corporate fig leaf than an honest statement.

One of the current examples is that of the global corporation Siemens. Despite the ongoing massive forest fires in coal-rich Australia, the company is holding on to the sale of a signaling system[12] for one of the world's largest coal-fired power plants. On the side, Kaeser, the chairman of the supervisory board, is still baiting Neubauer, a German climate activist, with a seat on Siemens' supervisory board. The man-made dangers, that habitat is destroyed by the consumption of water in the coal-fired power plant or that the unique Great Barrier Reef is further destroyed by the transport of coal over water, seem to be a minor matter for business and politics. This extremely short-sighted and misguided corporate and political mental vicious circle routine is certainly not altered by the fact that an estimated one billion living creatures have already fallen victim to the current fire in Australia.[13]

Siemens presents itself in advertising on Corporate Social Responsibility[14] (corporate social responsibility or voluntary sustainable development), with:

---

[12] Siemens- Adani-Carmichael-Project, on which Joe Kaeser, President and CEO of Siemen AG made the following statement on 12.01.2020: "After considering all the facts and consulting with third parties, it has been decided that we will fulfill our contractual obligation and supply signaling equipment for the Carmichael rail line." See statement in English: https://press.siemens.com/global/en/news/joe-kaeser-adani-carmichael-project (accessed 01.07.2020) The contract has been signed, but delivery is delayed due to the Corona pandemic, according to Siemens employee by phone (01.07.2020).

[13] Many current press outlets are reporting across continents on the climate catastrophe in Australia and the lack of action by the Australian government, headed by Prime Minister Scott Morrison, who is sympathetic to the coal lobby. See among others.

https://www.weser-kurier.de/deutschland-welt/deutschland-welt-wirtschaft_artikel,-siemens-haelt-an-auftrag-fuer-australisches-kohlebergwerk-fest-_arid,1889089.html (accessed 2020-01-13).

https://www.zeit.de/wissen/umwelt/2020-01/buschbraende-australien-feuer-naturkatastrophe-hitze-welle-tiere-3 (accessed 2020-01-13).

https://www.weser-kurier.de/deutschland-welt/deutschland-welt-wirtschaft_artikel,-siemens-haelt-an-auftrag-fuer-australisches-kohlebergwerk-fest-_arid,1889089.html (accessed 2020-01-13).

[14] https://new.siemens.com/global/de/unternehmen/nachhaltigkeit.html (accessed 2020-07-01).

> Sustainability at Siemens follows our values of responsible, excellent and inno-
> vative. At Siemens, we define sustainable development as the means to achieve
> profitable, long-term growth.

By growth is presumably meant less a sustainable growth in the sense of Carlowitz
and more an excessive growth in the sense of economic profit maximization. It
is inconceivable that Siemens, with this revelation and definition of its growth
philosophy – due to social pressure – will withdraw from its Adani-Carmichael-
Project. If it does, I'm sure other companies are standing by. The climate gas
disaster caused by CO_2 emissions continues in Australia and elsewhere.

Finally and unfortunately, it must also be stated that valuable forward-
looking practices in the sense of Carlowitzian sustainability, which could also
promote cross-border cooperation and thus also prevent conflict-prone risks
up to and including wars, have for a few years now been subject to massive
pressure from individual politicians who, obsessed with power, are pushing
ahead with a confrontational course in their policies, with little hope of quick,
resounding successes. Climate change is simply denied or accepted as natural
(Trump in the USA, Bolsonaro in Brazil, Morrison in Australia and many more).

Who will ultimately win the battle of nature against humanity is beyond
question. But the much more crucial question is, how long can the power-
holding simple minds with their simple solutions continue to ignore nature's
warning signals?

At the risk of repeating myself several times, which – due to the enormously important topic clearly expressed by the title of the book – is partly
deliberate, the course of Sect. 7.2 so far shows that a science like Systemic
Bionics is also deeply political, if we consider the anthropocene effects in our
societies, which we ourselves are to blame for.

Who, if not the survival genius that is nature, with its ingenious product
and process developments, its ingenious communication channels, its
resource-efficient processes, its perfectly coordinated energy conversion processes, could help us achieve the *strong* sustainability we need to help secure
the future of our children, grandchildren and grandchildren's children?

After the key terms mentioned above, the forestry of systemic bionics leads
us directly into the network of systemic bionics as the foundation of biotechnical or system bionic transfer processes, as Fig. 7.15 shows.

> Classical bionics and systems bionics differ in that the former takes the direct
> route from a natural *model* to technical *imitation* and *application,* while the lat-
> ter seeks the cybernetics of both biological and technical subjects/objects or their
> systemic networking as the starting point for application-ready and sustainable
> solutions.

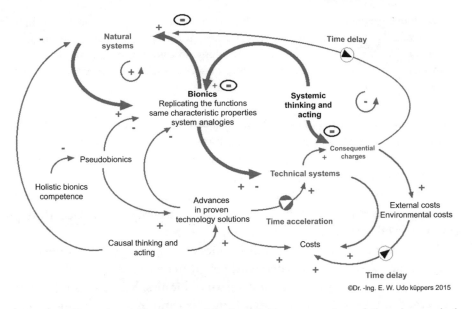

Fig. 7.15 Effect network of system bionic activities – meaning of the plus symbol: Effect between two system parameters is strengthened or weakened in the same direction. Meaning of the minus symbol: Effect between two system parameters is strengthened or weakened opposite – balancing – effect. A minus symbol in the three-quarter circle indicates a balancing circular feedback between system elements or for the system as a whole (is also called an angel circle). A plus symbol in the three-quarter circle indicates an increasingly reinforcing/weakening circular feedback between system elements or for the system as a whole (also called a vicious circle)

The impact relationships in Fig. 7.15 are largely self-explanatory. Nevertheless, a brief explanation is intended to provide an introduction to reading and understanding the effect network illustrations, which differ significantly from the usual representations (tables, lists, flow charts, rankings, etc.), but nevertheless depict the task complexes dealt with more realistically than the standardized and often linear relationship patterns are able to do. The aim of every network of effects – in nature, for example, the food webs of organisms – is to achieve a fault-tolerant overall stability that enables progress to be made without major consequential problems.

Nature has been applying (evolutionary) effect network strategies highly successfully for billions of years. Technology has been making impressive progress in detail for centuries with its goal-oriented development strategies. However, these have been accompanied by no less impressive regressions in the embedding of technological solutions in the natural and social environment. The search for natural models for bionics follows a feedback loop that can be interpreted as follows:

The more the secrets of nature are deciphered, the higher the probability that their valuable principles, functions, structures, forms, processes or organizations can be transferred into potential technology solutions through bionics. The more the bionic solutions have a beneficial effect in the technosphere, the stronger the desire to discover new secrets in nature.

Tapping into the source of natural models can be done in two ways, the inspirational epistemological and the pragmatic engineering. From a systemic point of view, it becomes clear that bionics makes use of the highly complex natural world and serves an equally complex technical environment. Concentrating on detailed bionic solutions means at the same time ignoring potential consequential problems that could certainly be avoided through careful consideration of real networked influences.

The network of effects of systemic bionics in Fig. 7.15, which is only reproduced in an excerpt, is associated with far greater structural complexity than is the case with a classical, more direct path of bionics. Via this systemic bionics path, one approaches reality – and thus also and in particular the capture of externalities, side effects and the anticipatory avoidance of possible consequential problems, etc. – much more quickly. Six highlighted impact relationships – representative of others – are to be briefly interpreted.

1. *Negative* effect of *consequential loads on natural systems.*

The greater the consequential burdens (of technical processes) on nature and the environment, the lower the services provided by nature (and further: the less bionics can profit from nature's role models).

2. *Negative* effect of *systemic thinking and acting on bionics.*

The stronger the impact of a new systemic view on conventional detailed bionics, the weaker its isolated development strategy becomes (or the more strongly conventional bionics is integrated into the real systemic environment).

3. *Negative* effect of *advances in proven technology solutions on bionics.*

The more purely technical and economic advances dominate social and natural habitats, the more difficult it is for bionics – with few exceptions – to survive sustainably in the competitive market. Experience has shown, however, that advances in technology can be integrated into an established market much more quickly than new bionic solutions, however efficient they may be (positive effect of advances in proven technological solutions on technical systems).

4. *Positive* effect of *natural systems on bionics.*

Bionic solutions live from knowledge processes about natural systems. The greater the biodiversity, the more intelligent natural solutions can be identified and used for bionics.

5. *Positive* impact of *pseudobionics on progress of proven technology solutions.*

The more pseudo-bionic solutions penetrate the market, whose long-term cost creation – not to be equated with value creation – is doubtful, the more the market focuses on proven technology solutions and their advances.

6. *Positive* effect of *consequential charges on external/environmental costs.*

The greater the consequential costs – from short-term profit-seeking in the market – the greater the impact of external costs or costs for reconstructive environmental damage on the portfolio and the greater the effective profit loss.

> Unlocking nature's secrets without blindly copying it calls for *systems thinking;* transferring its secrets into fault-tolerant sustainable applications calls for *systems action.* And the two are closely linked.

Quality controls are the be-all and end-all of any progress. This also applies to the holistic development strategy of systemic bionics. Its quality is based on the interplay of three interlinked quality features of the quality or optimisation space of systemic bionics shown in Fig. 7.16:

- Quality Coordinate 1 of Systemic Bionics: Degree of *Natural Example*
- Quality coordinate 2 of systemic bionics: degree of *technical (technospheric) implementation*
- Quality Coordinate 3 of Systemic Bionics: Degree of *Networked Sustainability*

Figure 7.17 shows the qualitative evaluation of the three coordinate axes of systemic bionics according to a self-explanatory evaluation scheme ranging from (–) very poor quality to (++) very good quality.

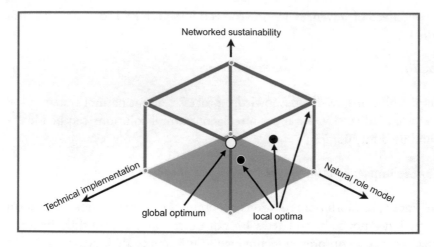

© Dr. -Ing. E. W. Udo küppers 2015

Fig. 7.16 Quality or optimization space of systemic bionics

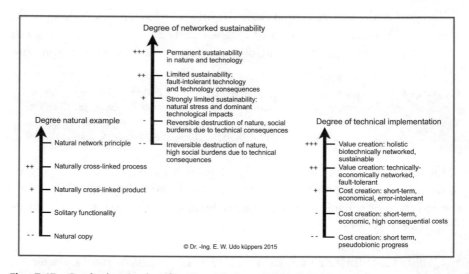

Fig. 7.17 Gradual axis classification of the criteria for measuring or comparative determination of system bionics solutions in the quality or optimisation space according to Fig. 7.16

The special feature of the "degree of interlinked sustainability" coordinate is that sustainability is assumed in the Carlowitzian sense, i.e. the interlinking of economic, ecological and social concerns is taken into account as a whole.

Figure 7.16 shows the so-called *systemic bionic quality or optimisation space,* which is spanned by the three quality axes mentioned above. The best of all results of a development of systemic bionics is achieved when the highest qualities on the three value axes coincide (global optimum).

Each approach to bionic development can be positioned in the quality space shown. The graduated classification on the axes of the quality cube in Fig. 7.17 can only be a first approach to the *holistic* recording or measurement of qualities of bionic results. Nevertheless, for the first time a realistic evaluation of the result of a bionic development seems possible, which takes into account both the networked influences in nature and technology and the quality criterion of sustainability.

For demonstration purposes, we pick out five of the large number of bionic solutions and assign them – after subjective selection of the quality criteria – to the three quality axes, not without pointing out that a realistic, well-founded quality assessment of bionic solutions would go beyond the scope of this work (Table 7.3).

Table 7.3 Five examples of bionic solutions with (subjectively) assigned axis positions of the quality or optimization space according to Fig. 7.16

Organism-principle, process, product, interconnected animate and inanimate world	Degree of natural model	Degree of networked sustainability	Degree of technical implementation
Tree *Principle:* Axiom of constant voltage	+++ Branching	-/- Use of substances that are harmful to nature, some of which are highly toxic	+ In the market
Natural flow systems *Principle:* Entropy minimization	+++ Free flowing waters of nature, biological pipe systems	+ Elements of technical pipe systems with partly natural pollutants	+ In advanced Development
Plant leaf *Principle:* Self-cleaning	+++ Textured blade surface	+ Technical elements with partly natural pollutants	+ In the market
Ostrich Egg *Product:* Eggshell, Composite material, breathing function	++/+ Shell structure, permeable gas transport	− Currently functional verification only with plastic film as carrier material	+ In advanced Development
Trunkfish *Product:* Body shape Fluid dynamics	+ Shape with low flow resistance	-/- Use of substances that are harmful to nature, some of which are highly toxic	− Museum model

7.3 Central Survival Function: The Circulation Principle – Minimization of Trivial Linear Processes by Efficient Low-Loss Circulation Processes

The nature principle of the circular processes is central to all technical material processes. The routine cycle in Fig. 7.18 instructs us to treat materials in such a way that they fulfil the required functions over their life cycle, guarantee qualities, avoid or minimise harmful side effects and lead to new cycles with low losses.

The economic dominance of the majority of all human-initiated processes shows its short-sightedness with regard to sustainable goals through causal linear thinking with exponentially increasing consequential problems. Over decades, perpetuated or new innovative areas of capitalist-driven human activities have emerged, whose revenue and profit creation has led to enormous prosperity, but at the same time to externalized problem and cost accumulation or waste of materials.

Three Examples Among Many Should Illustrate This

1. The energy industry, with a decades-long focus on nuclear and fossil fuels with simultaneously unsolved disposal problems of inevitably accumulat-

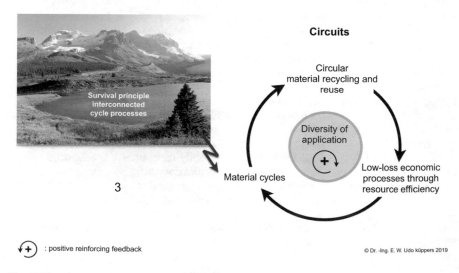

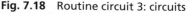 : positive reinforcing feedback

© Dr. -Ing. E. W. Udo küppers 2019

Fig. 7.18 Routine circuit 3: circuits

ing radioactive waste, with half-lives of decay, e.g. of iodine-129 with 17,000,000-seventeen million years.

At the same time, after the 2011 nuclear power plant accident in Fukushima, Japan, politicians in Germany decided to phase out nuclear fuel energy conversion, the practical implementation of which is still unparalleled in terms of political dilettantism, economic narrow-mindedness and civic intransigence in 2020.

2. The motor vehicle industry, which develops and manufactures new vehicle type classes "on the fly", with entire robotic production lines worth tens of millions of euros, being completely scrapped once the vehicle type is discontinued, despite still retaining a high functional value![15]

If at the same time at the Consumer Electronics Show – CES – in January 2020, the automotive company Daimler presents its science fiction concept car "Vision AVTR", in cooperation with the makers of the movie "Avatar", but at the same time its CEO Källenius literally declares: "Nature is and remains our best teacher",[16] then in the context of this book Källenius' statement is correct, but his inferential action afterwards is short-sighted and misguided.

3. The technically "highly equipped" agribusiness or industrial agriculture, whose companies occasionally sell industrially produced seeds and pesticides to farmers – preferably in developing and newly industrialized countries such as India, Bangladesh, and others. The consequences of this compulsory liability of many small farmers, but also of large landowners and their monoculture cultivation, are partly completely depleted infertile soils after a few years, mass poverty of small farmers in developing countries, increasing destruction of natural fertility cycles in food cultivation, acceptance of mass insect mortality and much more.

Even if – as already practiced in greenhouses in the Netherlands – "robot bees" pollinate tomatoes, peppers, cucumbers or strawberries, a monetary calculation (Heinrich-Böll-Stiftung et al. 2020, p. 45).

---

[15] Result of a personal conversation (January 2020) with an employee of a passenger car supplier of special devices for passenger car production lines.

[16] Weserkurier, Bremen, 08.01.2020, Page 17, Daimler: show car in Las Vegas, lawsuit in Germany, dpa.

For a fraction of the money that millions of pollination robots would cost in the future [...] maintain and strengthen today's ecosystems.

The Insect Atlas 2020 (Heinrich-Böll-Stiftung et al. 2020) shows with impressive facts misguided strategies in politics and agriculture, highlighting global insect mortality (ibid., pp. 14–17), pesticide use and insect death (ibid., pp. 18–19), showing higher feeding damage by insects due to climate change (ibid., pp. 22–25).

A fundamental variable of the cycle principle of nature is the *negative* or *balancing feedback* that maintains the overall system in a stable steady state through various stations of change and action. Due to their interconnectedness, the organisms of nature know the limits of their actions and thus of their ability to survive. Humans have largely freed themselves from this by their acquired causal thinking and acting. In the long term, they have thus created a condition for themselves which, with the consequence of increasingly incalculable conditions or deliberately ignored warning signs of nature, catapults them into their own geological age, that of the Anthropocene. For the first time, man is demonstrably the cause of environmental damage on the largest scale, whether as plastic waste in the sea or as micro-nano-particles in our food, whether as the destroyer of natural climate regulation, whether as the destroyer of essential life-supporting forests, whether as a creature that is in many respects unreasonable and saws at the branch of life on which it is itself sitting. Evolution is also inexorable with us as organisms when we believe that we have to place ourselves outside the rules of life of nature.

7.4 Nature's Growth Strategy: Regulated Input-Optimised Instead of Controlled Output-Maximised Growth

Growth is always present in nature. Year after year, in the growth phase of the seasons – in spring – new growth cycles take place, with a material wealth and unimaginably large amounts of material, far beyond that which humans produce or leave behind per year. And yet, year after year, nature regulates its growth cycles with material processing at the end of the growth periods, without leaving behind the slightest *"useless"* waste – according to human conception. A few figures may illustrate this.

In 2018, global plastics production reached a value of almost 360 million tonnes. The volumes sent for recycling increased by almost 12% to more than

9.4 million tonnes.[17] According to the report, 350.6 million tonnes of persistent plastic was not recycled and left to the environment and nature, where it develops life-destroying qualities.

The quantitative ratio becomes drastic if we compare the "waste" management, or better: the waste-free *management of material recycling* in nature. With its mature cybernetic structures, nature recycles around *170 billion tonnes of new biomass annually* (Gleich et al. 2000). Everything happens without any destructive effects on itself, on the environment and of course without affecting the health of us humans.

Obviously, nature-growth management works with only a handful of different starting material molecules (Küppers and Tributsch 2002, p. 14 ff.) and handles a cyclic material quantity per year that is about 450 times higher than human material processing of plastic, with more than 10,000 different material compound formulations and, as a consequence, huge environmental and natural catastrophes, far more efficiently.

The routine cycle in Fig. 7.19 shows how we can use growth management in the future without the enormous consequential problems we have caused by our narrow-minded development strategies. To systematically avoid, or at least significantly reduce, growth waste or worthless loss material on a human

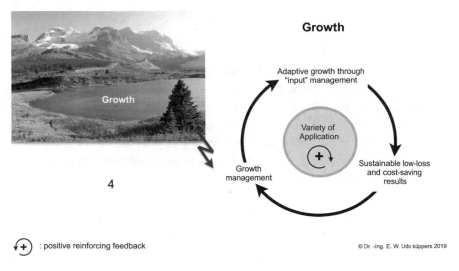

(+) : positive reinforcing feedback

© Dr. -Ing. E. W. Udo küppers 2019

Fig. 7.19 Routine circuit 4: growth

[17] https://newsroom.kunststoffverpackungen.de/2019/10/18/k-2019-plasticseurope-praesentiert-plastics-the-facts-2019 (accessed 2020-01-19).

scale requires a fundamental reversal of our output-fixated growth strategy, towards nature's growth principle, which is reflected in the following sentence:

> Adaptive sustainable growth in nature follows an "input management", with the consequence of a resource-optimized utilization with qualitatively highly efficient productions and complete avoidance of material losses through perfect recycling or reuse.

For our technical-economic production processes and huge waste mountains of "unproductive" materials, this goal is an enormous challenge, which is also accompanied or must be accompanied by completely new *systemic* organizational structures, beyond the old, encrusted hierarchies that still exist in many cases.

Generating growth, while at the same time creating value, but without *systemic* consequential problems – which include not only the material-technical waste side, but also ecological and social burdens – would be a general growth strategy that can develop future perspectives for sustainable solutions out of the anthropocene environmental burden.

It is – as so often – in the end less a question of technical ability, but rather a question of political-economic will! The power of people to change their previous mental models, from which short-sighted, misguided but economically successful strategies are derived, by new mental models for far-sighted and sustainable practices is required for a high degree of self-reflection and insight into the previous practices of misguided and less sustainable processes.

Upton Sinclair's personal impressions of working conditions in Chicago slaughterhouses around 1900 (Sinclair 1980), his description of business practices and working conditions in the American oil industry (Sinclair 1984), and other socially critical analyses probably led him to the well-known saying of universal validity, which also includes the present industrial perspectives of not a few decision-makers in society:

It is difficult to make a man understand something when his salary depends on his not understanding it.

7.5 Biospheric Zero Material Waste Strategy: Networked Cycles of Material Recovery or Recycling in Small-Scale Local/Regional Industrial, Commercial and Private Sectors

Nature shows us how to use materials or their costs in a highly economical or intelligent way without reducing functionality in the perfection of small-scale networked associations. Figure 7.20 shows us the principle circular process of how something like this can also be implemented in technical-economic terms, whereby the usual routine is replaced by an innovative routine of "small-scale heterogeneous material circulation processes".

At the end of a production cycle, all material that we produce and use is recycled or reused, sometimes to a small extent, but the majority is left to the geosphere and/or the biosphere as unproductive, worthless waste material. The example of plastic outlined above in Sect. 7.4 shows in a drastic way how a material that does not – as the name suggests – come from natural sources but exorbitantly damages them, whereby natural food webs that have grown over millions of years suddenly entail dangers to life and limb of organisms that were hitherto unknown.

The value-adding processes of substances or materials from natural sources and artificial production are predominantly carried out separately from each other. This is not surprising, because the production of natural materials and

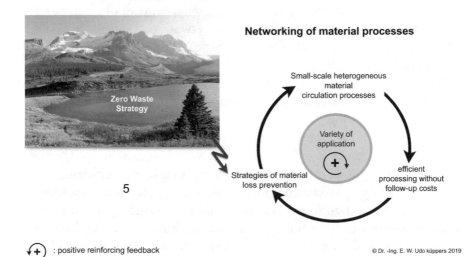

$\checkmark\!+$: positive reinforcing feedback

© Dr. -Ing. E. W. Udo küppers 2019

Fig. 7.20 Routine cycle 5: networking of material processes

materials obey completely different prerequisites, properties and goals than technical or artificial materials and materials do.

The temperature scales alone for the production of plastic formulations (approx. 200–300 °C), or technical iron melting processes (approx. 1500 °C), or special ceramics (>1600 °C) are far beyond the production temperatures of biomaterials and their composites (room temperature around 20 °C).

The extraordinarily high manufacturing temperatures are of course also associated with extremely high energy requirements, which in turn are preceded by more or less lossy conversion processes. However, energy is increasingly becoming a critical variable in both the technical and the social environment, as shown by the energy-consuming processes of the digitalization of the economy (Industry 4.0/5.0, etc., robotics) and society (Internet of Things), which are increasingly coming into focus.

The interplay of highly efficient properties of natural materials – in particular of composite materials whose individual layers have different, sometimes contradictory physical properties, such as hardness and elasticity – with technical materials of specific properties would be a new challenge for innovative sustainable material processes of the future. Resource efficiency, minimal energy consumption and a multifunctional range of applications would be goals of a biosphere-technosphere material future.

Highly efficient material processes, which can be launched much faster and more effectively than previous material cycles in a heterogeneous network of different companies, industries, services, etc., use – similar to the networked material processes of nature – small-scale circulation processes through linked "input-output" transports of transformed material waste of one network participant to recyclable materials of the other network participant. It would be ideal to use existing industrial or business parks for this purpose, with the one but essential target change:

Municipalities and communities as owners of industrial and business parks do not select their settling companies according to the maximization of business tax, but according to the minimization of technical and other waste materials or energy compound losses.

This would result in a differentiated WIN-WIN situation for all parties involved. Municipal tax revenues would be replaced by cost-efficient, significantly lower energy provision or global energy transports and a reduction in cost-intensive waste transports. The industrial and business park participants would be coordinated in their processes in such a way that new and expensive purchases of materials would be reduced and inevitable production and other costs of disposal could be minimized or would not be incurred at all. In addition, interesting synergies can arise through the networked – production-technically coordinated – cooperations, rather than among the classic industrial and business park participants.

> **Tip**
>
> Small-scale, coordinated alliances of heterogeneous material processes of cross-industry production companies, service providers, trades and other participants aim at material and cost efficiency or energy efficiency over the lifetime of their products through a WIN-WIN situation[18] from cooperative synergetic or symbiotic action.
>
> Municipalities as owners of industrial and business parks turn their purely economic objective of maximised tax revenue around into the sustainable objective of minimised cost expenditure with simultaneously adjusted tax revenue, which together also lead to a WIN-WIN situation.

7.6   Nature's Energy Flows: System-Adapted Energy Cascades of Minimal Energy Losses

The routine circuit in Fig. 7.21 shows the circular path of efficient energy processing for technical applications.

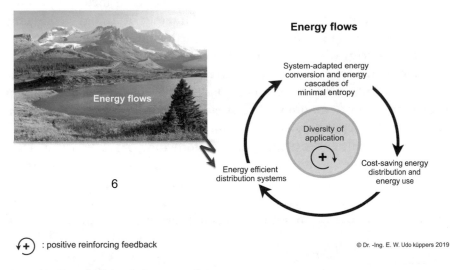

Energy flows

System-adapted energy conversion and energy cascades of minimal entropy

Diversity of application

$(+)$

Cost-saving energy distribution and energy use

Energy efficient distribution systems

6

 $(+)$: positive reinforcing feedback   © Dr. -Ing. E. W. Udo küppers 2019

Fig. 7.21   Routine circuit 6: energy flows

---

[18] All partners benefit from a WIN-WIN situation – in the example of the industrial and business park – through cooperative coordinated action; similar to what often occurs in nature through synergetic or symbiotic processes.

The only source of energy that is available to us on our earth and that sustains and develops life through varied processes of change is the sun. It is central for all organisms.

Humans have evolved through intelligence and creativity to devise energy systems that facilitate work, promote and empower progress, but also support convenience.

In contrast to the energy systems of nature, whose energy conversion systems, for example all organisms, including the vital oxygen producers trees or entire forests, convert, transport and use energy in an adapted, i.e. economical way, without any consequential problems, technical energy systems are often limited to individual isolated objects with sometimes enormously high risk potential. Examples are nuclear power plants, with radioactive, persistent fuels, power plants with fossil fuels (coal, oil, gas) and environmentally destructive resource procurement as well as massive influences on climate changes, central energy parks with regenerative wind energy and complex distribution systems, distributed solar energy conversion systems with partly environmentally harmful toxic photoelectric materials, individual hydroelectric power conversion systems, numerically insignificant geothermal energy conversion systems. The aim of the individual technical energy conversion systems is to supply industries, municipalities, private households, etc. with electricity and heat.

The industrial and social change taking place, which is increasingly influenced by digital processes ("Industry 4.0", "Internet of Things", constantly growing "digital means of communication", etc.), demands new qualities and quantities of energy without restraint. However, in order to achieve the exergy efficiency[19] of the conversion product electricity, which is the ultimately effective form of energy for many technical consumers of energy (machines, means of transport, robots), a number of utilization systems, such as steam power, hydropower, wind power, solar energy, fuel cells, etc., are upstream.

Total energy efficiencies along these transformation processes are therefore associated with not insignificant anergetic – work-loss – energy components. The lossy global transport of energy forms over long distances, including the existing infrastructural and socially detrimental side effects, is our way of dealing with energy. The enormous challenge of providing a complete industrial

---

[19] Exergy is that portion of energy that can be completely converted into work, while anergy, is the other portion of energy that cannot be converted into work. According to the first law of thermodynamics, energy = exergy + anergy = constant. Electricity is the form of energy that can do 100% of the work. "The conventional energy efficiency of converting electricity into heat is close to 100% [...]." (Fricke and Borst 1981, p. 22).

country like Germany with energy and also high safety standards has become more risky rather than less risky, especially after the politically initiated phase-out of nuclear power in 2011 and the digital attacks on the data or power grids that are taking place all over the world.

The question is allowed: Why are nature's functionally linked energy networks, which evolve evolutionarily and self-repair in the event of a local disruption, largely intact without having to fear the global threat of a biospheric "system-destroying" attack?

It is their decentralized and local energy systems and supply units that adaptively regulate their energy demand through clever networking with each other. This is not done with a maximized 100% exergy efficiency, but with a perfectly coordinated energy network in a cross-species manner. Processes such as the principle of symbiosis, growth adaptation to climatic conditions, specially optimised protective functions against energy loss (e.g. in mountain plants or desert plants in extreme environments) and many more natural energy tricks are nature's highly effective energy craft and repair tools.

It is true that organisms in nature also have the ability to generate electricity, such as elephant trunk fish, Nile pikes or electric eels. They use electricity for communication and defense against enemies, but not to maintain their own organism.

The social and industrial change that we are currently experiencing due to the increasing influence of digitalisation and thus also the increasing *"hunger for energy"*, with not inconsiderable stressful consequences, as indicated above, should be worth our while to look beyond the *"technospheric energy boundary"* into the *"biospheric energy world"* and its intelligent organismic handling of energy. It would be worthwhile, after all, nature is based on a progress process of millions of years, with energetically effective, adaptively developed principles, without comparable solutions in the technosphere, if we consider the complex reality in its entirety.

7.7 Meandering of Watercourses: Minimising Entropy and Natural Hazards

Straight ship channels, technical flow systems of high energy loss, etc.

The routine cycle in Fig. 7.22 intervenes in existing technical flow systems of man, which already carry the potential of high energy losses due to their constructive designs guided by economic goals (river canalization or river straightening, technical energy-loss-rich shaped parts of pipe transport

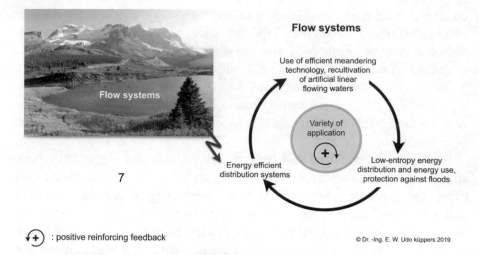

Fig. 7.22 Routine circuit 7: flow systems

systems with different transport goods or multiphase flows, as well as other technical flow systems in various products and processes), although there are clearly more efficient – because more cleverly designed – ways and forms of flow transport.

Why do free flow systems in nature, such as river courses, in plains without geological obstacles, rarely flow straight for more than 10–20 times their river width? Because random changes in complex dynamic environments cause rivers to meander. This happens as follows: Slight irregularities in the substrate caused by rocks, plants, or other objects are enough to cause the straight course of water to be diverted from its path. Water flows asymmetrically towards one bank, bounces off there and changes direction to the other bank. As water bounces off one bank, loose sands, clayey soils, stones, and other materials are removed from the bank and re-deposited by the flow process of the water downstream on the opposite bank. The water flows faster on the side of material removal than on the opposite side of material accumulation. This alternating process of material removal and accumulation, with the different flow velocities on both sides, leads to characteristics of the river course similar to a serpentine line. This results in the so-called energy-efficient meanders of flowing waters.

This principle of efficiency causes a minimization of entropy in the course of the river, as an integral part of a water cycle of nature.

Experimental tests on technical pipe systems with so-called Meander-Effect® pipe bends have proven the advantage of flow efficiency compared to circular pipe bends. (Küppers 2007a, b, c).

Why do natural flow systems in nature, after intervention by humans, flow as if pulled on a string, often transformed into artificial channels, for miles straight ahead? Because planners in politics and business want it that way for purely economic reasons!

The general efficiency principle of natural flowing waters is overridden and replaced by a short-sighted and misguided limited economy principle.

The increasing extreme weather conditions due to climatic changes in recent years have also caused small rivulets to grow into raging rivers, which develop exponentially increasing flow velocities in straightened river sections, canals and similar constructions, with enormous destructive force. The number of these consequential problems due to human – not very far-sighted – interventions in nature is increasing. The destruction of towns and cities close to rivers is in some cases assuming catastrophic proportions, which – if the predictions of credible climate experts are taken into account – will continue to increase. In many cases, the only solution people have to protect themselves against such self-inflicted natural events is to build barricades and barriers against floods. – It is not so much remedying or meaningfully acting on the causes of the problem as it is short-sighted costly repair behavior.

Besides this large destruction potential of water due to a relatively small technical intervention of man in nature, immature designs of technical flow systems in pipes used for air-conditioning and ventilation systems, existing in technical compressed air systems, pipe transports in plants and machines and buildings seem to be only a minor evil of energetic losses. However, if we calculate all the "hidden" energy losses in technical pipe flow systems together, the result is likely to be costs that are significant in economic terms.

Why does this very effective type of energy-efficient action by minimizing flow losses and thus minimizing the necessary electrical drive energy of the systems happen so rarely? Behind this lies – psychologically speaking – the problem of every operator of such systems: They would have to make these losses public! A negative effect for the customers, which must be avoided at all costs. It is preferable to place energy-wasting larger drive units in front of the flow systems, or to conceal the true extent of the energy losses. Even standardized cost models, based on flow-physical measurements or calculations, only work with circular-arc pipe body geometries of the high-loss baffle parts that have been standardized for decades.

7.8 Plant Communication: Communicative Cooperation Instead of Confrontation as an *Emergent* Survival Strategy in the Technosphere

Communication or networks for the exchange of information, together with energy flow and material processing, form the three fundamental transport processes of nature. They are also fundamental processes in the technosphere.

Figure 7.23 understands the "incentive" of systemically oriented communication as the *"routine" of "using efficient communication algorithms across sectors"*, the *"reward" of which is* shown as *"safeguarding and stability of progress through networked early warning systems"*.

Plants are a great role model in communication. With their at least twenty senses, including the five we use, plants, as stationary creatures, possess an impressive sensorium for communication or information exchange with other organisms and the environment. Another distinctive feature sets plants apart from animals (Mancuso 2018, pp. 10–13):

> [...] the decisive and hardly anyone conscious difference between plants and animals lies [...] in the contrast between decentralized distribution and concentration. While the bodily functions of animals – like those of humans, the author – are concentrated in certain organs, those of plants are distributed over the whole body. (ibid., p. 10)

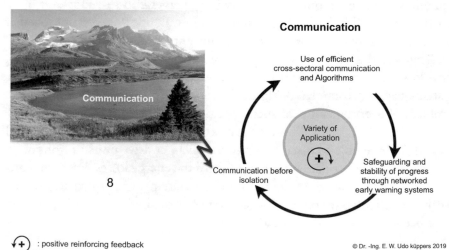

8

 : positive reinforcing feedback

© Dr. -Ing. E. W. Udo küppers 2019

Fig. 7.23 Routine circuit 8: communication

Such a construction was rarely a model for human developments, rather the bodily functions including communication of animals and of ourselves. The construction of computers up to the present day is an example of a blueprint for communication, in which a central processor – similar to a brain – is linked to working memory, mass storage, audio signal unit and more.

> Even our society is based on this archaic, hierarchical and centrally controlled model. Yet it offers only one advantage. It provides quick answers – but they do not necessarily have to be right. Otherwise it is fragile and anything but innovative. (ibid., p. 11)

Our administrative hierarchy, established in every public institution – from the smallest municipality to the federal government – and perpetuated over decades in terms of structure and communication, is evidence of the inflexibility and lack of innovation of this form of organization and the communication that takes place within it. This is particularly remarkable because there is a dynamic of complex processes around us to which a rigid administrative communication cannot contribute anything decisive to correct tasks and problem solutions!

> Plants, on the other hand, do not possess an organ like a central brain. Nevertheless, they can perceive their environment more sensitively than animals, they can actively fight for limited soil and air resources, reliably assess situations, perform sophisticated cost-benefit calculations and react appropriately to environmental stimuli. So they could be an alternative blueprint for us to seriously consider. All the more so now that we need to detect change ever more quickly and develop highly innovative solutions immediately. [...].
>
> The plant model is much more resilient and contemporary than that of animals. Plants are a living example of the fact that robustness and flexibility need not be contradictory. Their modular structure makes them the epitome of modernity: thanks to their cooperative, distributed architecture without a command centre, they remain functional even in disaster situations and adapt quickly to changing environmental conditions. The most important bodily functions of the complexly structured plants can rely on a highly developed sensory system that enables efficient exploration of the environment and a rapid response to threats. For example, plants have a sophisticated, continuously evolving network of root tips that actively explore the soil. It is no coincidence that the Internet, the epitome of our modern world, is also structured like a root network.
>
> Because plants have a much more contemporary structure than animals due to their unique evolutionary path, no one can match them in terms of robust-

ness and innovation. We would do well to take this into account when thinking about our own future. (ibid., pp. 11–13)

Robustness against changes in the environment and the innovative power to adopt new ways of communication instead of perpetuating old ones would be a promising path for building new adapted administrations, as suggested by Küppers, E. W. U. (2011a, b, c) and Küppers, J.-P. and Küppers, E. W. U. (2016), among others (see also Figs. 7.24, 7.25 and 7.26). What in plants are highly developed sensory systems, the aforementioned authors interpret as criteria for a communicative "high-mindedness" that goes far beyond the mental boundary of departmental thinking and acting, thereby understanding the dynamics of change as a challenging task, less as a problem. In this respect, communication under high regard is adaptive, forward-looking and fault-tolerant.

Figures 7.24 and 7.25 compare a strictly hierarchical with a networked mindful communication structure of public organisations, whose tasks and problem solutions lie in dynamic complex environments. It becomes clear that concentrated communication based on largely "isolated" and linear administrative and centralistic areas of responsibility also only performs "isolated" tasks, and can therefore also only develop "isolated" solutions to problems (Fig. 7.24). Incidentally, industrial companies have similar organisational structures, with centralised and linear communication processes and areas of responsibility, to those of public administrations.

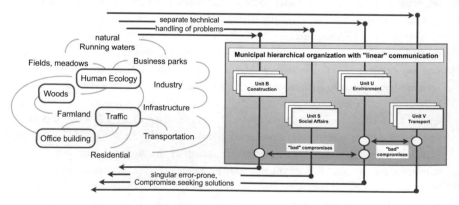

© Dr. -Ing. E. W. Udo küppers 2019

Fig. 7.24 Hierarchical management structure with largely "isolated" communication elaborates specific solutions to problems in networked dynamic environment without taking into account their feedbacks

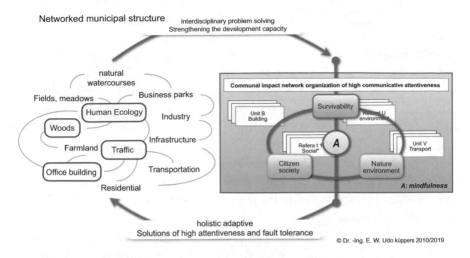

Fig. 7.25 Holistically managed organizational structure of public administrations with extensively networked communication for adaptive problem solving in dynamic complex environment. M stands for mindfulness

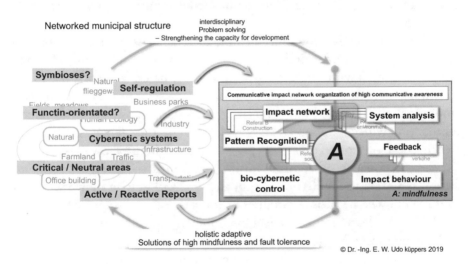

Fig. 7.26 System understanding strengthens communicative attentiveness, marked M in the figure. (Template after F. Vester 1991a, b, Suppl., 19)

In contrast, Fig. 7.25 shows an administrative organization that perceives tasks in a dynamically complex environment holistically through networked communication, thereby solving the strengths and weaknesses of the interrelated tasks and problems far more effectively and efficiently. The mindfulness quality of the communication system is thereby significantly increased because the "sensor system of the communication system" reacts adaptively to changes in the network and is not focused on "isolated" individual areas of the specific system environment.

This makes the communication and environmental system more robust and innovative, as is inherent in the communication system of plants.

Municipal or regional tasks are processed and solved from within the individual office areas, but largely to the neglect of real networked complexity. This is fragmented. Statutory goals are achieved in the short to medium term, accompanied by delayed – often surprising – consequential problems (with consequential costs), which not infrequently reveal the lack of manageability and predictability of the natural interrelationships. Communication across departmental boundaries only leads to "lazy" compromises, rarely if ever to sustainable solutions.

Municipal or regional tasks are processed and solved from networked departments or functional organisational units. In doing so, real complex interrelationships are taken into account and great attention is paid to the smallest errors, which means that consequential problems (with consequential costs) can be largely avoided. Communication is robust and innovative in the face of changes.

The communicative and thus also organizational structural approach of approaching the complex problems of the environment and nature adaptively and solving them with foresight, as indicated by the example of a networked communal structure, in Fig. 7.25, can only succeed if our mental model of (mono)causally practiced ways of solving problems is transformed into a mental model in which an *understanding of* the *system* is the prerequisite for adaptive practical solutions. Figure 7.26 provides an insight into how we strengthen the central concept of communicative *mindfulness* – or *high mindfulness* (Küppers and Küppers 2016) – within an organization with networked communication, whether it is communes or companies.

The more I learn about the real environment and its interconnectedness and communicate this, for example by recognizing and using symbioses, self-regulating systems, or cybernetic systems (in Fig. 7.26 on the left), and the more intensively I treat these insights cooperatively through systemic tools such as impact network analyses, pattern recognition, or feedback processes (in Fig. 7.26 right), the more strongly I support the central mindfulness within a system, its robustness and ability to progress, as can be seen in Fig. 7.26 in state hierarchies, in Fig. 7.27 in industrial hierarchies.

Communication of high mindfulness requires networked organisational structures in order to be able to react adaptively to constantly occurring dynamic changes in the environment. Communication of high mindfulness is robust and innovative, it strengthens error prevention and the view for foresighted, error-tolerant action.

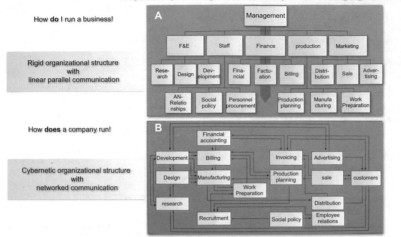

Comparison of practice Organizational Hierarchy - A - and Learning Organization - B

according to Gomez/Probst 1999 | ©Dr.-Ing. E. W. Udo Küppers 2010/2019

Fig. 7.27 Communication in rigid hierarchical and dynamically networked learning business organisations (after Gomez and Probst 1999, pp. 70–71). Whereas organisational structure A, with its rigid processes and pre-programmed communication channels, has to make considerable extra efforts if unforeseen events or planned innovations occur, organisational structure B is designed from the outset as a learning structure. In the event of unforeseen events or impending innovations, it adapts to the new tasks and communication channels in a time- and cost-efficient manner

7.9 Sequoias: Earth-Historical Experience in Protection Against Fire – Bionic Technology of Fire Protection

Several thousand years old and full of life experience and fire fighting techniques. That's what you could call the *sequoiadendrons,* or redwoods, that grow on the west coast of California and in the Sierra Nevada. Their experience in harnessing "naturally occurring" fires is impressive. On the one hand, in that their seed pods are first blasted by fire to release the seeds for new generational growth. On the other hand, the sequoia trunks, with heights of over 100 m, trunk diameters at ground level of more than 15 m, and trunk circumferences of more than 30 m, are surrounded by a light, airy bark up to ½ m thick. This protects the giant organism not only against predators but also against total destruction by fire.

Using millennium-old and therefore long-term proven redwood fire techniques as bionic fire protection, as suggested in Fig. 7.28, has hardly arrived in relevant branches of industry yet. The dominance of plastics as fire-retardant

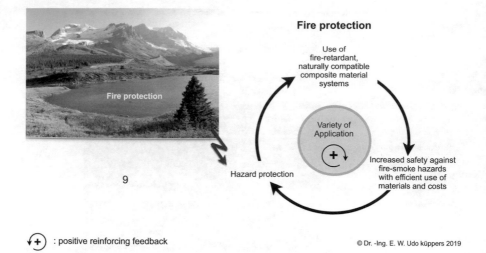

Fig. 7.28 Routine circuit 9 fire protection based on biological model of long-term proven technology in dealing with fire

materials (e.g. polystyrene) is unmistakable in the market, despite existing risks, criticism and counter-criticism of the use of these non-natural materials.

Figure 7.29 shows the model of a bionic fire protection product based on the bark of the redwood tree.

Past and current fire accidents with fatal outcomes for people have shown and continue to show that, despite technical progress in fire prevention measures, there are still many technical problems, especially with regard to materials. The airport fire in Düsseldorf in 1996, the cable car fire accident in Kaprun, Switzerland in 2000, the railway accident in Nancy, France in 2002, or current fire disasters involving walls insulated with polystyrene (PS), such as the fire in the "Grenfell Tower" residential tower block in London in 2017, in which apparently safely tested fire retardant materials, in conjunction with other technical and human influences, led to catastrophes on a large scale, are just a few examples. In all of these disasters, insufficient knowledge about the complex behaviour of material composites with high proportions of plastics and thus highly toxic gas mixtures in the event of fire was involved.

The more the anthropocenic effects of self-inflicted causes spread in nature and technology and affect our living and working space, the more highly proven natural techniques come into the focus of technical applications or should attract intensive attention. Carrying on as before only promotes the encrusted standard routines in our mental models and the short-sighted technical-economic, often misguided progress.

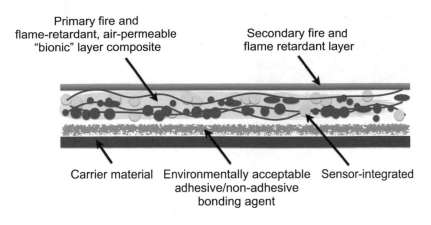

Primary fire and
flame-retardant, air-permeable
"bionic" layer composite

Secondary fire and
flame retardant layer

Carrier material   Environmentally acceptable   Sensor-integrated
adhesive/non-adhesive
bonding agent

© Dr. -Ing. E. W. Udo küppers 2003/2019

Fig. 7.29 Structural model of a bionic fire-retardant protection, as seen here, could be the start of a new development for nature-compatible composite material technology that avoids previous fire-retardant techniques – but also fire hazards – caused by plastics and their dangerous additives, such as the toxic flame retardant hexabromocyclododecane (HBCD)

7.10   Self-Cleaning Surfaces: Multiple Efficiency Approaches for Self-Maintained, Dirt-Repellent Clean Surfaces of Technical Products

The bionic routine circuit in Fig. 7.30 refers to a prime example of bionics, the transfer of optimal structures of dirt-repellent, self-cleaning leaves to technically functional surfaces. Ingenious models in nature are plants, such as lotus flower, lady's mantle, cabbage leaves, columbine, but also animal organisms with a dirt-repelling effect, as found in many insect wings.

Technical applications are already being tested in various research and development facilities and are also already in practice. The breadth of the technical application is shown by the random selection of product examples (without claim to completeness), which realize the lotus effect or advertise with it:

- Nanotol Sanitary 2in1 Cleaner + Protector with lotus effect sealing[20]
- Facade paint Lotusan with patented Lotus-Effect[21]

---

[20] https://www.nanotol.de/sanitaer/2in1/reiniger-mit-nanoversiegelung?number=NS21-5&gclid=EAIaIQobChMIh-GdyqOj5wIVk-R3Ch12XQgeEAQYASABEgK0_PD_BwE (accessed 2020-01-27).
[21] https://www.farbenbote.de/produktdetail.cfm?pro_id=1101&gclid=EAIaIQobChMIh-GdyqO-j5wIVk-R3Ch12XQgeEAQYAiABEgLoN_D_BwE (accessed 2020-01-27).

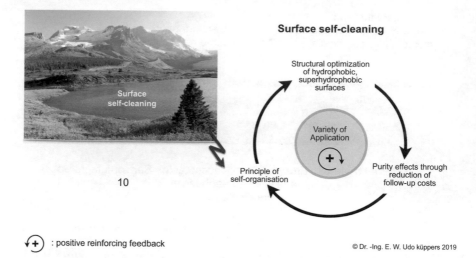

Fig. 7.30 Routine circuit 10 surface self-cleaning

- Professional Glass & Ceramic Nano Sealing Set from HEUREKA with lotus effect[22]
- **Nano sealing** with Nansolid™ with optimized lotus effect[23]
- PreConnect® Lotus: Self-cleaning fiber optic contacts from Rosenberger[24]
- Erlus Lotus – the world's first self-cleaning clay roof[25]
- Erlus Ergoldsbacher E 58 S Lotus Air®, a further development of the lotus effect[26]

7.11   Lightweight Constructions: Treasure Trove for Technically Stable, Economical and Aesthetic Building Constructions

Figure 7.31 shows the bionic routine circuit, based on ingenious lightweight constructions of nature.

What do ray echinoderms, diatoms, brown algae, bulrushes, water lilies, quiver trees, diatoms, honeycombs, and the skeleton of a bird, to name just a handful of organisms, have in common?

---

[22] https://www.amazon.de/stores/node/4945862031?_encoding=UTF8&field-lbr_brands_browse-bin=HEUREKA&ref_=bl_dp_s_web_4945862031 (accessed 2020-01-27).

[23] https://www.xpertco.de/nansolid.html (accessed 2020-01-27).

[24] https://osi.rosenberger.com/de/produkte-services/lwl-verkabelung/preconnect-lotus/ (accessed 2020-01-27).

[25] https://www.innovations-report.de/html/berichte/medizin-gesundheit/bericht-29970.html (accessed 2020-01-27).

[26] https://www.erlus.com/produktdetails?pid=43004 (accessed 2020-01-27).

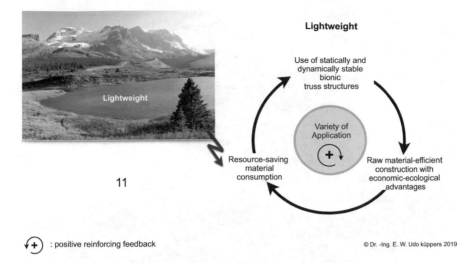

Fig. 7.31   Routine lightweight construction cycle

They are all lightweight constructions of nature in perfection! The principle of material efficiency (minimum use of material with correct positioning) is followed as well as an adapted maximum of stability or breaking load.

Lightweight constructions of nature can be found in all organisms to a greater or lesser extent, which is why we can also speak of a general principle of nature.

The lightness of bird flight, for example, is also due to so-called pneumatized bone structures. The bones of the modern robust bird skeleton are of "[…] phenomenal lightness […] and interspersed with air-filled cavities" (Hickman et al. 2008, p. 867).

The principle of robust lightweight construction is therefore present in the technosphere – not least due to the examples of natural lightweight constructions – whether in aircraft construction, vehicle construction, shipbuilding, as well as in individual bionic-excellent constructions in the field of tent roof construction.[27] Other areas of technology, such as house and apartment construction, still rely in many cases on conventional – often manual – methods of building construction, although nature also shines in this exposed and socially highly relevant area with quite a few models for energy- and material-efficient construction techniques (Becker and Braun 2001; Küppers 2001). New experiments with digitized planning and glass fiber (GFRP) or carbon fiber (CFRP) as a building material for spatial lightweight structures (Fig. 7.32), which are also inspired by nature's role models, are still in the

---

[27] Imposing example, using a biological trick of spiders, for loop-like secure suspension of the curved roof surfaces, is the Olympic tent roof in Munich, developed and constructed by Günther Behnisch and Frei Otto, between 1968 and 1972.

Fig. 7.32 Bionic lightweight construction in technical development or application: carbon and glass fibre construction of a spatial truss at BUGA-Wiesbaden, 2019. (Courtesy of ©ICD/ITKE University of Stuttgart and ©Roland Halbe)

development stage This special type of material construction, has a weight per unit area between 7 and 8 kg/m² and is 1/5 as heavy as a comparable steel structure (Kurmann 2020, pp. 8–9). However, the "ecological" question remains open as to what happens to the CFRP-GFRP composite material after its service life. Does a material process for recycling exist for these plastic composites? If not, how and where does the disposal take place?

Aircraft that are intended to be particularly robust and light in weight – because they save fuel – use lightweight honeycomb composite constructions in their wings, for example, as Fig. 7.33 shows.

For passenger cars, the company EDAG-Engineering GmbH developed the lightweight body structure of the concept model EDAG Light Cocoon, which, in addition to material efficiency for weight saving (approx. 25% lighter than a conventional aluminium construction) and robustness, also took into account traffic engineering criteria such as impact behaviour, etc. Figure 7.34 shows the Cocoon body and pictures of the concept model EDAG GENESIS, with the biotechnical lightweight construction analogy to the protective shield of a turtle.

Notes

The quantity and quality of nature's ingenious lightweight constructions is immeasurable. – In contrast, the quantity and quality of bionic applications is still in its infancy. The recognition of ingenious lightweight construction

Fig. 7.33 Bionic lightweight construction in aircraft construction

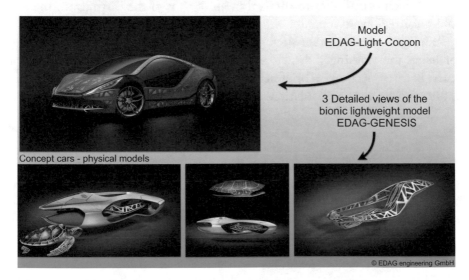

Fig. 7.34 Bionic lightweight construction for automotive engineering. EDAG concept model Cocoon, top; EDAG concept model GENESIS, bottom. (With the kind permission of EDAG-Engineering GmbH, Fulda)

principles – as well as other principles – of nature is only the first step of a bionic development towards beneficial application. Bringing together different boundary conditions, development strategies, scales, etc. from both spheres, the biosphere and the technosphere, requires systems thinking. Nature evolves its "products and processes" in a highly complex environment with the best adaptive progress in each case, while technology develops its "products and processes" in a goal-oriented and dominant, economically-focused manner. The guiding orientation of people for both spheres is sustainability in the forestry sense (see Sect. 7.2). Only when this is recognised and practised will there be a chance, in my view, of *sustainably* coping with the enormous tasks and risks that a new geological era, the *Anthropocene, has* presented us with, or at least of reducing the risks involved.

7.12 Composite Structures: Thermoregulation with Energy Efficiency

Figure 7.35 shows the routine circuit composite structures for climate control.

Nature has mastered the technique of thermoregulation perfectly through its varied composite layer structures, often together with other regulatory principles. The biological example of concentrically arranged onion skins (Sect. 4.6) is only one of many examples of organismic adaptive thermoregulation, which contributes to survival in all regions of the earth, from the ice-cold polar regions to temperate zones and hot deserts.

People have realized that such product technologies can also be beneficial for us, as the clothing industry, among others, has been demonstrating for years. But first, let's take a brief networked look at another potential technical application for thermoregulation.

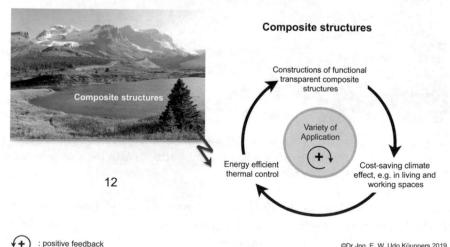

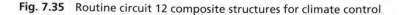

Fig. 7.35 Routine circuit 12 composite structures for climate control

Not much actually needs to be said about the broad relevance of packaging technology with regard to our human packaging technology and the associated packaging materials. Packaging is not limited to cardboard folding boxes, plastic films or tin cans for food. If you look beyond this, you will see that, in the broadest sense, house walls, roofs, means of transport and so on can also be defined as stationary or mobile wrappings or packaging. Why should they not also be able to make use of the packaging advantages of the universal onion packaging, i.e. its physical and chemical properties in terms of energy efficiency, heat storage capacity or resistance to microorganisms? For example, through composite forms of flexible, light-permeable and heat-insulating membranes, or multi-layered house wall constructions, with the physical property of balancing out climate fluctuations in an energy-saving way, in the sense of uniform, pleasant living throughout all seasons.

The systems thinking postulated in Sect. 7.11 is also the driving force behind these suggestions for "new" bionic developments for practical application, for innovative, sustainably efficient solutions.

The textile and clothing industry has long recognized the effectiveness of the *onion-skin principle* for itself: the textile industry in general and the "outdoor" textile industry in particular. In an article from 2013, this is discussed in detail.[28]

> The onion principle refers to the layering of clothing during outdoor activities. However, in my opinion, the traditionally known and taught onion-skin principle is not optimally adapted to outdoor activities that are more demanding than walking.
>
> Choosing the right clothes is basically about protection from the elements like rain, snow and wind. Clothing should be selected according to this principle. The second principle is especially important when exercising: the physical effect of evaporative cooling. When water evaporates on a surface, energy is released. This leads to a cooling effect. It is now important to use this effect intelligently. In summer, moisture or its evaporation can help to cool the body to the optimum operating temperature. In winter, of course, this cooling effect should be avoided at all costs. In short: In summer, sweat can and should cool the body, but in winter this can have dire consequences. With the onion principle you can control these effects. [...]
>
> The conventional onion-skin principle is based on the idea that several layers of functional clothing are worn on top of each other. Adapted to the climate and the weather, different layer thicknesses and lengths are combined, i.e. from the

---

[28] https://www.aufundab.eu/outdoor-kleidung-das-optimierte-zwiebelprinzip (accessed 2020-01-28).

inside out sweat-wicking underwear (baselayer), on top of this an insulation layer if necessary, and then on top of this a breathable and/or waterproof layer (hardshell or softshell – depending on the weather either with a waterproof membrane or only windproof material).

In principle, this layered look is a good basis for walking. However, it needs to be considerably fleshed out and carefully adapted for sweaty outdoor applications such as hiking, mountaineering, running or trail running. In this article, we address other important points to consider when choosing functional clothing. Often too much clothing and too warm clothing is taken along. This quickly adds up to kilos that are never worn anyway or worn incorrectly. Who doesn't know them: the perfectly equipped, the summit stormers wrapped in fancy black softshells, who struggle through the midday sun with their heads full of red and sweating. I confess, I tend(ed) to do this myself. You pack your greatest fears and even sacrifice a certain exhilaration and naturalness to these fears along the way.

This bionic example of textile thermoregulation over the seasons proves once again that a systemic approach is needed to make optimal and adaptive use of the advantageous principles of nature – in this case, energy-efficient thermoregulation for organisms with different purpose-bound uses. After all, polar bears in the Arctic and desert jackals in the Namib also engage in thermoregulation, but each in their own adapted way.

7.13   Shells and Casings: Spatial Technospheric Construction Principle Free of Supporting Elements

Figure 7.36 shows the routine cycle 13 Routine packaging trays and sleeves.

Packaging in nature and technology can be described without great modesty as universal cross-sectional products of the biosphere and technosphere. There is hardly an organism or a technical product that can exist without some kind of packaging as an interface to the environment, or that is used by us over a longer period of time. In Sect. 7.12, we briefly discussed packaging in relation to thermoregulation. Here we want to take a closer look at packaging shells and wrappings as load-bearing, space-spanning natural models for technical construction solutions, without being able to even begin to show the wide range of natural models.

The idea of using hazelnut shells as bionic models for small and large rooms or halls, without additional supporting structural elements, is based on their

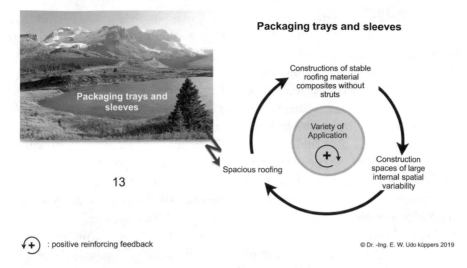

Fig. 7.36 Routine circuit 13 packaging trays and sleeves

adaptively loadable structures and geometries, which save material in the structure where it does not necessarily contribute to the shell's stability, but may take on other functions, such as in the hazelnut the peripheral circular cavities through which nutrients flow. On the other hand, SEM images[29] of the cross-section of a hazelnut shell show the optimal distribution of stresses by alignment of the cell structure. It is unidirectional where the life inside the shell needs to be protected, transitions to a mixed cell orientation in the middle of the shell, and then multidirectionally aligned at the shell periphery, where the nutrient guidance channels worth protecting are located, to optimally ensure nutrient transport. In addition, there is the design feature of an adaptive shape, which is constructed as a functional interface to the environment.

The whole is – as everywhere in nature – system-optimized, adaptively tailored to the respective survival purpose (Küppers and Heyser 2004/2005).

Technical cross-room building constructions do not have such detail optimizations from a holistic point of view. Individual trades work hand in hand. Nevertheless, systemically not considered risks show suddenly occurring consequences, with partly small and large catastrophic effects, such as the collapse of the ice rink roof in Bad Reichenhall in 2006.[30]

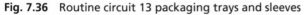

[29] SEM = Scanning electron microscopy imaging with nanometer-scale resolutions ($<10^{-9}$ m).

[30] https://www.sueddeutsche.de/bayern/eishalle-in-bad-reichenhall-das-begleitet-einen-immer-1.2802588 (accessed 2020-01-29).

Many structural defects only become apparent after the accident has occurred (aftercare principle). A necessary systemic, networked construction analysis in advance (precautionary principle) often fails because of money. The consequences are well known.

Shells and casings of organisms are perfectly adapted to their survival. Their functionality and shaping adapts with environmental changes in habitats. We currently find in nature the best results of natural techniques and qualities that have emerged over time. Given our inadequacies in dealing with nature technology, which includes frightening facts of Anthropocene impacts, the call for alternative and sustainable solutions for our survival is growing louder. Nature is an incomprehensibly vast treasure trove of new sustainable solutions and *Systemic Bionics* can help put them into action for our progress.

Finally, direct your attention to the two shells of and macadamia and hazelnut in Fig. 7.37, which are of almost the same spherical shape but differ in compressive loading by a factor of 5! What structural secrets does the macadamia nutshell hold that it comes up with five times the stresses of a hazelnut shell?

With a view to bionic applications, it could be concluded that arch structures or similar building structures, which use analogous structural secrets of the macadamia shell or other natural packaging, could lead to significant material savings at comparable component loads compared to conventional components.

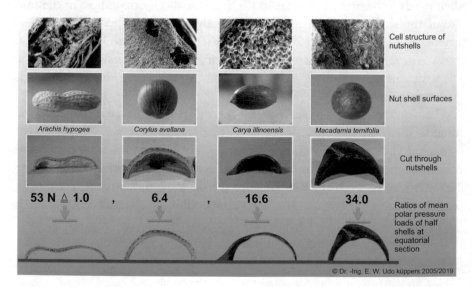

Fig. 7.37 Four biological packaging shells as models for bionic robust shell designs

In addition to this obvious bionic perspective for hall components with shell geometries, a completely different field of application in the social environment of accommodation for millions of asylum seekers and displaced persons from Africa and the Near and Far East could also enter the field of application.

For the majority in the refugee camps on the Greek islands, such as the Moria camp on Lesbos, catastrophic living and working conditions have persisted for years, despite international aid, through e.g. overcrowded, leaky, uninsulated tents as makeshift accommodation, patched-together unstable wooden pole grids with makeshift tarpaulins as roofed rain protection (Papadopoulos 2020, p. 10) and the like. Stable shell structures, with the help of the new manufacturing technology of *additive manufacturing*, could be used to relatively quickly and easily mass-produce stationary living or working areas that are more than makeshift shelter and that maintain the minimum level of human dignity. The stability and functionality of shell housing shelters, as indicated by Fig. 7.38, which can be combined in any way, is not so much the problem as their political and economic provision and organization for such quarters, as can be seen at present on the Greek island of Lesbos.[31] (see also "Greece is sealing itself off",[32] Küppers and Küppers 2015).

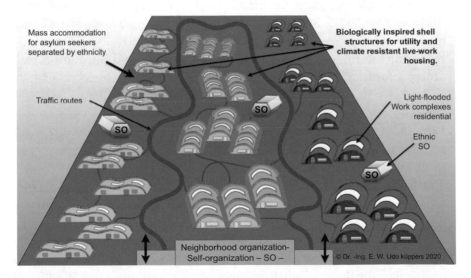

Fig. 7.38 Vision of temporary bionic shell housing in mass accommodation for asylum seekers

[31] https://www.spiegel.de/fotostrecke/lesbos-lager-moria-nach-dem-feuer-fotostrecke-141244.html (accessed 2020-01-29).

[32] https://www.dw.com/de/griechenland-schottet-sich-ab/a-52195054 (accessed 2020-01-31).

Practicable solutions developed from scientific disciplines, however useful they may appear to society, still require the approval of politics and science. Anyone who therefore believes in the independence of science from politics and business is thinking very short-sightedly.

7.14   Low-Friction Slideways: Routine Circuit 14 Low-Friction Slideways

The journeys in urban subways are accompanied by sometimes deafening noise, especially when the subway railcar together with the trailer car drives in curves. This noise is generated by tribological forces,[33] which act on the vehicle's wheel-rail systems. Enormous energy consumption is also coupled to frictional forces that occur. Holmberg and co-authors studied energy consumption from four societal sectors – energy production, manufacturing, construction and transportation, with the following summary result (Holmberg and Erdemir 2017, pp. 263–284):

- In total, ~ 23% (119 EJ) of the world's total energy consumption originates from tribological contacts. Of that 20% (103 EJ) is used to overcome friction and 3% (16 EJ) is used to remanufacture worn parts and spare equipment due to wear and wear-related failures.
- By taking advantage of the new surface, materials, and lubrication technologies for friction reduction and wear protection in vehicles, machinery and other equipment worldwide, energy losses due to friction and wear could potentially be reduced by 40% in the long term (15 years) and by 18% in the short term (8 years). On global scale, these savings would amount to 1.4% of the GDP annually and 8.7% of the total energy consumption in the long term.
- The largest short term energy savings are envisioned in transportation (25%) and in the power generation (20%) while the potential savings in the manufacturing and residential sectors are estimated to be ~10%. In the longer terms, the savings would be 55%, 40%, 25%, and 20%, respectively.
- Implementing advanced tribological technologies can also reduce the CO_2 emissions globally by as much as 1460 MtCO_2 and result in 450,000 million Euros cost savings in the short term. In the longer term, the reduction can be 3140 MtCO_2 and the cost savings 970,000 million Euros.

---

[33] Tribology is the science of friction, wear and lubrication in bodies moving against each other, as is the case with moving subways on rails.

- (Comment by the author: "EJ" = exajoule, energy unit, 1 EJ = 10^{18} joules or 10^3 petajoules (101^5 J), for comparison: primary energy consumption in Germany in 2018[34] was 12,900 petajoules, GDP equals GDP, gross domestic product).

The essential statements from the study by Holmberg and Erdemir are:

- Tribological contacts account for around 23% of global energy consumption.
- Wear-reducing measures can reduce energy loss by 40% in the long term (over 15 years) and by 8% in the short term (8 years). Transport can contribute 25%, energy production 20% and construction 10%.
- Advanced tribology techniques can also significantly reduce global CO_2 emissions.

The savings effect in the long term is in the upper three-digit million euro range.

This is where looking to nature and its friction or friction-minimized surfaces comes into play. Figure 7.39 shows the routine circuit 14 Low-friction slideways.

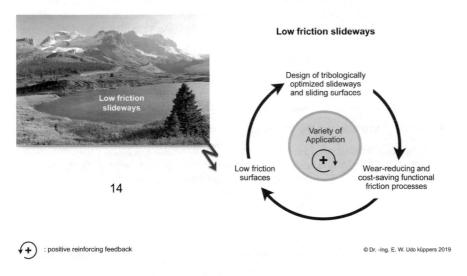

(+) : positive reinforcing feedback   © Dr. -Ing. E. W. Udo küppers 2019

Fig. 7.39 Routine circuit 14 low friction slideways

---

[34] https://www.umweltbundesamt.de/daten/energie/primaerenergieverbrauch#entwicklung-und-ziele (accessed 2020-01-29).

Moving a body against another body, or moving a body against air, water, or through earth or sand requires force. This is no different for organisms than for technical bodies such as vehicles, aircraft or ships. We know from organisms that in the course of evolution they have adapted by special techniques to the friction or frictional force that is directly effective during movement and acts in the opposite direction to the body's direction of movement, not least in order to save energy.

Four examples from nature and their bionic solutions will illustrate this friction-reducing energy-saving effect.

Example 1 Surface of Sharks (Selachii)

Sharks have a scale-like structure on the surface of their skin that feels like sandpaper when brushed over. These special tooth-shaped structures reduce the occurring and resistive current turbulence on the body surface in the fast-swimming fish. So-called "riblet" structures are able to reduce the frictional resistance in turbulent boundary layers by about 10% compared to smooth surfaces. Riblet structures are U-shaped grooves that run in the direction of flow. First investigations on artificial sharkskin structures were carried out at the TU-Berlin (Bechert et al. 1986; Bechert and Reif 1985).

In principle, the application of bionic structured films to aircraft wings, modelled on sharkskin, to reduce the frictional resistance of an aircraft works well. However, the completely different boundary conditions of flying, such as temperature resistance (from +30 to +40 degrees near the ground and – 60 degrees at altitudes of 10,000 m) as well as UV radiation exposure, still hinder the permanent use of friction-reducing films.

Example 2 Surface of Dolphins (Delphinidae)

Like sharks, dolphins are fast swimmers in turbulent currents. Their special trick to minimize the frictional resistance of the body in an energy-saving way is that the skin structure has a so-called vortex damping system. The plasticity of the skin causes a reduction in vortex energy, which in turn saves propulsive energy. This energy saving effect was first discovered by Kramer, in the mid 1950s (Kramer 1960). Heinert (1976, p. 74) describes Kramer's technical investigations at that time on an artificial dolphin skin as follows:

In 1956, Kramer conducted experimental investigations on underwater vehicles and cylindrical underwater bodies, using an artificial dolphin skin as the outer shell of the underwater bodies in order to transfer the biological phenomenon of boundary layer stabilization to technical objects. Kramer did this without directly copying the structure of the dolphin skin, but attempted to technically implement active principles of turbulence damping. The result of comparative model tests without the artificial dolphin skin envelope was a 50% reduction in frictional resistance with this artificial skin.

Example 3 Surface of Sandfish (Apothecary Skink, (Scincus Scincus))

With its scale-like surface of the up to 20 cm long body and its pointed snout, the slender reptile slithers through the desert sand at lightning speed, as if frictional resistance did not matter. In fact, however, the sandfish has developed an energy-saving method of locomotion in its environment that is unparalleled in technology. On its scale-like surface, the sandfish has micrometer-sized transverse grooves that are able to dissipate the electrostatic charges on the crystal-hard scales that occur when it wiggles through the sand (Rechenberg et al. 2009, p. 14, BMBF project). More significant, however, is the comparative result of sliding friction measurements on an inclined plane on sandfish and technical materials, such as glass, polished steel, aluminium, nylon and Teflon (ibid. pp. 5–9). On average, the critical sand friction angle for the sandfish was 21 degrees, while all engineering materials were above this. Here, too, the special organism surface masterfully demonstrates its ability compared to technically comparable surfaces.

Example 4 Surface of Pitcher Plants *(Nepenthes)*

Pitcher plants love meat, mainly from insects (see Sect. 4.8). Attracted by the sweet nectar, the insects sit on the edge of the plant, which is native to the tropics, and not infrequently slide down along a slippery wax-layer sliding surface, where they are decomposed and digested. This sliding zone of pitcher plants is considered to be at least one of the most slippery or low-friction surfaces in the plant kingdom.

A fifth and certainly more examples from the plant kingdom of low-friction surfaces are present, including that of the Lotus Effect® mentioned earlier in Sect. 4.3.

Like so many surfaces in the animal and plant kingdoms, the pitcher plant has structures perfectly adapted to its survival strategy, often on a micrometer and nanometer scale. It would be a mistake to believe that low-friction surfaces in nature are extremely hard and smooth as glass.

Technology knows various means to reduce friction and thus wear on technical, often metallic surfaces that move against each other, by greasing, oiling, polishing, coating. However, the systematic manufacturing technology of micro- and nanostructuring of surfaces to reduce tribological effects is still largely in its infancy. It must also be taken into account that in a bionic solution, the diverse biological techniques of energy-saving friction reduction are often only effective in one direction, while e.g. technical sliding and rolling friction, or a combination of both, also perform alternating directions of movement. If, for example, the running surface of a microstructured rail track, as in the case of a U-track, should want to effectively reduce friction losses, the microstructures would have to be equally effective when the track moves forwards and backwards. However, this change of direction is not provided for in nature – at least for the four examples mentioned above – or there was no evolutionary pressure for it. However, the lotus leaf surface shows universal, direction-independent rolling friction with the effect of surface self-cleaning by rolling water drops and other partly highly adhesive substances.

Another – not insignificant – aspect of bionic, micro- or nanostructured surfaces for friction or wear reduction is that of effective pressure forces on the friction or sliding pair. Under certain circumstances, bodies weighing several tons move against each other or travel over stationary rails. Where are the limits of micro- or nanostructured strength under different dynamic load cases before failure occurs?

So there is still plenty of potential for development, both for new knowledge from nature and for technically efficient applications of natural principles.

7.15 Specialised Dynamic Folded Constructions: Functional Technical Folded Bodies with Space-Saving Effect

How does a thumb-sized plant bud develop into a 4–6 times palm-sized rhubarb leaf? Even more impressive is the size ratio between banana plant bud and adult banana leaf, which reaches a length of up to 3 m and a width of up to 0.6 m. Finally, through the size ratio between plant bud and adult leaf of the aquatic plant *victoria amazonica*, with a diameter of up to 3 m and a carrying capacity of up to 80 kg, nature shows us its great leaf masterpiece among all the smaller masterpieces of plant leaves.

Figure 7.40 shows the routine circuit associated with bionic folding techniques.

Special folding techniques are involved in all the design developments of these plants, from the small bud to the fully grown leaf surface. For years, folding technology has been using these natural models for new folding constructions, especially where space and weight-saving constructions are required, such as in satellite technology or space travel.

But also terrestrial applications use folding techniques as a means of hiding large areas in small envelopes, in order to let these unfold again into

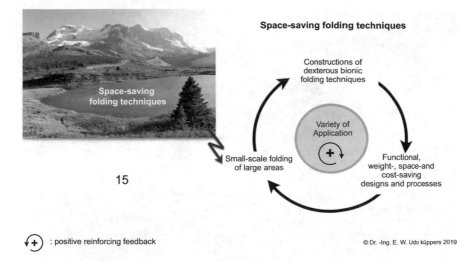

Fig. 7.40 Routine circuit 15 space-saving folding techniques

large-scale constructions when used for a specific purpose. Meter-sized flexible folding constructions can be found, for example, in the area of trade fair construction, or special folds in residential construction or furnishings. The uncomplicated folding and unfolding of street maps also goes back to the so-called Miura folding technique.

Recall the spatial geometry of the folds of rhubarb leaves as seen in Fig. 4.15 in Sect. 4.9. The spatially stable folding mountain of the leaf results when three mountain edges and one valley edge interact, which was very nicely seen in Fig. 4.15.

It was the Japanese origami artist Koryo Miura who used nature's folding technique for engineering purposes with his origami diamond structure, also known as Miura folding or Miura-ori, for example for retracting and extending solar sails on satellites, as shown in Fig. 7.41.[35]

It is little surprise to bionists that folding techniques spread not only in animate but also in inanimate nature and that all types of folding show similar structures. Figure 4.15 shows the folding of a rhubarb leaf, and Fig. 7.42 shows Miura folding compared to chaotic folding on a paper leaf, whose mountain-like structures are the spontaneous result of a leaf being randomly crumpled up and unfolded again. All three folding results (rhubarb leaf, Miura fold, chaotic fold) show the same typical geometries required for spatial folding: three mountain edges and one valley edge. Evolutionary folding is one result, sophisticated engineering folding is the second, geometric folding

---

[35] https://web.archive.org/web/20051125174630/ http://www.isas.jaxa.jp/e/enterp/missions/complate/sfu/2dsa.shtml (accessed 2020-02-02).

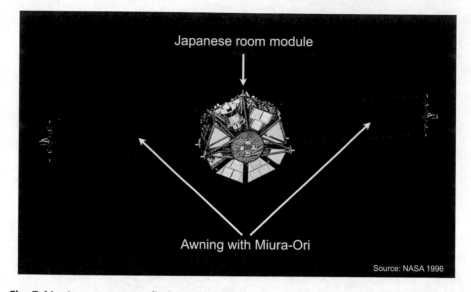

Fig. 7.41 Japanese space flight module with solar sails using a Miura fold. Image file in the public domain, description: The Japanese Space Flyer Unit photographed from the approaching space shuttle Endeavour during mission STS-72, date: January 1996. (Source: National Aeronautics and Space Administration – NASA)

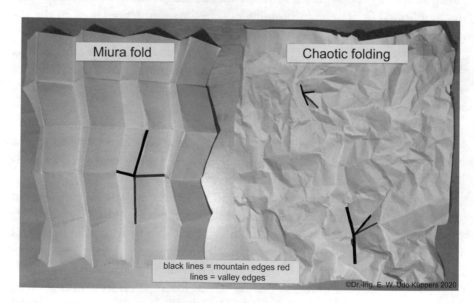

Fig. 7.42 Miura folding compared to chaotic folding

laws arising spontaneously from chaos are the third. Obviously, spatial foldings – wherever they take place – are subject to universal laws.

Another type of folding that spirals around technical bodies and for which organisms such as state jellyfish are models for the bionic solution of volume control is shown in Fig. 7.43. State jellyfish belong to the class of hydrozoans. The species shown in Fig. 7.43 belongs to the physophore group. Its constructive structure is remarkable. Like many jellyfish species, physophores move forward by special pumping techniques. At the "head end" of the jellyfish (in the picture on the left) there is a so-called pneumatophore, a bladder filled with gas. This in turn is connected to a system of interconnected, transparent tubes, the so-called swimming bell.

The gas from the pneumatophore is admitted to the floating bell and returned to the "bubble reservoir" to fill and empty the floating bell, which alternately expands and contracts. By reducing the internal volume of the floating bell, a recoil is generated which provides the propulsion.

In terms of folding technology, of interest in the case of packaging is, in addition to the tear resistance of the transparent underwater envelope, in particular the aerodynamic control technology for growing and shrinking packaging volumes. See Küppers (2002a, pp. 33–68 and Küppers 2002b, p. 75 f.).

However, chamber packaging which, for example, adaptively reduces its volume after the removal of packaged goods and thus independently generates a minimum transport volume on the way to disposal or recycling, are still dreams of the future for the dominating throw-away mentality of our technical packaging industry and packaging disposal, which in the meantime leaves its "calling card" in six huge sea whirlpools of packaging plastics! How much longer?

Source: Dr.-Ing. E. W. Udo Küppers 2002, own archive

Fig. 7.43 State jellyfish with pneumatically controlled packing volume. (Source: Own picture archive)

Combined bionic dynamic printing and folding techniques can be used to develop innovative and sustainable packaging and packaging techniques that were previously unknown and which give hope that packaging problems with further consequential problems will be solved better than has been done to date. Their acceptance and application is not exclusively a problem of material technology or machine technology. Last but not least, it is also a social problem, which on the one hand involves problem-preventing (here: Transport volume and packaging weight reducing as well as waste avoiding) innovations and on the other hand the avoidance of a further increase of our packaging waste volume. It is indisputable that plastic packaging plays a major role in this anthropocene merry-go-round of substances and materials.

Figure 7.44 shows two bionic variants of variable volume control of plastic packaging, on the one hand by organismic folding technology on a microscopic scale, for example in protein folding, and on the other hand by clever combination of pneumatic pressure and folding, as in state jellyfish.

The abundance of biological models as long-term proven and ingenious experts of folding technology is immeasurable. Architecture has long benefited from this through fancy spatial body constructions using nature's folding principles (Nachtigall and Pohl 2013, pp. 173–178, among others).

An initial résumé for bionic folding and packaging, whereby the concept of packaging goes far beyond the technical teaching and research and

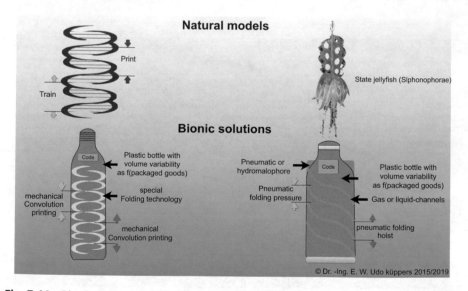

Fig. 7.44 Bionic pressure and folding technology of volume-controlled plastic packaging, left: volume-controlled dynamics of folding (biological model protein folding), right: volume-controlled dynamics of folding (biological model state jellyfish)

development discipline of packaging a wide variety of packaged goods, could include the following points:

1. The production technique of natural folding is embedded in complex life processes of organisms. Therefore, biological folding structures cannot be classified as pure construction features or construction types following the model of technically standardized classes.
2. Biological organisms live, they grow and shrink and the associated folds with them. Technical packaging is manufactured for specific volumes of packaged goods. In the case of relatively rigid packaging, the volume remains the same even after the packaged goods have been emptied. This is, by the way, a highly disadvantageous consequence-cost-intensive feature in the disposal of packaging.
3. From 2 it follows therefore: Foldings in organisms are not necessarily guided. Due to the elastic material properties, the folds adapt optimally to the local loads. For technical packaging, this built-in trick would be of enormous advantage. A striking example is the so-called "crumple zone" of passenger cars, which elastically absorb the enormous pressures in a collision in a fold-like manner.
4. Folds connect large surfaces with the smallest space packages. Important for volume-saving processes such as in aerospace technology.
5. Folds stabilize. Unlike unfolded, flat structures of the same size and shape, they absorb greater external forces (pressure, tension, bending, shear). The striking folded roof – affectionately known as the lemon squeezer – of the Dresden Art Academy is one of many examples here of the building art of folded roof structures.
6. Protect folds. Skillful folding around growing offspring protects their life and thereby ensures propagation and preservation of the species. For technical folding, see also point 2.
7. Folding processes are influenced by external stimuli. In nature, folding processes are sensor-controlled and part of biocybernetic control circuits. Dynamic volume-controlled/regulated folding is – for whatever applications – still a challenge for engineers.
8. Irregularly regular folding structures often arise in nature with the participation of self-organised processes. For processes of self-organization in the technosphere in general and in folding technology in particular (see Fig. 7.42), a wide field of research, development and application is still available.

9. Mechanical folding in nature is often associated with another chemical or physical technique, e.g. aero- or hydrodynamic pumping or suction. Target for innovative technospheric solutions.

As we have seen, folding fulfils a wide variety of tasks in nature. Folding is also not new to our packaging technology and beyond its disciplinary boundaries. We know millions of folded (folding) boxes with different contents from our daily dealings, e.g. with technical packaging for food. And yet nature still uses and conceals many secrets of folding, often in combination with other proven packaging techniques or principles of material-saving handling or energy-efficient production. Perhaps it is precisely the clever combination of individual packaging techniques that holds the key to new, bionic innovations!

7.16 Window Plants: Stable Living and Working Places in Local Environments Influenced by Extreme Climates

Figure 7.45 shows the circular path of a routine process that proposes bionic protective measures to provide more precautionary protection to populations in areas of extreme weather against the destructive forces of nature.

From Sect. 4.10 we know the ingenious tricks of desert plants to protect themselves against the sun's merciless rays and desiccation by hiding most of their bodies below the surface and only receiving light through small dome-shaped openings. In doing so, plants have developed special water-storage techniques, as already mentioned (see Sect. 4.10). These organisms are well adapted to the dry – arid – environment. Light is collected through the small curved window openings, which act like *Winston collectors* (internally mirrored paraboloid funnels) and therefore do not have to follow the course of the sun. Helmut Tributsch carried out solar experiments with the window plant *Fritia pulchra* and proposed as a bionic building concept an underground multi-storey desert building, which has not been realised to date and which is modelled on the working principle of the window plant (Tributsch 1995). Figure 7.46 shows on the left the biological model and its technical imitation.

A new functional approach to the construction of underground living and working spaces powered by solar energy, which can also use the technology of the Winston collector to build self-sustaining units for the population, is the protection against the destructive forces of storms and floods in arid and tropical areas. The consequences of these extreme weather longs are evident in eastern Africa (Mozambique, Malawi, Ghana, Madagascar), or in the tropics

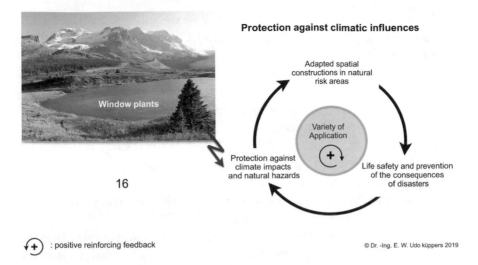

Fig. 7.45 Routine circuit 16 protection against climatic influences

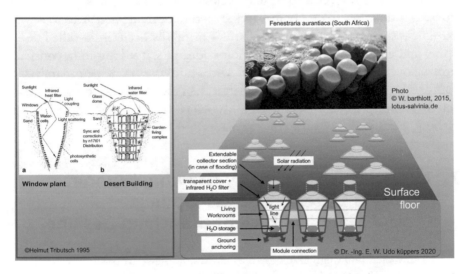

Fig. 7.46 Window plants as a biological model for underground living complexes, left, and as a protection against extreme weather and its potential destruction of aboveground living and working areas, right

(such as Haiti,[36] which has been hit by cyclones several times), with total destruction of mainly light huts and other unstable dwellings. The Climate Risk Index from 2016 by Germanwatch[37] warns: "Weather extremes have claimed 530,000 lives and trillions in property damage worldwide since 1996."

---

[36] https://www.spiegel.de/gesundheit/diagnose/hurrikan-matthew-in-haiti-ein-jahr-nach-dem-tropen-sturm-a-1171156.html (accessed 2020-02-03).

[37] https://germanwatch.org/de/13028 (accessed 2020-02-03).

What would be more obvious than to follow the biological path of a window plant, whose bionic functional characteristics would be, with the billions of temporary aids for the reconstruction of similar unstable buildings, which are always announced after a destruction, but flow tenaciously:

- Protection against destruction by storms or hurricanes, because only so-called dome-shaped (if necessary telescopically extendable in case of floods) light windows are in contact with the surface of the earth
- Temporary self-supporting living and working units
- hydrophobic construction materials
- Underground interconnected systems and small-scale living and working units, similar to Fenestraria (Figs. 4.16 and 4.17)
- Stable building or shell constructions through new additive manufacturing techniques on site, ideally with a high proportion of renewable raw materials

Instead of being subjected to regular cycles of expensive reconstruction measures, as is the case after every major destruction of residential areas following extreme weather situations, which – still – primarily occur in so-called developing and emerging countries, the bionic idea of stable long-term construction below the earth's surface could withstand the immediate dangers of approaching climate extremes much better than the often unstable and endangered dwellings in the relevant regions, which are directly exposed to the extreme weather.

7.17 Dynamic Stability by Growing Without Stress Extremes: Stress-Optimized and Material-Efficient Branching Designs

Half-timbered buildings, also known as trusses, frameworks or studs, are skeleton constructions in which the entire loads and forces are taken over by load-bearing timber, while wall parts that close off the room but are not load-bearing form the wall end.

Understanding the development of timber frame structures and timber frame construction requires knowledge of the basic wall constructions with wood. Taking all preliminary stages into account, there are only three different types of wall construction in timber construction worldwide: block construction, stick construction, half-timbering.

All variations and combinations of these basic wall structures, from post-and-beam construction to modern timber-frame construction, are based on one of the three basic constructions.

This is how Manfred Görner begins his description of half-timbered buildings in the chapter "Fachwerkgefüge und Fachwerkentwicklung in Deutschland" (Gerner 2007, p. 9).

A modern truss is a skeleton structure made of standardized, machined wooden beams. Horizontal, vertical and diagonal beams are connected to each other in such a way that they act as load-bearing elements providing the building with basic stability against wind forces. The free, non-load-bearing areas between the wooden skeleton construction are filled with various materials in order to make interior spaces thermally insulating or comfortable against external dynamic weather influences.

Trees are stationary organisms whose trunks are supported by a network of roots in the soil and unfold upwards to form a sprawling biological framework. From a mighty round trunk near the ground, a multitude of different branch ramifications and varying branch thicknesses lead upwards and, together with leaves, form the canopy of a tree.

The supporting branches of the tree are arranged in such a way as to avoid, according to the principle of energy efficiency, any superfluous tension or stress peaks that could lead to breakage.

Claus Mattheck has analysed tree growth strategies in several studies (see Sect. 4.11) and worked out the *axion of constant stress*. According to this, the tree organism adapts in such a way that where potential fracture lines or high stress peaks could arise, e.g. in branch forks, more tree material is settled for stability and where there is no danger of excessive stress, tree material is saved.

The organism tree seems to be permanently searching for the best stability criteria in a dynamic environment.

This permanent search for the best energy-saving structures against overloading cannot, of course, be performed by a technical framework once it has been created. But new technical constructions can learn from the dynamics of trees with their stress-optimal branching, size ratios of branched branches and much more. Figure 7.47 gives the routine cycle for such a process flow, which can be mapped to a variety of applications.

The tree growth strategy of the trees mentioned in Sect. 4.11 and the axiom of constant stress derived from it were transferred into a *Computer Aided Optimization method – CAO –* a simulated adaptive growth in the computer, so to speak.

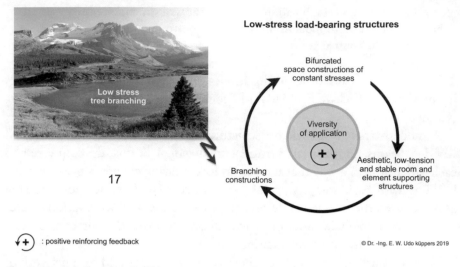

Fig. 7.47   Routine circuit 17 low-stress supporting structures

With the CAO method, additional material is deposited at points of the computer-generated tree model that are subjected to higher loads, thereby reducing stress peaks and increasing the load-bearing capacity and service life of the component to be optimized – in this case a fork branch (Fig. 7.48, left). Figure 7.48 shows the CAO method on the right, using the example of a pedicle screw (self-tapping screw for use in spinal surgery) that is sensitive to stress and cracks. This practical example of surgical procedures shows an enormous increase in the durability of the screw and thus the service life of the implanted product as a result of the CAO method (Mattheck 2010, pp. 24–26).

The CAO method was supplemented by the *soft kill option* method – SKO – which simulates technical lightweight constructions based on the model of bone growth. Where material is subjected to low stress, material is reduced until it is removed. Starting from a loaded initial component as a design proposal, a lightweight truss structure made of tension and compression bars of adaptive stress profiles finally develops (Fig. 7.49, left). Figure 7.49, right, shows the SKO progression using an industrial rocker arm as an example. Both results show a significant peak stress reduction, which increases the service life of the lightweight structure (Mattheck 2010, pp. 27–28).

Finally, both methods of *CAO* and *SKO* are further extended by the *CAIO method,* which calculates the optimal fiber flow in technical components and thereby increases the load capacity (Mattheck 2010, pp. 29–30). See also further literature (Mattheck 2017).

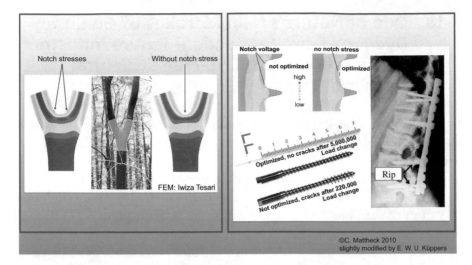

Fig. 7.48 CAO method in use. (With kind permission of C. Mattheck)

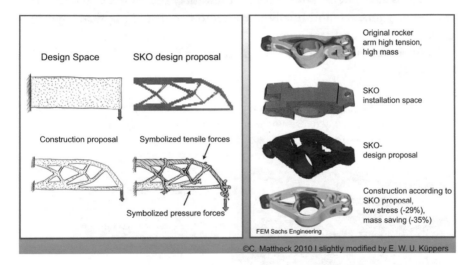

Fig. 7.49 SKO method in use. (With kind permission of C. Mattheck)

All the above-mentioned simulation methods based on biological models have been used for years in various industrial sectors, mainly in automotive engineering (Harzheim 2008). As bionic methods, the adaptive optimizations not only show their ability in technical lightweight construction. They thus prove not least the long-term proven ingenious design principles of nature, which are driven by energy- and material-efficient solutions, but also by sustainable nature-compatible material utilization. The latter is still a huge problem in technology, economy and society. Of course, nature also has suitable solutions for this, which still need to be recognised and adaptively applied in a bionic way.

7.18 One of Nature's Many Strokes of Genius: Dye-Free Paint Production! Future Industry of Material-Saving Dye-Free Structural Dye Production

Let's construct a thought experiment with the following question: How do we manage if all material colors we know no longer exist, but we are still able to perceive colors with our eyes?

The solution is of physical origin and nature – who else – has included it in its rich repertoire of long-proven principles. The trick of the morpho butterflies to shine steel-blue without the aid of the smallest pigment has already been explained in Sect. 4.11.

Figure 7.50 shows the circular path to effective and efficient color design through light refraction, rather than pigmentation.

But let us also take a digression into the variety of colours in nature, which uses colour pigments, and finally move on to the area of physical colour production in nature and technology.

Excursus on natural color production (Küppers and Tributsch 2002, pp. 73–74).

Technical products are printed with a myriad of dyes, many of which have questionable properties, such as synthetic dyes that cause significant damage to nature and the environment.

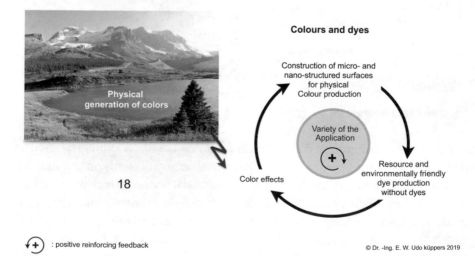

© Dr. -Ing. E. W. Udo küppers 2019

Fig. 7.50 Routine cycle 18 colors without dyes

Nature shows us that we could do without this to a large extent. It gets by with just a few environmentally friendly organic dyes (pigments) and achieves a variety of shades as well as the brightest colours via a purely physical structuring of the base material or a combination of structuring and pigmentation.

The pigments of living creatures are usually in the form of tiny droplets or granules incorporated into the skin, feathers or fur hairs. The most common pigment is melanin. It is responsible for black, brown, as well as many reddish and yellow colors. Examples are the red hair of humans or the yellow feather fluff of small chicken chicks. Magnificent red, orange or yellow is produced by carotenoid dyes. They get their name from the carrot, in which they also occur. Pink flamingos also adorn themselves with them. They absorb it from algae, which is part of their diet. Pterins are also used as dyes in nature. Pterins are molecules derived from uric acid. They are produced by metabolism and used as dyes in nature. They got their name from butterfly wings (pteros = wings) because that is where the molecules were first found. Carotenoids bound to proteins produce many blue and green colors. The dark blue of living lobsters is also produced in this way. If you cook them, the carotenoid splits off and determines the color itself. The green of many caterpillars and grasshoppers is produced by a mixture of blue and yellow pigments, so it has nothing to do with the plant pigment chlorophyll. Sometimes mixed colors are obtained by point distribution and mixing of two different pigment colors.

However, many colors on natural bodies are not produced by pigments, but by purely physical effects in the microstructured body shell. Small particles scatter short-wave, blue light more strongly than long-wave, red light. This is why the sky is blue and the setting sun is often red. The blue eye color in animals and humans is thus produced by light scattering off small protein particles against a dark melanin background. The phenomenon is called *Tyndall scattering*[.].[38] It is responsible for the piercing blue plumage of many birds including kingfishers and parrots. The brilliant green of many birds, snakes, lizards and frogs is produced by Tyndall scattered light modified by a yellow filter. So you need three layers for this. At the very back a dark melanin layer, in front the Tyndall scattering layer and at the very front the yellow filter. If everything works, the color is a bright green. If the yellow filter is missing, which sometimes fades away in the light, the color is blue. If the melanin background is missing, the color is again yellow. If the melanin layer is missing and the yellow filter layer is missing, the color is white. In some birds, there are all these color variations. In Malaysia and Indonesia lives a magpie, which is perfectly adapted to the forest with its green color. But if it moves its territory to the savannah, its plumage turns blue. This is because the sun bleaches its yellow filter. The white color in nature is

---

[38] The Tyndall effect, named after the Irish naturalist John Tyndall (1820–1893) describes the scattering of light by microscopic particles, of similar size to the wavelength of light, suspended in liquids or gases, i.e., floating in them.

caused by light scattering, when the scattering particles are too large to produce blue light. White hairs use cavities as scattering centers, and white butterflies use the highly structured surface of light-reflecting wings. Nature often produces shades of gray by mixing in melanin. Gray bird feathers often have white and black barbules, feather hairs, mixed in.

Iridescent colors in nature are created by interference of sunlight of different wavelengths on highly structured optical layers. A simple example are the colors of a gasoline film on water or the colors of soap bubbles. Depending on the thickness of the film and the length of the light wavelength, interference may or may not occur. Tiny platelets, of organic crystals like calcium oxalate, or of keratin, chitin, in the outer skin, in hair, in feathers produce the same phenomenon. The magnificent iridescent colors of hummingbirds, pheasants, sunbirds, butterflies, and beetles are produced in this way. Other iridescent colors, for example those of pearl shells, are produced by diffraction phenomena, much like the rainbow.

This brief overview shows how nature can produce its wealth of color design with a small palette of recyclable pigments, no heavy metals, but a wide range of optical artifices that require no chemistry.

Technical colours by physical effects instead of environmentally harmful dyes and additives already exist in various applications. Figure 7.51 shows some examples of these.

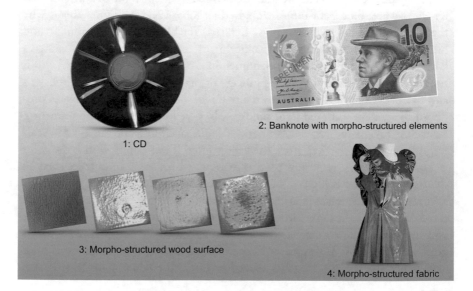

2: Banknote with morpho-structured elements

1: CD

3: Morpho-structured wood surface

4: Morpho-structured fabric

Fig. 7.51 Technical examples of "morpho" surface-structured products. (To 1: © Dr.-Ing. E-. W. U. Küppers 2020; to 2: In agreement with the Reserve Bank of Australia's guidelines (including clause 1); to 3: © Yi Liu et al. (2020, p. 7); to 4: © Donna Sgro, Morphotex Dress, https://www.uts.edu.au/about/faculty-design-architecture-and-building/staff-showcase/morphotex-dress and http://www.donnasgro.com/Morphotex-Dress (accessed 10.02.2020))

7.19 Refined Bonding in All Situations: Bionic Joining by Adhesive Bonding – The Economical Functional Manufacturing Technique

In Sect. 4.15 we saw how organisms in all regions of life have succeeded in developing special joining techniques that allow them to come into immediate contact, temporarily and permanently, with surfaces, however smooth or fissured, watery, oily or dust-dry they may be, without prior "surface cleaning". This is by no means a matter of course in the technosphere on this organismic scale!

Adhesive bonding is considered to be the most economical manufacturing technique, especially where there is the long-standing competition of welding, soldering, riveting, seaming and other joints (Fritz and Schulze 2015). The general goal in many engineering cases is the individual durability of joints of materials with different properties. In the automotive industry, whose products aim at durability and weight reduction at the same time, and in other industries as well as application areas, such as medical technology, adhesive bonding technologies are already in proven use as useful joining techniques (Industrieverband Klebstoffe e. v. 2016; Rasche 2012; Onusseit 2008; Brockmann et al. 2005; Habenicht 2005). The basis for the systemic circular process of bonding is shown in Fig. 7.52.

However, macromolecular plastics are still the dominant raw material for adhesive bonding in technology to this day. The wish of former Henkel employee Dr. Ramón Barcadit, 2009 Vice President Research of Henkel AG & Co. KGaA, Adhesive Technologies, Düsseldorf, a:

Bonding and debonding on command bonding technique, i.e. bonding and debonding on command, with easy machining! (Brickmann 2009, p. 32)

The development of sustainable, marketable adhesives seems to be a particular challenge up to the present and probably also in the near future. After all, Fraunhofer-IFAM is working on new techniques for debonding on command.[39]

---

[39] https://www.ifam.fraunhofer.de/content/dam/ifam/de/documents/Klebtechnik_Oberflaechen/Klebstoffe_Polymerchemie/entkleben_knopfdruck_fraunhofer_ifam.pdf (accessed 2020-02-06).

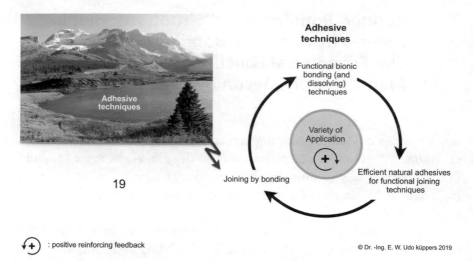

Fig. 7.52 Routine circuit 19 bonding

Organisms are known to have mastered this joining technique for millions of years and have constantly improved and extended it.

Another challenge is no less great, because the majority of all component adhesives as plastic formulations are hostile to nature because they cannot be disposed of in a sustainable and environmentally compatible manner.

Often, even the bonded joints between two connecting parts cannot be removed without leaving residues after use, so that they can be used again as joining material through a recycling cycle. As with any technical material, technical adhesives also have a number of advantages and disadvantages, which should only be pointed out here.[40]

Adhesive bonding technology has indeed made enormous progress in recent decades, in terms of the adhesive and the bonding technology itself and the material combinations to be joined.

However, from a holistic point of view, from the elegant cohesive bonding and unbonding of geckos to the permanent bonding in aqueous environments of barnacles, the superiority of natural adhesive bonds is still demonstrated in an impressive manner worthy of imitation, and all the more so as the raw material and material consequential problems affect the Anthropocene present and future (see Fig. 7.53).

---

[40] https://de.wikipedia.org/wiki/Kleben#cite_note-4 (accessed 2020-02-06).

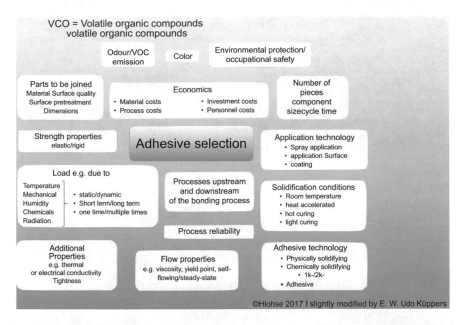

VCO = Volatile organic compounds
volatile organic compounds

Fig. 7.53 Adhesive selection. (Author Hlohse, created 18.10.2017, CC BY-SA 4.0. Source: https://de.wikipedia.org/wiki/Datei:Wichtige,_bei_der_Klebstoffauswahl_zu_berücksichtigende_Kriterien.jpg#/media/Datei:Wichtige,_bei_der_Klebstoffauswahl_zu_berücksichtigende_Kriterien.jpg (accessed 2020-02-06))

7.20 *Open* Packaging as a "Long-Term" Freshness Store: Packaging Trick Extends the Life of Packaged Foodstuffs

Figure 7.54 introduces the routine cycle of packaging, which should transfer the secrets of natural packaging in a beneficial way to the technical packaging industry. This is not an easy but urgently necessary path if we take into account the consequences of technical packaging for nature and the environment.

Section 4.16 made us marvel at what is probably nature's most universal and varied packaging and the packaging techniques associated with it: The egg. Similarly, rich natural packaging was revealed in fruits and nuts, which take on spherical shapes.

All of nature's packaging seems to exist in round shapes. – Evolution does not seem to have created any developmental pressure for angular wrappings, which are often found in technical packaging – there was simply no need for them!

Natural packaging, as biomass, is part of a cross-species network of inter-connected, material-processing cycles. No matter which packaging, which

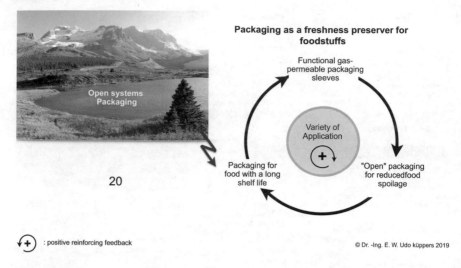

Fig. 7.54 Routine circuit 20 open packages

special materials are processed, they are all 100% returned to the natural cycle, where they are available again on a molecular level as packaging and for the life of new generations to be protected.

In this respect, packaging technology is still lagging far behind – in spite of the acknowledgeable progress made in detail (see Küppers and Tributsch 2002).

In the technical packaging sphere, other boundary conditions are present than in nature. Nevertheless, there are clear weaknesses in the course of technical packaging developments, starting from the raw materials through to disposal or recycling. The reverse side of consumer goods (e.g. food, cosmetics) and capital goods (e.g. cars, machines, mobile phones) advertised on the shelves of retail outlets, together with their elegant packaging, can easily be seen on the ramps and in the containers in the backyards of retail outlets. In addition to the well-known travel tourism, *waste tourism* – across all continents – has become a common term. The consequences of *waste tourism* are not only frightening in terms of material and can be observed daily. Further explanations along this path would clearly go beyond the scope of the book.

However, I would like to point out a general feature in the comparison between natural and technical packaging. It is hidden in the nanometre and micrometre fine structure and surface of packaging, which is not visible to the naked eye, but which makes the difference:

Biological packaging is an open system to the environment. Technical packaging is a closed system to the environment.

Packaging that provides functional protection for organisms over their lifetime is in permanent contact with the environment. This is already necessary in order to supply growing life with the necessary oxygen in the early stages of its development and at the same time to dispose of the exhaled carbon dioxide. This open packaging system can be seen particularly clearly in egg shells, whose calcareous hard shell ensures the respiratory activity of the life inside through direct connecting tubules to the external atmosphere, as Figs. 4.33 and 4.34 show.

According to the functional model of the ostrich shell were therefore:

1. Developed a flexible functional packaging film that is open to the environment.
2. Bionic experiments were conducted to confirm the principle of open packaging of nature, while preserving "life" – the technical goal was: long-term preservation of edibility or edibility of various fruits under general environmental conditions or room temperature conditions.

Figure 7.55 shows a scanning electron micrograph – SEM – of a bionic film surface and Fig. 7.56 shows an early specimen (2000s) from a series of PE film treatments as a packaging material open to the environment.

With the aid of these, various so-called "spoilage" tests were carried out, a selection of which can be seen in Figs. 7.57 and 7.58.

Details of the fabrication of the bionic flexible packaging film are described in Rojas-Chapana et al. (2005) and Fink et al. (2006).

The following two figures show start-finish states of two practical tests to determine time limits for the edibility or spoilage of fruit under different packaging. In the first test in Fig. 7.57, neutral glass containers were sealed against the environment by various films; in the second test, the same thing happened, except that one glass container was open to the environment.

- *Keep-fresh test or spoilage test 1*

Initial experience in keeping tomatoes fresh was gained in several tests, including at an international fruit and vegetable trading company in Bremen. Figures 7.57 and 7.58 show the results of these practical tests.

The boundary conditions for the first field investigation under summer room climate conditions (without air conditioning) in June 2004 (Fig. 7.57):

Start tomatoes fresh from the cold storage of a fruit trader:

Fig. 7.55 Bionic flexible packaging film made of functionally treated polyethylene – SEM image

Fig. 7.56 First generation bionic packaging film – functional PE film

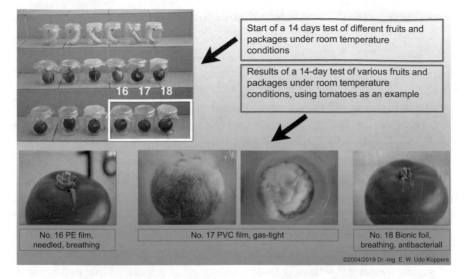

Fig. 7.57   Bionic packaging film in the spoilage test with tomatoes, apples and bananas

Fig. 7.58   Bionic packaging film in the spoilage test with plums

Tomato #16: PE film, needled, breathing
Tomato #17: PVC film, gas-tight
Tomato #18: Bionic film

Objective Result of the first fruit test with tomato (Fig. 7.57):

After 14 test days (24 h/d), slight fungal growth is visible under the PE film (No. 16) and clear fungal growth under the gas-tight PVC film (No. 17). The tomato under the bionic film (No. 18) shows no fungal growth at all, only dehydrated petals.

The bionic film in the practical test shows the expected result and confirms the prediction of long-term protection – here for tomatoes – under room temperatures. Likewise, the effect of repelling microorganisms – very clearly in contrast to No. 17 (PVC film) – is confirmed in the best sense. In addition, the period of consumption is important for the storage of food, even without refrigeration. After 14 days, the consumption of tomatoes packed with the bionic film still seems possible.

• *Keep-fresh test or spoilage test 2*

A further packaging test with the bionic film was carried out in 2006 by an external company, also under room conditions and a test period of 14 days throughout. This time it was fresh fruit from a plum tree that was externally undamaged. The boundary conditions for the external field test under room climate conditions (without air conditioning) in October 2006 (Fig. 7.58):

Start plums freshly picked:

Plum A (left): Glass container covered with gas-tight PVC foil
Plum B (center): Glass container covered with bionic foil
Plum C (right): Glass container without cover

Objective Result of the external fruit test with plums: after 14 test days (24 h/day), plum A shows clear thread-like fungal attack under the gas-tight PVC film due to lack of oxygen. The fruit in the right-hand jar (plum C, without cover) also shows severe fungal attack after intensive dehydration of the fruit.

Only plum B in the middle jar with the bionic film as a cover shows no change in shape or fungal infestation after 14 days of testing, but only a slight dehydration, which can be seen from the wrinkled skin of the fruit.

A nice proof that the principle of millions of natural packagings, as systems open towards the environmental atmosphere, also for technical packagings in experiments close to reality, leads to a clearly increased life time resp. edibility of the packaged goods compared to closed technical packagings.

However, the systematic introduction of this significant freshness retention advantage into packaging technology and the packaging industry is still in progress – despite enormous long-term benefits for industry, consumers and the environment – although reliable preliminary results have been available for more than 1.5 decades. The clinging to decades of entrenched packaging techniques in combination with relatively inexpensive, but in part environmentally destructive plastic packaging materials is nothing other than the misguided clinging to old, outdated routines in the minds of decision-makers, which we know enough about!

7.21 State-Forming Insects as Masters of Climate Control: Function-Optimized Climate Control Without Additional Energy!

In Sect. 4.17, we described the ingenious climate regulation of a termite burrow open to the environment, housing millions of individuals, and visualized it by Fig. 4.36. Figure 7.59 describes the routine cycle for creating bionic buildings modeled on the air-conditioned termite mounds.

In 1996 opened the first high-rise building, whose natural air conditioning without artificial air conditioning and was built by the architect Mick Pearce on the model of the air conditioning of termite mounds. About the design and passive cooling of the building it says (original English, translated into German):[41]

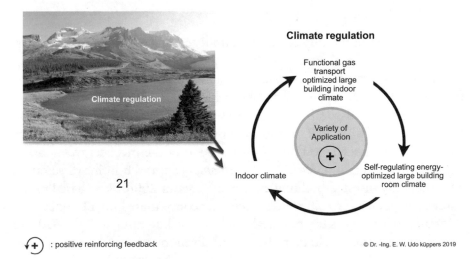

Fig. 7.59 Routine circuit 21: climate regulation

[41] https://en.wikipedia.org/wiki/Eastgate_Centre,_Harare (accessed 2020-02-10).

The *design of the Eastgate Centre* is a deliberate departure from the 'big glass block'. Maintaining a comfortable temperature in glass office blocks tends to be expensive, requiring significant heating in winter and cooling in summer. They tend to recycle air to keep the expensive conditioned atmosphere inside, resulting in high levels of air pollution in the building. Artificial air conditioners are high maintenance and Zimbabwe has the added problem of having to import the original system and most spare parts, wasting foreign exchange reserves.

Architect Mick Pearce therefore took an alternative approach. Due to its altitude, Harare has a temperate climate despite being in the tropics, and the typical daily temperature variation is 10 to 14 °C. […] This makes a mechanical or passive cooling system a viable alternative to artificial air conditioning.

Passive cooling stores heat during the day and ventilates it at night when temperatures drop.

- Daytime: The building is cool. Daytime: Machines and people generate heat and the sun is shining. The heat is absorbed by the building fabric, which has a high heat capacity, so that the indoor temperature does not rise much, however.
- Evening: Outside temperatures drop. The warm indoor air is vented through chimneys supported by fans. However, it also rises naturally as it is less dense, drawing in denser cool air at the bottom of the building.
- Night: This process continues as cold air flows through cavities in the floor slabs until the building fabric reaches the ideal temperature to start the next day. Passively cooled, Eastgate consumes only 10% of the energy required for a similar conventionally cooled building (formatted by the author) (see Fig. 7.60).

A few years later, in the City of London, under completely different – temperate – climatic environmental conditions than those prevailing in Harare in southern Africa, the *Portcullis House* was completed in 2001, modelled on the Eastgate building there (architects Michael Hopkins & Partners). Figure 7.61 shows two views of the house on whose façade several flow channels are attached, four of which connect to a chimney-like structure on the roof, they are part of a natural convection – flow transport of thermal energy.

Braun pursued an interesting – *multibionic* – approach in his dissertation on "Bionically inspired building envelopes" (Braun 2008). It is based on 25 natural phenomena, which were successively concentrated on 11 conceptual integrations for a possible practical application, but until today, 2020, no concrete application could be realized yet.[42] Even more recently, a number of

---

[42] Statement by Braun (RWTH-Aachen) during a telephone call on 11.02.2020, who also attributes the delay to material-technical problems in the bionic implementation. (Nature is just more sophisticated than it seems at first glance, the author).

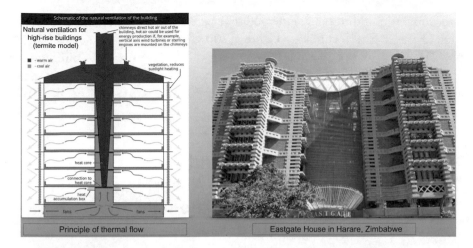

Fig. 7.60 Natural climate control at Eastgate Centre of Harare, Zimbabwe, Natural flow-thermal transport, left, ©3Natural_ventilation_high-rise_buildings.JPG: KVDP – CC BY-SA 3.0 – https://en.wikipedia.org/wiki/Eastgate_Centre,_Harare#/media/ File:Natural_ventilation_high-rise_buildings.svg (accessed 10-02-2020). Front view, right, ©3-MOB.COM, 19-05-2018, https://www.3-mob.com/technology/watch-this-video-on-the-innovation-around-how-eastgate-cools-itself-is-life/#.XkFkNi1oTow (accessed 10-02-2020)

Fig. 7.61 Natural climate control in Portcullis House in London, UK. Left: Westminster station building Portcullis House, London GB. ©Sunil060902, 18.01. 2009 CC BY-SA I 3.0. https://de.wikipedia.org/wiki/Portcullis_House#/media/Datei:Westminster_station_ building_Portcullis_House.JPG (accessed 02/10/2020). Right: Side front of Portcullis House with prominent ventilation shafts, Creative Commons Attribution 2.5. http:// www.photoeverywhere.co.uk/britain/london/slides/government5329.htm (accessed 02/10/2020)

concepts and buildings have emerged that take natural phenomena as models, such as the Erfurt Technology Center (2002, see Nachtigall and Pohl 2013, pp. 212–213) with a thermal flow system modeled on termite-building air conditioning, as well as work on energy-efficient buildings with bionic green architecture (Nazareth 2018; Yuan et al. 2017; Zuazua-Ros et al. 2017; Hammer 2016, etc.).

7.22 Massive Hammering Without Headaches! Technology for Energy-Absorbing Impact Protection and for Functionally Efficient Loading Techniques

Figure 7.62 shows at the beginning of the chapter again the routine cycle for the research and development of innovative bionic damping systems.

As small as these birds are, their ability to withstand the load of more than 1000 g (1 g = 9.81 m/s^2 acceleration due to gravity) without damage is great. In comparison, pilots or astronauts must withstand 3–4 times – in extreme cases 10 times – the acceleration due to gravity.[43] In extreme speed tests with racing cars, one driver was even subjected to a measured 214 g when his car

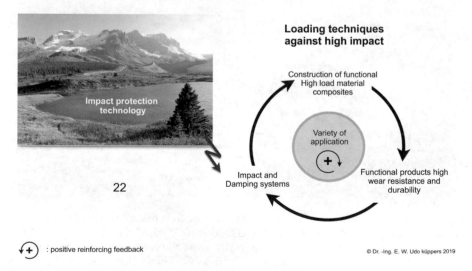

Fig. 7.62 Routine circuit 22 load-damping

---

[43] https://www.ds.mpg.de/131983/18 (accessed 2020-02-11).

hit a steel bollard of the safety fence. The driver survived this impact alive except for a few broken bones.[44]

If we humans had the ability to hammer against tree trunks with our beaks like woodpeckers, then possible head injuries from falling, head collisions in football or American football, head blows in boxing or a car collision, etc., would be... – without any external damping aids – simply none! Two examples from the US Football League:

- While some players suffered mild craniocerebral trauma from a centrifugal force of 60 g, one of the active players survived even the force of 168 g without damage. (online Focus 2007)[45]
- According to the US National Football League, concussions occur in players over 80 g, (McKittrick 2020).[46]

Neurology also deals intensively with diseases of the human brain, which is exposed to considerable stress through strong impacts or blows. Athletes of various sports, as mentioned above, are particularly affected by this – with possible late effects such as *Alzheimer's disease.*

It is therefore not surprising that for quite a few years, the invulnerability of woodpeckers and their injury-free hammering under high head loads beyond 1000 g has become the bionic focus of medicine, with the aim of researching and practically applying effective and efficient protection techniques against head impacts (including Farah et al. 2018; Smoliga 2018; Wallace et al. 2018).

Although a large number of head protection products exist in the form of helmets or protective covers that inflate in the event of a fall, in a wide variety of materials and shapes, the secrets of the invulnerable woodpecker head during hammering on the one hand, and their bionic imitation on the other, have not yet been satisfactorily solved.

Knowledge of the biomechanism of the woodpecker's ability to resist impact injury the head and the material properties and distribution of shock-absorbing spongy bone could be incorporated into the design for new protective helmets, sports products and other devices to effectively resist impact. In addition, a better understanding of biomechanism, biomechanics, and

---

[44] https://web.archive.org/web/20141019234856/http://www.kennybrack.com/pages/personal-info/2003.html (accessed 2020-02-11).

[45] https://www.focus.de/gesundheit/ratgeber/gehirn/news/sportmedizin_aid_228705.html (accessed 2020-02-11).

[46] http://theconversation.com/how-do-woodpeckers-avoid-brain-injury-120489 (accessed 2020-02-11).

suitable new structures of shock-absorbing composite materials is still needed before a woodpecker-type head protection technology can become marketable (see Wang et al. 2013, p. 718). Despite all efforts to unlock the secrets of nature – in this case the extraordinarily perfect head protection mechanism of woodpeckers – and to bring them into application bionically, the different boundary conditions that a bionic transfer entails should always be taken into account – and this requires a *systemic approach* to the problem and the solution.

But one bionic result is to be presented. The spongy bone structure of woodpeckers inspired Anirudha Surabhi,[47] CEO of Quintessential Dallas, Texas to design a lightweight bicycle helmet (Fig. 7.63). The material is textured cardboard. This particular design is also used to protect black boxes in aircraft and for a new generation of shock absorbers.[48] An all-encompassing test with results of the protective function is still pending.

Figure 7.64 compares the g-values of woodpeckers with those of humans in various situations, some of which are risky and hazardous to health.

Technical industrial damping systems or systems for vibration damping for impact, shock and other loads are still far removed from the damping technology of woodpeckers, if one compares the corresponding calculation values of mass, velocity, displacement and g-value.

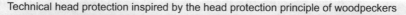

Technical head protection inspired by the head protection principle of woodpeckers

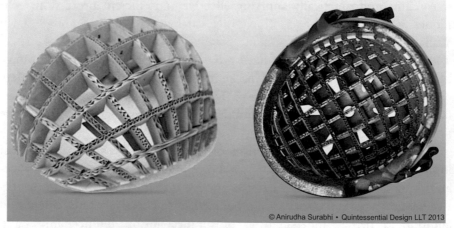

© Anirudha Surabhi · Quintessential Design LLT 2013

Fig. 7.63 Technical head protection according to the head protection principle of woodpeckers

---

[47] https://anirao.com/project_helmets/project_helmets.php (accessed 2020-02-14).

[48] However, scientific studies or experiments on the bioanalogous effectiveness of this particularly light head protection or as impact protection of a "black box" in airplane crashes or their simulations, etc., are not to be found in any technical publication.

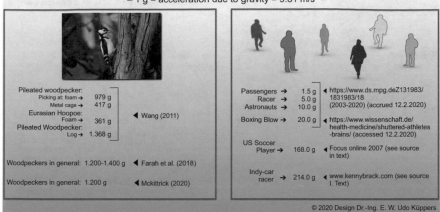

Fig. 7.64 Comparison of MAX g-values in woodpeckers and humans. (Great spotted woodpecker: Source: Pixabay Licence free download, https://pixabay.com/photos/woodpecker-tree-bird-animal-world-4761644/ (accessed 10.02.2020))

In individual cases, however, technical damping systems reach three-digit g-sizes. According to statements by various companies (Weforma, Zimmer, etc.) that manufacture damping elements for various industrial and other purposes, g-values of 500 are not unusual. However, the determination of the g-value is not primary for users of damping elements, rather the values for mass, number of strokes and number of shock absorbers, position of the damping elements, installation length and other technically relevant data more.

As imposingly as nature – through the woodpeckers – shows us its technical advantage, it seems all the more difficult to apply this ingenious protective principle – whose importance is beyond question, not only in the medical field – in a bionically effective and sustainable way.

7.23 Functional Dynamic Bonding and Debonding Without Adhesive: Adhesive-Free, Temporary Redetachable Bonding Techniques on Various Technical Surfaces

Figure 7.65 is again the first to introduce the routine circuit for a particular type of connection technique.

The fantastic world of the special technique of walking, sticking, releasing and walking on geckos, in literally all spatial positions, described in Sect.

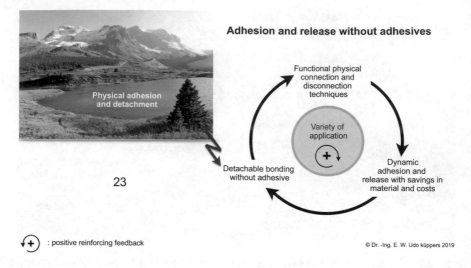

Adhesion and release without adhesives

$\overset{\downarrow}{+}$: positive reinforcing feedback

Fig. 7.65 Routine circuit 23 bonding without adhesive

4.19, does not remain hidden to sticking and gluing experts in the technosphere, with a view to industrial applications.

- *Industrial application*

Researchers have taken this principle as a model to reproduce the gecko effect from synthetic polymers in the form of microscopically small columns. One research focus in the development of these photolithographically produced, fibrillar microstructures lies in the geometric variation of the contact surfaces. For hard, smooth substrates, for example, an enormous increase in adhesive force is shown when the ends of the columns are widened similar to the spatulae of the gecko. For industrial applications, the adhesion principle of the gecko's foot is also so interesting because the adhesion does not require an additional energy source. With a suction cup principle, air consumed under energy must be displaced; a negative pressure must be applied and maintained for a certain time. The Gecko principle does not require this additional energy source and can also be implemented in a vacuum. Another point that is playing an increasing role in many production chains is the residue-free removal of the gripping systems. Typical products in automotive, semiconductor and display technology are characterized by highly sensitive surfaces. At the INM – Leibniz Institute for New Materials – the potential of the gecko principle was recognized early on and thus the applicability in robotics could be demonstrated with the establishment of the gecomer technology. An industrial robot equipped with gecko structures, the Gecobot [...], can achieve adhesive forces of 1 N/cm² on smooth, flat surfaces. By manipulating the structures, the adhesion can also be controlled specifically and thus switched on and off with pinpoint accuracy [...]. Thus, it is possible to move sensitive parts very precisely even in a vacuum within a production chain without leaving residues on these parts. At the same time, the adhesive system works almost without wear; in the test run, the adhesive force of the "Gecko structures" remains intact even after up to 100,000 "Pick&Place" runs. The technology thus offers great potential that can be used in a wide range of production processes.

The working group "Functional Microstructures" of the INM-Leibniz Institute for New Materials developed a fibrillar adhesive structure (Fig. 7.66, left), which, however, still turns out extremely coarse compared to the gecko foot. The INM[49] writes about this:

> The program area *Functional Microstructures* (of INM, the author) conducts pioneering research on novel functionally structured surfaces. Special mechanical, optical, thermal and haptic functionalities are generated by micro/nano-structuring and targeted material selection. The design of the structures and their functions are modelled on concepts from living nature, which are transferred to artificial systems. The focus is on researching switchable adhesion to

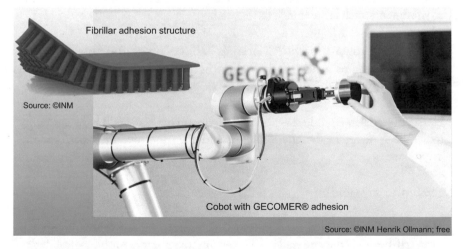

Fig. 7.66 Cobot with fibrillar adhesion. (Photos: ©INM Leibniz Institute for New Materials, photos freely available)

---

[49] https://www.leibniz-inm.de/forschung/grenzflaechenmaterialien/funktionelle-mikrostrukturen/ (accessed 2020-02-11).

various surfaces using our patented Gecomer® structures and transferring them to applications. These range from technical gripping systems to protective systems and medical surfaces.

A current practical example with the Gecomer® structure adhesion technology is the so-called *"Soft Cobot"*. According to INM, it is the first implementation of Gecomer® technology in a collaborative robot (Fig. 7.66, right). To this end, the INM:[50]

> Collaborative robots are a new generation of robots for direct cooperation with humans, even without safety distance and protective cages. Scientists at INM – Leibniz Institute for New Materials now present for the first time a cobot equipped with microstructured surfaces for handling objects. Since these structures are very soft and have no sharp corners or edges, the risk of injury to humans is further reduced. [...]. "Our innovative gripping systems are made of a highly elastic material. The gripping and detaching of objects is influenced by intelligent surface structures. This allows us to do without pointed grippers or tweezers," [...]. As a result, objects can be transported and deposited in the production process without causing injury to people or damage to the objects. The adhesive structures are particularly suitable for sensitive parts, such as devices for the automotive, semiconductor and display industries. "Our innovation can be extended to all processes where cobots are used, even in challenging environments such as vacuum". [...].

With the new program "Living Materials" 2020, "programmable living materials", INM initiates a new path for bionic techniques. The head of INM's "Dynamic Biomaterials" program area says:[51]

> Unlike materials we synthesize in the lab, living materials would have the ability to self-heal, adapt to the environment, and even improve their performance during use.

It goes on to say (ibid):

> The applications already known cover a broad spectrum: self-ventilating sportswear, self-healing concrete walls, self-renewing membranes for biological detoxification, biosensory tattoos for the detection of harmful substances on the skin or therapeutic use for the long-term and personalized delivery of medical agents into the body in chronic diseases are just a few examples.

---

[50] https://www.leibniz-inm.de/pressemitteilung/hannover-messe-weicher-cobot-erstmals-implementierung-der-gecomer-technologie-in-einem-kollaborativen-roboter/ (accessed 2020-02-11).

[51] https://www.leibniz-inm.de/pressemitteilung/living-materials-2020-programmierbare-lebende-materialien-eroeffnen-neue-moeglichkeiten/ (accessed 2020-02-11).

C. J. Meier (2017) of the Max Planck Society describes a multipurpose gripper that sticks like a gecko in an MPI article of 26 May 2017:

> An elastic membrane with tiny nubs coupled with negative pressure gives a new gripping system high adhesive force even on curved surfaces. Robots usually need a gripper arm that adapts to three-dimensional surfaces. Such a gripper should be as soft as possible so that it can adapt to a wide variety of shapes, but not too soft either, because otherwise it would detach too easily and not be able to support weights for long. Researchers led by Metin Sitti at the Max Planck Institute for Intelligent Systems in Stuttgart have now produced a membrane with microscopic nubs modelled on the fine hairs of a gecko's foot and attached it to a suction cup-like body. By means of negative pressure, this flexible gripper adapts so perfectly to a wide variety of surfaces that the load is distributed evenly over the entire contact area. In this way, the researchers avoid the stress concentrations that otherwise occur at the edges due to the load, which lead to the connection detaching. The gripper can achieve 14 times the holding force of grippers without this load distribution. (end of quote).

In Fig. 7.67, the author, the multipurpose gripper is seen in action: it playfully holds various objects such as a round Erlenmeyer flask (a) filled with 200 ml of liquid (total weight: 307 g), various coffee cups (b–d), a pair of cherry tomatoes (e), or a 139-gram plastic bag. (Size order, 10 cm) © PNAS, see Song et al. (2017) Fig. 1, P. E434.

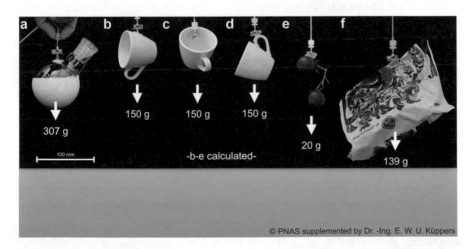

Fig. 7.67 Multi-purpose gripper with "Gecko" holding technique. (©PNAS, see Song et al. (2017) Fig. 1, P. E4345)

Learning from nature means opening up new innovative fields of application with new technologies. If the sustainability of bionic solutions is also taken into account, this would be a promising path for future environmentally compatible products and processes.

7.24   Aero-, Hydro- and Lithospheric Dynamic Low-Drag Locomotion: Wide Field of Development for New Technically Efficient and Energy-Saving Designs of Flying and Floating Transport Bodies

Birds of prey such as condors; ospreys, kites, hawks and others dominate their airspace with imposing flight maneuvers. Characteristic are the broad wing surfaces, which run out in individually small hand-fittings. Gliding for hours in thermal updrafts (condors), daring but always controlled flight manoeuvres over water (ospreys), enormous manoeuvrability between trees (goshawks) as well as perfect take-off and landing manoeuvres are some of the flight characteristics of the world champions of flying, largely over land.

Royal albatrosses, gulls, petrels and terns are no less gifted flight artists over water. Their narrow, tapered wing tips are perfectly adapted for flight maneuvers over the sea.

Like birds in the air, sharks, dolphins, seals, penguins and countless other swimmers move in the water, elegantly and gracefully, as if there is no resistance to overcome.

In the earth, here especially in the desert sand, rattlesnakes and horned vipers show their particularly adapted way of locomotion on and in the "sea" of billions of small stones that are constantly in motion. The mastery of locomotion to dive through the sand at lightning speed, however, remains reserved for the sandfish.

Air, water and earth or sand are habitats full of perfectly adapted organisms that have all developed one thing in common for their locomotion: To use energy as sparingly as possible.

It is precisely this criterion of energy efficiency that has also long driven engineers in aviation and maritime shipping. Before we discuss a few basic properties and recent developments from both industries, Fig. 7.68 introduces the chapter through the routine bionic cycle on the subject.

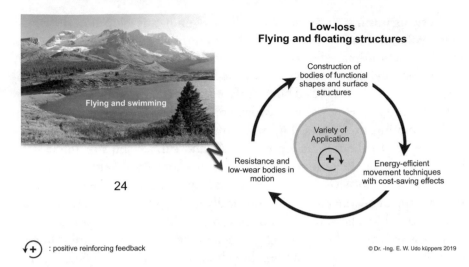

Fig. 7.68   Routine circuit 24 flying and swimming

Whoever moves through the atmosphere (gaseous earth shell), hydrosphere or aquasphere (water or liquid water) or lithosphere[52] (stony earth shell, stony crust and outer part of the earth mantle) as a subject or object or in general as a "resistance body", has to overcome the resistance applied by him. This resistance, also called flow resistance, is physically divided into form resistance, surface resistance, interference resistance, wave resistance and induced resistance.[53] Force is required to overcome it.

> The secret of every organism and every technical flying, swimming or diving body is how skilfully the resistance body and the impulse force generated to overcome the resistance body in a fluid (in this case air or water) interact optimally in order to move with the least amount of energy.

---

[52] The lithosphere is left out of this context because it is – compared to the other two spheres – still relatively little explored. The same applies to the deep sea (several 1000 m below the sea surface).

[53] The resistor components are:

1. form or pressure resistance: pressure distribution around the resistance body.
2. frictional resistance: occurring shear stresses on the body surface.
3. interference resistance occurs between two closely spaced resistive bodies.
4. characteristic impedance: occurs in the range of supersonic and transonic speed (range of simultaneous supersonic and subsonic speed).
5. induced resistance: generation of flow in a fluid by a resistive body in the fluid.

Every organism with the ability to fly or swim has developed in the course of its evolution a unit of resistance body and impulse propulsion or buoyancy, which has an optimal effect on its energy balance.

Technical flying or floating bodies have followed different paths of development, whose body shape and body surface results are linked up to the present with enormous energy wastage, which comes about not only through flying and swimming themselves, but also through the interlinked development processes up to the production of aircraft and surface ships or underwater boats. Of course, the technical boundary conditions are different from those in evolution – but nevertheless, considerable weaknesses and deficiencies still become apparent when analysed holistically and from the perspective of sustainability, which, however, will not be explored further at this point.

Thrust systems and drag bodies are separate in aircraft and ships – in contrast to the analogies in the animal kingdom. In aircraft, jet engines are usually suspended in nacelles below the wings; ship propulsion systems have propellers as thrust generators, usually at the stern, whereby the thrust generated is transferred to the ship's front drag body.

Significant losses in speed and momentum are therefore unavoidable in aircraft and ships due to this separation of drag and propulsion mechanisms.

A pioneer in aviation bionics in the 1960s was the aeronautical engineer Heinrich Hertel (1901–1982). He headed the Institute of Aircraft Engineering at the Technical University of Berlin from 1955 to 1970. His exemplary ideas on bionics, which included areas of structures, cybernetics, bird flight, insect flight, fish – form and motion, superior natural processes, and others, are summarized in his book Structure – Form – Motion (Hertel 1963). From the wealth of his findings and experimental results, one significant criterion will be highlighted as it relates to recent developments in aircraft design, shown in Fig. 7.69.

Starting with the observation of swimming fish and recognizing the coincidence of momentum loss and momentum supply, with later mapping to flight technology, Hertel (1963, pp. 201–202) writes:

In general, one also expects and looks for a water jet behind a fish swimming at constant speed, which is directed backwards in relation to the still water and whose impulse current Js is opposed to the flow resistance of the body. When the fish stops swimming and lets its orbital motion run out, the flow resistance due to surface friction and detachments shows up in a "wake" behind the fish. The "wake" represents a forward directed "jet" with velocities (vo-v_{xW}) relative to the still water. The momentum of this jet is Jw. When swimming at constant

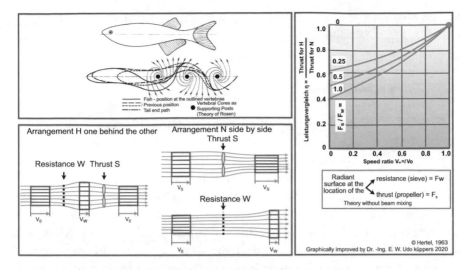

Fig. 7.69 Top left: Vortex field with counter-rotating vortices following the body-tail beat of swimming fish, top right: power comparison of arrangements of drag and thrust behind and beside each other. Theory without jet-mixing, below left resistance-thrust-arrangements. (©Heinrich Hertel, taken from Hertel (1963, pp. 201–202), graphically improved by the author)

velocity v_o, the loss of momentum Js (jet forward) from drag is balanced by the addition of momentum Js (jet rearward) from propulsion (fin stroke).

In the flow close to the hull, the decelerations (loss of momentum) and accelerations (supply of momentum) of the water can now follow each other in such a way that behind the fish the momentum is completely balanced, i.e. in relation to the still water a flow is neither forward nor is left behind. Behind a fish swimming fast at constant speed, the water can thus remain completely at rest, apart from eddies and turbulence "standing" at the place. The fish will approach the flow processes to such a balance behind the stern, as this achieves a particularly high propulsion efficiency. The power that the swimming fish uses to overcome the resistance, apart from being converted into heat lost in the water, is then to be found in the eddies and turbulence in the otherwise unmoving water behind the fish.

From experiments (Rosen 1959) to make the flow visible at and behind a swimming fish, of whose results [...] (Fig. 7.68) gives an example, it can be seen that the vortices behind the fish remain at the place of detachment from the body. It follows further from the above that a current between the vertebrae to the rear is also not necessary. Thus it is clear that in the case of fast swimmers the flow processes caused by the surface friction and those caused by the propulsive movement are not to be considered separately, but belong directly together and lead to the fact that only vortices and turbulence, but no changes in impulse current in the direction of travel, appear.

Hertel (1963, p. 202) writes about the arrangement of the resistor body and the thrust unit in flight technology, as well as about the mathematical proof of the power reduction by "interlocking":

- *To unity of airframe and engine*

In aviation technology, the principle of the "unity of airframe and engine" developed with the high-performance aircraft, according to which the engine is to be installed in the airframe and merged with it in the overall form in such a way that minimum overall drag, best engine efficiency and minimum weight are achieved. This principle is clearly expressed in the author's airplane design (1946), which [...] (Fig. 7.69, the author) reproduces. However, this unity of airframe and propulsion only becomes complete in engineering when what seems possible in fish is realized, that the impulse losses and impulse feeds follow each other in such a way that no change of impulse results behind the airplane or boat.

- *For the mathematical proof of the power reduction by "interlocking"*

The "interlocking" shall be explained by an example in [...] (Fig. 7.69, the author): A sieve by itself in the airflow V_o causes the resistance W and the velocity decrease to v_w, a propeller by itself in the airflow causes the thrust S and the velocity increase to v_s – If one wants to overcome the resistance W by the thrust S of the propeller, the jet power is required to be $L_N = W (v_o + v_s)/2$ when switched in parallel. However, if the propeller is switched into the decelerated jet behind the sieve, it accelerates the jet back to the face velocity v_o. Behind this sequence of drag and drive, there is no longer any change in velocity, but v_o prevails everywhere. The required power is now $L_H = W (v_o + v_w)/2$. But since V_w is less than V_s, L_H becomes less than L_N hence the power requirement L_H when connected in series becomes less than L_N when connected in parallel. The size of the difference between L_H and L_N is shown by the diagram in [...] (Fig. 7.69, right, the author). In the limiting case $v_w / v_o = 0$ and $F_s / F_w = 1$, the power required for series connection is only 40% of that for shunt connection.

Hertel points out in advance that in aviation technology, high-performance aircraft – or high-speed aircraft – are committed to the principle of "airframe and engine unity". That was the case in the 1960s and remains so today.

But why has the sensible, energy-efficient back-to-back connection of missile and propulsion in passenger aircraft been omitted to this day? It is true that – from a bionic point of view – a new design has been created by shape variations on wings, such as those created by a single upstanding winglet at the wingtips, or surface structures modelled on the shark surface have been tested. The first shape design leads only to minimal flow advantage with lower induced drag, but at the same time also to higher interference resistance! The surface experiment to reduce the frictional resistance is still far away from systematic use on passenger aircraft.

The fact that a simple change in the position of engines on the aircraft, which currently hang *next to* the fuselage under the wings, and thus – by their arrangement alone – produce enormous energy losses, can turn these energy losses into energy-saving effects, is apparently not a target for aircraft manufacturers.

After all, Boeing and Airbus present current designs of aircraft, as shown in Fig. 7.70, which resemble the body of a ray, with a distinctive change in the position of the engines in the rear fuselage area, but are nevertheless arranged as a *"side-by-side" unit of "airframe and engine"*.

The really amazing thing, however, is the comparison of the two new aircraft types with an aircraft design by Heinrich Hertel from 1946, which has amazing similarities in the body shapes, but Hertel's design corresponds to the ingenious energy efficiency principle of the nature of fish, through the perfect *"one behind the other" unity* of *"airframe and engine"*!

Modern simulation techniques (see Huhn et al. 2015) through quantitative flow analysis of swimming dynamics with coherence to Lagrangian vortices (all field theories can be described with mathematical formula introduced to classical mechanics by Italian physicist Joseph-Louis Lagrange in 1788), as

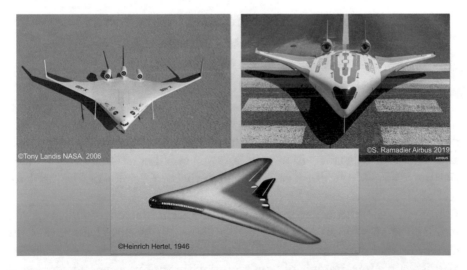

Fig. 7.70 New Boeing and Airbus aircraft design compared to a flying wing transoceanic aircraft, Project HTL 1946 by Heinrich Hertel (©Hertel 1963, p. 201), Boeing X-48: ©Tony Landis for NASA, Photo ID: ED06-0198-62, Public Domain https://en.wikipedia.org/wiki/Boeing_X-48#/media/File:X-48B_from_above.jpg (accessed 12.02.2020). Airbus Mavaric Demonstrator: ©Airbus S. A. S. 2020 https://www.airbus.com/newsroom/stories/Imagine-travelling-in-this-blended-wing-body-aircraft.html (accessed 12/02/2020)

well as modern visualization techniques, show that aquatic animals can modify their fluid environment to increase swimming and food gathering efficiency (Fish and Lauder 2013). Not surprisingly, both authors note that Leonardo da Vinci already recognized the benefits of streamlined bodies of fish based on the surrounding current. Further, Fish and Lauder state:

> Understanding how animals can control flow has immense implications not only for understanding the evolution of aquatic species, but also for the development of biologically inspired machines and even for the elucidation of global climate change. [...].
>
> The generation of vorticity is an inevitable consequence of propulsion in a fluid medium. The vortex flow detached in the wake represents a significant loss of energy for swimming animals. However, animals can swim more efficiently if they are able to extract energy from the swirling vortices. Because energy is a limited resource that can affect an animal's survival and ability to reproduce, the presence of hydrodynamic mechanisms that recover energy destined for wake entropy provides adaptive advantages.

As you can guess, dear readers, research into the secrets of swimming and flying animals and their transfer to technically energy-efficient ships and aircraft still holds great potential for development.

7.25   With Power and Speed to Catch Prey: Perspective for New Technical Impact Materials of High Speed and/or Resistance Strength

Section 4.21 described the impressive and sophisticated techniques of the mantis shrimp's structured material structure (spiral Bouligand architecture) and visual ability (spatial vision with only one eye). Let us focus here on material properties of the crustacean and how their principles can be used for technical products.

In Sect. 7.22, the injury-free head butting technique of woodpeckers was highlighted as a model for technical applications. In this chapter, it is the extraordinary injury-free head butting technique of a crayfish.

It is again a fine example of how evolution, in various ways and adapted to the respective regional conditions, has found ways to tame enormous forces by organisms – beyond human capabilities – in such a way that they pose no risk to the organism itself.

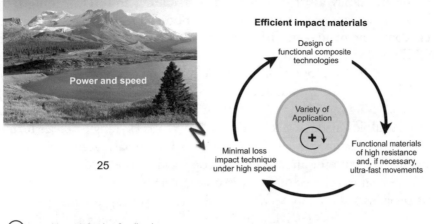

Efficient impact materials

(+) : positive reinforcing feedback   © Dr. -Ing. E. W. Udo küppers 2019

Fig. 7.71 Routine circuit 25 efficient impact materials

Figure 7.71 provides the pathway into bionic applications of these materials engineering services.

Scientists at the University of Southern California (Yang et al. 2017) developed a new additive manufacturing technique that technically mimics the resilient bouligand structure (or twisted plywood structure, see also Sect. 4.21) found in crabs, crab claws, peacocks, mantis shrimp, shrimp, beetle wings, and lobsters.

The bouligand structure with ordered collagen or chitin fibers in one layer and heterogeneous between different layers, has been extensively studied and shown to contribute greatly to the reinforcement of stopping initiated cracks in the material (Rabe et al. 2006; Al Sawalmih et al. 2008; Daly et al. 2016; Zheng 2019).

This fracture-resistant material effect occurs because the crack does not find a straight path to spread further due to the twisted structural layers. Consequently, energy dissipation and impact resistance are increased.

Yang and colleagues use the additive manufacturing technique, among other things, to mimic the fiber alignment in the human meniscus in order to create reinforced artificial meniscus replicas. The reason is that tears or damage to the meniscus (meniscus ruptures or lesions) are among the most common diseases of the knee joint. The diagnosis is made by clinical examination and can be confirmed by MRI examination.[54] Yang et al. (2017), speak of more than 1.5 million people in the US and Europe affected by this condition.

---

[54] https://orthinform.de/lexikon/meniskusverletzungen (accessed 2020-02-17).

Using electrically assisted additive manufacturing and 3D printing technology, a reinforcement architecture is fabricated with anisotropic material layers consisting of aligned surface-modified multi-walled carbon nanotubes (MWCNT-S).

For the technical production of bionic meniscus parts, Yang et al. see the material made of carbon nanotubes as more advantageous than the common practice of using miniskene substitutes made of polymers (polyurethane) or directed scaffolds made of silk (ibid., pp. 1605750, 1). Yang et al. see further areas of application for the nanostructured materials according to the Bouligand structure in aircraft construction and in the field of mechanical and tissue engineering techniques (tissue-engineering).

As previously shown in Sect. 4.21, the mantis shrimp reaches a flapping velocity of over 20 m/s (>72 km/h) within a few milliseconds of flapping and an occurring acceleration of over 10^5 m², in the original:

> The subject of the present study, the 'smasher' peacock mantis shrimp *Odontodactylus scyllarus,* can deliver strikes lasting only a few milliseconds, with accelerations of over 10^5-m-s^{-2} and speeds of over 20-m-s^{-1} (Patek and Caldwell 2005, p. 3655), q.v. (Patek 2016, 2019).

The acceleration value a = 10^5 m s^{-2} is equal to 100,000 m s^{-2} or 10,000 g!

Could such shatterproof materials as are possessed by the mantis shrimp and other animals with the special Bouligand material structure also be useful for protection against strong blows to the head of a footballer, or boxer, or football player? Section 7.22 has already addressed this issue in a similar way, on the biological basis of head protection in woodpeckers. However, long-term tests with reliable results for bionic products of this type are still lacking.

7.26   Filigree Material-Efficient Architecture of Body Skeletons: Building Constructions Under New Stable and Aesthetic Structure and Form Design

Building owners would like to build material-efficiently because it saves costs; have a stable building shell because it should be safe against increasing storms and hurricanes; use adapted, energy-efficient climatic conditions in the rooms; also feel a new living feeling by creating room-variable new living and working units and use other useful properties of a house. Living according to one's own needs would be very congenial to the organism of the inhabitant.

Or would you rather be housed in mass-produced quarters made of reinforced concrete, with standardised housing units like those from an assembly line and bureaucratically standardised building regulations – albeit halfway bearable and still affordable?

The flexibility of comfortable living and working spaces in buildings is – so it seems – still undermined (apart from necessary safety criteria) by decades-old practiced building standards, recognizable especially in urban agglomerations.

It is not new knowledge, but decades of practice that the well-being of people and thus their health *also* depends on the spatial design in which they live and work. School classes that are taught in metal containers for years due to a lack of space, hospital rooms with bare white corridors, alienated high-rise building atmospheres and other conditions show only one thing: the dominance of economic goals to the neglect of people and their feelings.

> The overwhelming architecture of building – with very few exceptions – focuses less on people and more on using them as a means. POINT.

Figure 7.72 introduces the circular sequence for material-efficient building, which is ultimately part of an overarching *"Evolutionary Architecture"*.

If we looking back from the new 3rd Millennium after our era, at the extremely sophisticated building techniques and building materials of peoples such as the Assyrians, Egyptians, Chinese or Romans, partly millennia before

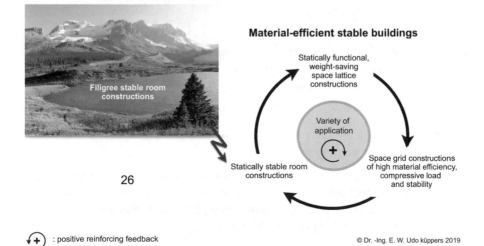

Fig. 7.72 Routine circuit 26 material-efficient stable buildings

our era, house walls are still set stone on stone with largely manual craftsman-ship, then on the one hand this is an incomparable building anachronism, on the other hand – in view of our current housing culture – we have learned little or nothing from earlier master builders thousands of years ago, and cer-tainly nothing from the ingenious master builders of nature (Küppers 2001).

Few engineers and architects use the perfect energy and material adapted living and working spaces of organisms in flora and fauna to offer us humans what animals and plants have perfectly mastered for millions of years: Building according to the purpose of their needs. Examples of this can be found in Jodidio and Gössel (eds., on Santiago Calatrava) (2016), Aldersey-Williams (2003), Becker and Braun (2001), Vélez (2000), Tsui (1999), Schaur (1995), Frey Otto (1988), Wunderlich and Gloede (1977), R. Buckminster Fuller and Applewhite (1975), Patzelt (1972), among others. This series of authors necessarily remains incomplete. Nevertheless, it provides an impressive insight into the architecture of natural or bionic building.

Tsui has compiled 12 learning principles for his building design construc-tions, which he has worked out from years of research (Tsui 1999, pp. 84–85):

1. Use material sparingly
2. Maximize structural strength
3. Maximize the enclosed volume
4. Produce an extremely high strength to weight ratio
5. Use of tension and stretch as a basis for structural efficiency
6. Designing an energy-efficient, well-insulated, comfortable environment, (largely, the author.) without external power supply
7. Shape design that improves air circulation
8. Use of local building materials
9. Use of waveforms that disperse or resolve multidirectional forces
10. Integration of aerodynamic efficiency into the structural form
11. Do not produce anything that is harmful or toxic to the environment.
12. Designing structures that can be built by a single organism

Further, Tsui writes (ibid):

> The basic meaning of the word *research* comes from the French, "re-again" and "chercher" – to seek again. [...]. And that is what drives us to look again at the world we live in, at the hidden universe below the surface, that is where we find the solutions to our widespread problems.

It is regrettable that research and development do not together determine the practice of architecture. Is it any wonder, then, that fundamental design development has been stuck for more than a century?

Werner Sobek, a successor of Frey Otto at today's Institute for Lightweight Design and Construction (Ilek) at the University of Stuttgart, expressed himself in the same direction – more than 20 years later – when he stated in an interview in the magazine VDI Nachrichten (VDI-N., 14. 02. 2020, pp. 8–9):

> Our construction industry is the only industry of significant scale where design, production and operations are systematically separated.

The Ilek contribution to the first and second learning principle according to Tsui is the so-called lightweight construction made of *graded*[55] concrete, which requires 40% less mass than a solid shell for the same load-bearing capacity (VDI-N. ibid.). However, at present only so-called *"art objects"* made of graded concrete exist as showpieces. It will probably be years before this technology can be introduced into the practice of residential construction and systematically used. The same applies to so-called self-healing concrete,[56] which is a composite material designed to reseal cracks that occur. Two of many construction techniques of mankind, which still find their world champion in the *"construction industry" of* nature for a long time.

The tent roof construction of the Munich Olympic Stadium (1968–1972), by Frey Otto and Günter Behnisch, is also incomparable. The lightweight construction was designed and realized using ropes and shells, whereby a spider with the "loop trick" in its web helped, as it were as a third partner of the two architects, to avoid occurring stress cracks in the tent roof (see Küppers 2001, p. 92).

One architect of biomorphic design is particularly noteworthy: Santiago Calatrava. His round, nature-like forms adorn buildings such as are typical for him and can be seen in Fig. 7.73 in a small selection.

Comparable biomorphic constructions, like those of Calatrava fill the book of Aldersey-Williams with the title "Zoomorphic – new animal architecture".

---

[55] Graded concrete changes its internal material structure in such a way that a dense material composition occurs at high load, and a porous material composition occurs at low load, resulting in a more material-efficient use for the same total load compared to a uniform solid concrete structure.

[56] https://www.baustoffwissen.de/baustoffe/baustoffknowhow/grundstoffe-des-bauens/selbstheilender-beton-bakterien-gegen-risse-betonschaeden-hendrik-jonkers-tonpellets-bio-beton/ (accessed 2020-02-18).

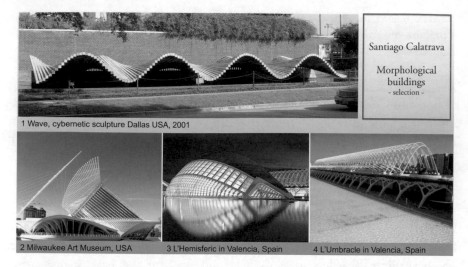

Fig. 7.73 A selection of morphological buildings by Santiago Calatrava. (No. 1: Photo: Andreas Praefcke – own photography, image free, https://de.wikipedia.org/w/index. php?curid=3942119. No. 2: Photo: Andrew C. from Flagstaff, USA CC BY 2.0 https:// de.wikipedia.org/wiki/Datei:Milwakee_Art_Museum.jpg. No. 3: Photo: Author: Diliff CC BY 3.0 https://de.wikipedia.org/wiki/Datei:Hemispheric_Twilight_-_Valencia,_ Spain_-_Jan_2007.jpg. No. 4: Photo: Author: Chameleon, public domain https://de.wiki-pedia.org/wiki/Datei:Ciutat_de_les_Arts_i_les_Ciències_-_L%27Umbracle.jpg)

Biomorphic buildings constructed as individual objects bear the signature of architects, some of whom are world-renowned. However, the question is allowed: What is the state of affairs of biomorphic apartment buildings in cities or metropolitan areas? About such approaches of building and well-being, of intelligent building with aspects of a different building culture than the existing one, which also takes into account the space with its infrastructure as a networked part of a whole, authoritative strategies and actions were also developed and presented from the then Frey-Otto Institute for Lightweight Planar Structures at the University of Stuttgart. *"Self-organizing path networks"* were as much a topic as *"Learning and research to build for the poor"*, "Earthquake-proof houses", "Adaptable building", *"Biocybernetic, system-compatible settlement structures"*, last contribution by the author (Küppers 1995) and others more (Schaur 1995).

With *absolute certainty*, the *"architects of nature"*, whether they live in small families or mass quarters, still possess their energetic and material secrets. If they are used by us humans for our purposes in a bionic way, we should – despite recognizable functional building principles in detail – not lose sight of the overall context, which still happens too often and is not infrequently punished with massive consequential problems.

Nature's master builders know their environment very well and orient themselves adaptively and sustainably with their functional solutions. Humans are still often overtaxed with an increasingly necessary *"thinking in systems"* instead of a *"thinking in products"* – also in the field of architecture – and are on the way to finding it, although the solutions of tasks in complex dynamic environments virtually challenge this *"systemic thinking"*.

If only it were possible to see the abolition of the separation of research and practice, of planning, production and operation, which has been called for several times before, in a holistic way again, much would certainly be gained for sustainable processes in architecture and the building industry.

7.27   Perfect Light Collectors and "Light Pathfinders": Solar Optical Techniques for Energy-Efficient Feel-Good Room Areas of Appropriate Climate

Polar bears are perfectly adapted to survive in their icy environment, with several physical principles contributing to this. One of them is the absorption of energy-rich UV radiation in the Arctic. The outer fur hairs, which are very close together, are transparent and hollow. It conducts the light to the black surface of the body, where it is absorbed and, with further protective mechanisms, maintains the internal body temperature of about 38 °C.

Figure 7.74 shows the routine circuit for bionic thermoregulation, while Fig. 7.75 shows the principle of "light gathering" by the polar bear.

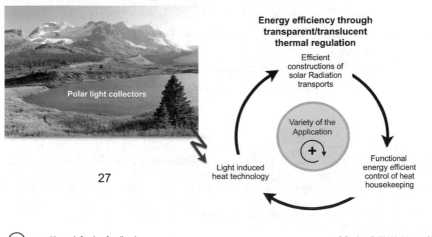

: positive reinforcing feedback   © Dr. -Ing. E. W. Udo küppers 2019

Fig. 7.74   Routine circuit 27 bionic thermal regulation

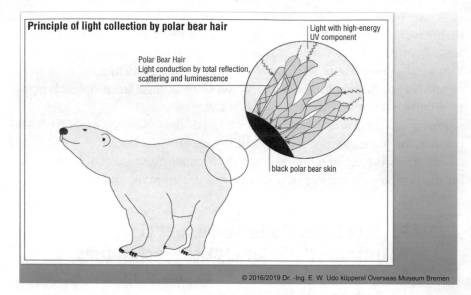

Fig. 7.75   Principle image of the polar bear as a light collector

The solar thermal principle of the polar bear (Fig. 7.75) has now been successfully tested several times on house facades (Fig. 7.76). The abbreviation "TWD" stands for "Transparent/translucent thermal insulation."[57] It consists of a heat-absorbing black absorber layer with an overlying transparent capillary sheet (capillaries are small tubes analogous to polar bear hairs) and an outer translucent weatherproof protective layer. Compared to an opaque (i.e. non-transparent) thermal insulation layer of the same thickness on the same masonry, TWD allows significantly more heat to be conducted through the masonry into the interior of the house. This is a very effective, because energy-saving application from building bionics for civil engineers and architects. TWD has been used successfully for years in both single-family homes and apartment buildings.

The following Fig. 7.77 shows the principle sketch of a translucent thermal insulation and associated light-diffusing materials from Wacotech in Herford,

---

[57] There is a clear difference between *transparent* and *translucent* from a physical point of view, although the term: *"Transparent thermal insulation"* for the abbreviation TWD has been used for years.

Transparent materials are completely translucent, while translucent materials only ensure partial light transmission. Light has a higher degree of scattering in translucent materials, which is why they appear milky in buildings with translucent facades. Incidentally, our skin is just as translucent as the light-processing green leaves of trees and shrubs.

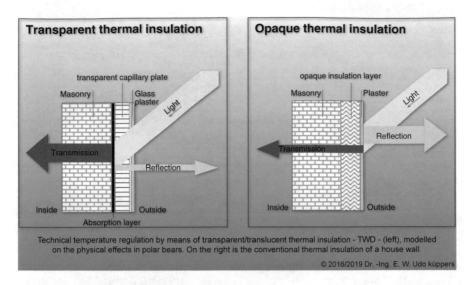

Technical temperature regulation by means of transparent/translucent thermal insulation - TWD - (left), modelled on the physical effects in polar bears. On the right is the conventional thermal insulation of a house wall.

© 2016/2019 Dr. -Ing. E. W. Udo küppers

Fig. 7.76 Principle of a solar thermal wall construction including absorber layer – TWD – compared to an opaque wall construction

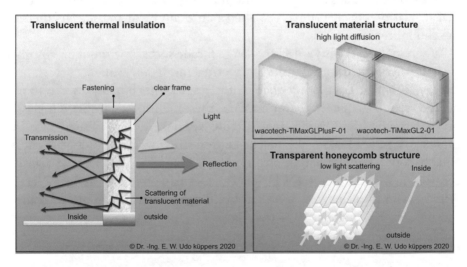

Fig. 7.77 Principle sketch of translucent thermal insulation, left. Right: associated materials, in comparison transparent grid modules, as TWD elements in combination with masonry and absorber layer. (Photos of the materials courtesy of Wacotech Herford, ©Christoph Frömming)

Fig. 7.78 Example of a building with translucent thermal insulation. (Photo sources: Courtesy of "Simon Sommer Fotografie", Architects: ZOLL Architekten Stadtplaner GmbH, Project participants: Wacotech GmbH & Co.KG)

Germany. In comparison, transparent materials can also be seen which, in combination with absorber panels and masonry (see Fig. 7.76), create a thermal insulation effect. Figure 7.78 shows an example of translucent thermal insulation in practice.

Where stable masonry is lacking, large façade surfaces without an absorber layer are often fitted with translucent thermal insulation elements, which produce a light scattering effect that is not unpleasant, but makes the rooms behind them opaque. In warm seasons, these translucent insulating glasses prevent rooms from overheating; in cold seasons, heat is released into the rooms. The milky-looking thermal insulation structure also avoids glare effects and shadow-free illumination of the rooms. One of the pioneers of translucent thermal insulation – TWD – is the German architect Thomas Herzog (1995), who implemented TWD in buildings such as the youth education centre Kloster Windberg, 1991 (see Fig. 7.79), the Wikhahn production halls Eimbeckhausen, 1992 and many others.

Youth Education Centre Windberg, «House of bedrooms» and «House of bedrooms1»

Photos: ©Youth Education Centre Windberg, 94336 Windberg

Fig. 7.79 Example of a building with translucent thermal insulation. (Photo sources: With the kind permission of the youth education centre Windberg, 94336 Windberg)

7.28   Functional Diversity: Optimised Controlled Multiple Use of Products and Processes

Figure 7.80 again points to the bionic circular process of using the thematised ingenious principle of nature in a technically sustainable way.

Biocyberneticist Frederic Vester has described the principle of multiple use as the fifth of eight basic rules of biocybernetics (Vester 1991a, b, pp. 66–86, first published, 1983):

Survivable systems develop products and processes that kill several birds with one stone – basically a variation of the Jiu-Jitsu principle. If possible, not what we create or do, if possible, no product and no process should be applicable for only one purpose. […].

The principle corresponds to the energy and effort saving way of working in nature, where the pollination of the flowers is coupled with the nutrition of the insects, where the earthworm not only serves as food for the birds, but at the same time aerates the soil, where the leaves regulate the humidity between plants and air, but also take care of the photosynthesis.

We find the same rational principle in the coupling of sexuality with social tasks such as partner cohesion and aggression inhibition, where the release of sex hormones not only serves reproduction, but at the same time strengthens the immune system and disease defenses, protecting us from infections and cancerous diseases.

This means that we should also try to kill several birds with one stone in everything we create, do and produce. A building does not only have its func-

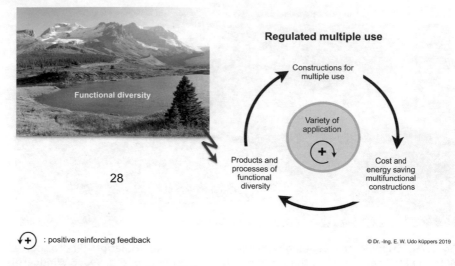

Regulated multiple use

Constructions for multiple use

Variety of application

Products and processes of functional diversity

Cost and energy saving multifunctional constructions

Functional diversity

28

(+) : positive reinforcing feedback

© Dr. -Ing. E. W. Udo küppers 2019

Fig. 7.80 Routine circuit 28 multiple use

tion for the inhabitants, because at the same time it can also give security to the cityscape and absorb noise, provide for heat-storing alleys in winter, for cool-releasing gases in summer; one can let fresh air corridors develop with its help instead of building them up; or one uses, where there is a roof anyway, huge collecting surfaces for solar energy or combines this with green spaces on the roof, which moderate the temperature as biological storage machines.

The same is true in many other areas: In schools, where learning and experience could be coupled, in sports, where relaxation, health and social contacts flow together, in technology, where a machine, in addition to its production function, simultaneously supplies waste heat for heating or steam for power generation. One example: companies such as paint shops or large bakeries, instead of polluting the environment, can even profit from it by installing a heat recovery system: energy costs are reduced. The amount saved simultaneously finances the plant installed on a hire-purchase basis, the environment is relieved of waste heat and exhaust fumes, raw materials are conserved, production costs fall, jobs are created and other companies earn – without anyone paying on top.

Some of these sustainable processes described in the 1970s and 1980s are already standard today, others are still far removed from Vester's visions. All of them carry an efficient multiple use, if they are recognized through *systemic thinking and action*:

- systematically used combined heat and power plants,
- multifunctional tool use (example in Fig. 7.81),

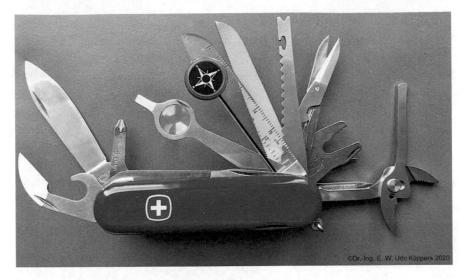

Fig. 7.81 Multifunction Swiss Army Knife

- sustainably used raw materials – without massively destroying nature,
- Schools designed to be learning and "living" in their architecture,
- Systematically understand sport and artistic subjects as part of holistic learning,
- Making cities and their infrastructure "alive",
- Recognize buildings systematically as relevant to the system and not just as individual isolated housing units to mass quarters.

But the truth is also: The environment is increasingly and sustainably relieved of waste heat and exhaust gases is still a fairy tale according to today's knowledge and climate facts!

Technical products, processes or organisational structures that have left open in their development process the ability to react adaptively to environmental changes or new task or goal directions in a timely manner not only strengthen a purpose-bound functional diversity, but also reduce completely new beginnings of developments with high consequential costs and last but not least also strengthen the stability of a system that does not have to be an entrepreneurial one.

Instead of offering, for example, five different, related or divergent functions in five different products, it is economically closer to integrate five functions in one product or process.

The famous Swiss Army knife, shown in Fig. 7.81 in a large version, is probably the prototype of a product with functional diversity. The *natural principle of multiple use* has been realized almost perfectly here.

However, as the incomplete list of urgently needed tasks presented above shows, the human inventive spirit will be challenged for years to decades to come, in view of the recognizable anthropocene state and the still often uncertain to unknown development of digital processes in our living and working environment!

A special form of multifunctionality among organisms is also described in Sect. 7.29. below, the *principle of symbiosis.*

7.29 Symbioses: Rich Effective and Efficient Solutions for the Benefit of All Partners

Forging alliances to the opposite advantage saves effort, energy and material. In Sect. 5.4.7 we have already gone into detail about the ingenious natural principle of symbiosis. What plant and animal organisms have managed perfectly for millions of years should actually also be of mutual advantage to humans, regardless of the development for which the natural principle is used. The routine cycle in Fig. 7.82 shows the way there.

If we look at the – with creativity and intelligence – created and further developed state of technospheric processes, with the indisputable, but from a

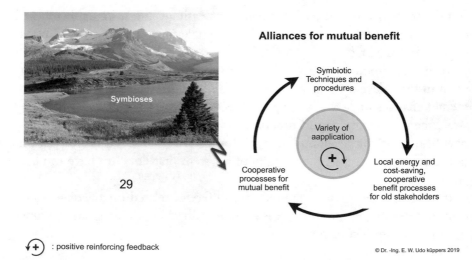

Fig. 7.82 Routine circuit 29 symbioses – alliances for the opposite advantage

systemic point of view short-sighted misguided economic engine of profit maximization, but also include the inevitable consequences of this development strategy for society, environment and nature, we can speak of many things, only not of a symbiotic behavior of mutual benefit, rather of a parasitic behavior.

Technical, economic or social "symbioses" are certainly present in detail, such as those between product manufacturers and product suppliers or product buyers; between energy providers and energy consumers; between care institutions and those in need of care; between homeowners and tenants, and so on. A closer look at these kinds of alliances, however, often reveals a dependent inequality between mutual give and take that throws the very purpose of an equally weighted symbiosis into disarray.

Symbiotic behaviour between humans is clearly different from symbioses between animals and plants, because the former, thanks to foresighted and planning processes, are not subject to strict evolutionary processes – in this case for mutual benefit – but take place temporarily, depending on possibility, need, purpose and goal, and are often dissolved again for selfish reasons.

But if symbioses serve organisms – we ourselves are realistically also only symbionts with an army of billions of bacterial colonies in our stomachs – as a beneficial universal principle of nature, together with many other principles of nature that we know, why has our economy so far not managed to establish an *ecosystem of economy that is* obvious and beneficial for all?

Many causes lie in the increasing alienation from nature, which is revealed in educational areas and work processes through performance constraints. Vester described it more than 30 years ago, today it is more topical than ever, thus (Vester 1991a, b, p. 26):

> Instead of symbiosis we find competition, helping becomes "sacrifice", letting oneself be helped becomes a burden. The result is an increasing stress load, coupled with a reduced ability to learn and thus an increasing inability to understand and cope with the complex problems of our world. [...] This development affects all areas of interpersonal communication.

The clear lack of perception and sustainable problem solving of complex problems of the present – even though it has been mentioned repeatedly and still is – is shown by the self-inflicted anthropocene consequences, but also by the incalculable intrusion of a digitalizing environment into our lives and work.

We are in the midst of a global crisis that is clearly emerging, through:

- Earth-wide climate change, or extreme weather,
- Population growth,
- Migration waves of refugees from wars and climate extremes,
- enormous shortage of raw materials,
- increasing contamination and destruction of our nature and environment – with which we should live in symbiosis,
- the occurrence of epidemics (such as currently in 2020 the spreading Covid-19 epidemic, first spreading in China), with already enormously burdening consequences for the world economy etc.

Where should new decision-making aids for sustainable developments come from, which steer the obviously wrong path of an insufficient ecological economy into a sustainable nature-compatible economy, in which there is also room for exemplary networked principles such as that of symbiosis?

This question is not new and yet more topical than decades ago!

Frederic Vester (1991a, b, pp. 124–135) an important pioneer of systemic thinking and cybernetic functional networking, in which a number of natural principles, including symbiosis, are integrated, has already proven for decades through various projects, including the leisure pueblo, the eco-land concept, the land workshop project, the Swissair project cabin, the Ford study, etc., that ecological economy is feasible – with common benefits for all involved.

Regarding his seventh biocybernetic rule, the principle of *symbiosis*, Vester (1991a, b, pp. 80–81) writes:

> Symbiosis always leads to considerable savings in raw materials, energy and transport and thus to multiplied, mostly free benefits for all the links involved. The more diverse these are, the more possibilities there are for symbiosis. It is thus favoured by diversity in a small space.
>
> Large uniform areas, central energy supply, pure dormitory towns, monocultures (also as far as industries and product manufacturing are concerned) must therefore do without the advantages of symbiotic relationships – and thus also their stabilising effect; relationships which in and of themselves would be possible multiplied by a different distribution. The use of symbioses therefore means small-scale planning, but also sensible coupling of existing facilities, for example in the industrial sector.
>
> Far beyond the function of "waste exchanges", a kind of *ecosystem of the economy* can be formed: Plants of the metal industry working together with those of the paper industry, [...] food industry linked with water purification and waste recycling and so on – where necessary also with cleverly selected *new* plants as links. In this way, further symbioses could be added: silent metal extraction from the most diverse ores through the cultivation of special microorganisms (as

already works today, (1980s, the author) a large part of the copper extraction). Photosynthesis plants for the extraction of energy, oxygen or algae proteins and similar biotechnologies. In such cybernetic composite solutions, in any case, a tremendous potential of our economy still lies almost completely untapped.

Similar to what was described earlier on the subject of architecture in Sect. 7.26, the further development of the last decades has already realized some of Vester's approaches expressed at that time, but others are still waiting to be put into practice more sustainably than before with cybernetic strategies, also using the symbiotic principle.

The extraordinary difficulty of integrating, in the environment of our evolved technosphere, long-term proven principles from a survival system of the highest "entrepreneurial" quality, proven over billions of years, is obvious. The development strategies in the two spheres, the biosphere and the technosphere, are too different.

But we have no other choice! The still acceptable limits of our technospheric growth have long since been exceeded and are beginning to show worrying consequences.

The question posed above, where the decision-making aids are to come from that will lead us into a future with the prospect of not accumulating even more consequential problems that are produced by our present actions, some of which are reckless towards ourselves, can only be answered:

> We must win back nature, its development strategy and principles, as a cooperative partner. This is primarily a communication task and requires new mental models; only secondarily is it an application-oriented task.

7.30 Design and Form: Blasting Sterile, Misappropriated and Pathogenic Technospheric Structures Through New Constructions Taking into Account Biological Rhythms

Section 5.4.8 has already addressed the issue of biological design, in the context of Vester's eight biocybernetic principles. How can technology benefit from this? Figure 7.83 shows the way there.

Some constructive approaches to biological design and body geometry worthy of imitation in technical terms have already been addressed above in

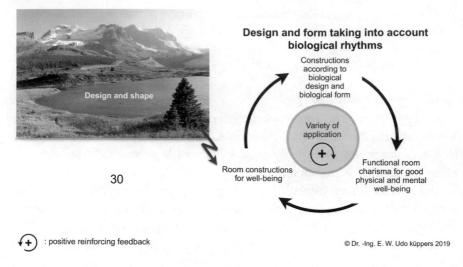

Fig. 7.83 Routine circuit 30 design and shape

Sect. 7.26. Here we want to explore principles and discuss with Vester the principled adherence to a basic biological design (Vester 1991a, b, pp. 82–83):

> Every product, every function and every organization should be compatible with the biology of humans and nature. This is not only an ecological demand, but increasingly also an economic one. The health of humans and nature is, after all, closely intertwined with the entire national economy via social costs and environmental impact. [...]. According to this principle, structures, functions and forms of organisation, if they are to lead to subsystems capable of survival, must conform to the laws of the biosphere.
>
> For example, a typical biological design principle is the irregularity in regularity mentioned in Sect. 5.4.8.

No bird's nest within a species is the same as another, although all serve their structural and form purpose.

The leaves of a tree all have similar shapes, but differ in details and are therefore never congruent, or, to put it technically: off the shelf.

The fur-patterns of zebras only look similar, is never completely same.

No matter what organisms are considered, the principle of irregularity in regularity is present in all.

> This also corresponds to the structure of our own body cells. The perception of such shapes therefore creates resonance with one's own pattern, i.e. recognition, familiarity and a sense of well-being – and thus not least also increased acceptance and thus, for example, increased sales opportunities on the market. Here, as so often, ecology also means economy.

The same is true of the attractiveness of house backdrops and their either stressful uniformity or homely irregularity, as we find them in the harbour of St. Tropez or in the differently inclined houses on the canals of Amsterdam or the clay and adobe forms of the natural building style of anonymous architecture, which radiate a deep sense of security and can be found from Africa to Persia to the pueblos of New Mexico, but also increasingly in modern buildings.

Planning and design should therefore never take place in isolation, but in feedback with the living environment. This extends to a new form of citizen participation, for example in state or regional planning, and a far greater inclusion of social compatibility in organisational as well as technological development (ibid.).

The theory is forward-looking – only the implementation leaves much to be desired. Citizen cooperatives are a ray of hope in the construction of housing, which aims at well-being, common ground and mutual help and support.

School rooms that often resemble the inner design of a sickly whitewashed cube more than a room design that stimulates all the senses of the pupils during learning are still the rule. The room has to be square, rectangular and practical, technically optimised instead of biologically adapted to the well-being and learning motivation of children and young people.

Open schools for holistic learning, in which the building structure and the spatial – organismic – design also have a considerable influence on the learning success of the children and young people, was realised, among other things, by the foundation of the Hanover Glockseeschule as an all-day school in 1972, Fig. 7.84. Decisively involved in this, in the tradition of German reform pedagogy, was the Hanover sociologist and social philosopher Oskar Negt (2004) as well as the educationalists Albert Illien and Thomas Ziehe.

The artist Friedensreich Hundertwasser,[58] who died in 2000, realized the principle of irregularity within regularity through his architectural design, as Fig. 7.85 shows. What a difference from the linearized house fronts of other school buildings! And what feeling arises from the contemplation of this irregular façade as compared with one that is monotonously smooth and gridded according to standards?

When in 2008 a German politician[59] publicly proclaims the so-called *"educational republic of Germany"* with great euphoria, and 12 years later metal

---

[58] https://www.hundertwasser.com/architektur (accessed 2020-02-20).
[59] https://www.faz.net/aktuell/politik/inland/nationaler-bildungsbericht-merkel-ruft-bildungsrepublik-aus-h1545858.html (accessed 2020-02-23).

Building of the Glockseeschule in Hannover, Germany

Founding building   current building

Fig. 7.84 All-day school Glockseeschule in Hanover. Left: Founding school seat in the former Fuhramt, today: Independent Youth Center Glocksee. (Photo source: AxelHH CC BY-SA 4.0, https://de.wikipedia.org/wiki/Datei:Jugendzentrum_Glocksee_seitlich. jpg). Right: Glockseeschule Am Lindenhofe, Doehren, Hannover. (Photo source: Christian A. Schröder CC BY-SA 4.0 https://de.wikipedia.org/wiki/Glockseeschule#/media/Datei:Glockseeschule_school_Am_Lindenhofe_Doehren_Hannover_Germany_01.jpg)

Luther Melanchthon Grammar School, Hundertwasser School, Lutherstadt Wittenberg

South side   North

Fig. 7.85 Friedensreich Hundertwasser: Building design and form of the Luther Melanchthon High School in Wittenberg. Left: Luther Melanchthon High School, Hundertwasser School, Lutherstadt Wittenberg, south side 2006. (Source: Doris Antony, CC BY-SA 3.0, https://commons.wikimedia.org/w/index.php?curid=539494). Right: Hundertwasser School in Lutherstadt Wittenberg, north side, 2010. (Source: Tnemtsoni CC BY-SA 3.0 https://commons.wikimedia.org/wiki/File:HW-Schule.jpg)

containers become the permanent provisional solution for German school classes, something must have gone considerably wrong mentally and practically in politics.

Design and form have not yet arrived in the architectural design and construction of educational institutions to the extent that children, young people and adults, as living organisms, are supported or encouraged in their learning, also and especially by their spatial environment, if lasting education is to endure as a central means of progress in a society.

7.31 Concluding Remarks

Section 7.30 ends the short voyage of discovery through biospheric-technospheric transformations, which of course could only "scratch the surface" of possible bionic products and processes, but which nevertheless – as the author hopes – could provide sufficient insight into the still often hidden genius of the natural *"creativity and intelligence" of* many organisms.

I also hope to have made it clear that the biological-technical transfer not only draws from the immeasurable reservoir of natural ingenious principles, but also has an effect on the whole range of technospheric applications. In this respect, the term *technology* or *technical* stands as a synonym for a multitude of differentiated, manufacturing, mechanical, infrastructural, structural, sociological, cultural, educational, economic, traffic-oriented, mobile and other areas of activity.

The strength of nature lies in its cybernetic decentralization, which allows organisms to develop peak performance for their survival. The principles behind this are the key to sustainable further development for ourselves and our habitat Earth.

Despite all the current trends, which also occupy the book market, imagining, among others, "The world without us", (Weisman 2008), describing a "Life 3.0" (Tegmark 2017), or even Jørgen Randers' (2012) global forecast for the next 40 years, entitled "2052", probably underestimate the power of nature, to which we will have to bow with insight sooner rather than later.

References

Aldersey-Williams, H. (2003) zoomorphic. new animal architecture. Harper Design International, New York

Al Sawalmih, A. et al. (2008) Microtexture and Chitin/Calcite Orientation Relationship in the Mineralized Exoskeleton of the American Lobster. Advanced Functional Materials, Volume 18, Issue 20, Pages 3307–3314

Bechert, D. W.; Barlenwerfer, M.; Hoppe, G.; Reif, W.-E. (1986) Drag reduction mechanism derived from shark skin. ICAS, Congress, 15th, London, England, September 7–12, 1986, Proceedings. Volume 2 (A86-48976 24-01). New York, American Institute of Aeronautics and Astronautics, Inc., 1986, S. 1044–1068.

Bechert, D. W.; Reif, W.-E. (1985) On the drag reduction of the shark skin. 23rd Aerospace sciences meeting, 1985 – arc.aiaa.org

Becker, P.-R.; Braun, H. (Hrsg.) (2001) nestWerk. Architektur und Lebewesen. Isensee, Oldenburg

Braun, D. H. (2008) Bionisch inspirierte Gebäudehüllen. Diss., Fakultät Architektur und Stadtplanung der Universität Stuttgart (D93)

Brickmann, J. (2009) Kleben und Entkleben auf Kommando. Interview mit Dr. Ramón Barcardit, Henkel AG & Co KGaA. In: labor & more, 1/09, S. 32

Brockmann, W. et al. (2005) Klebstoffe, Anwendungen und Verfahren. Wiley-VCH, Weinheim

Daly, I. M. et al. (2016) Dynamic polarization vision in mantis shrimps. Nature Communications 7, Article number: 12140 (2016)

Duhigg, Ch. (2014) Die Macht der Gewohnheit. Piper, München, Zürich

Duhigg, Ch. (2012) The Power of Habit. Random House Group, London

Farah, G.; Siwek, D.; Cummings, P. (2018) Tau accumulatins in the brains of woodpeckers. PLoS ONE 13(2): e0191526

Fink, D. et al. (2006) The "artificial ostrich eggshell" project: Sterilizing polymer foils for food industry and medicine. Solar Energy Materials and Solar Cells. vol. 90, Issue 10, pages 1458–1470

Fish, F. E.; Lauder, G. (2013) not just Going with the Flow. Modern visualization techniques show that aquatic animals can modify their fluid environment to increase the efficiency of swimming and food collection. American Scientist, March-April 2013, Vol 101, Number 2, S. 114

Fogel, D. B. (1999). An overview of evolutionary programming, in: (Davis et al., 1999), S. 89–109

Fogel, D. B.; Chellapilla, K. (1998). Revisiting evolutionary programming, in: Rogers S. K.; Fogel, D.B.; Bezdek, J. C.; Bosacchi, B. (Hrsg.), Applications and Science of Computational Intelligence, S. 2–11, SPIE, Bellingham, WA.

Fogel, D. B. (1988). An evolutionary approach to travelling salesman problem, Biological Cybernetics, 6(2), S. 139–144.

Fogel L. J.; Owens A. J.; Walsh, M. J. (1965). Artificial intelligence through a simulation of evolution, in: Maxfield, M.; Callahan, a.; Fogel, L. J. (Hrsg.), Biophysics and Cybernetic Systems: Proc. of the 2nd Cybernetic Sciences Symposium, S. 131–155, Spartan Books, Washington, D.C.

Forrester, V. (1997) Der Terror der Ökonomie. Wien: Zsolnay.

Fricke, J.; Borst, W. L. (1981) Energie. Ein Lehrbuch der physikalischen Grundlagen. Oldenbourg, München, Wien

Fritz, A. H.; Schulze, G. (Hrsg.) (2015) Fertigungstechnik, 11. Aufl., Springer Vieweg, Berlin, Heidelberg

Füsslein (2013) Zum großen Atem der Nachhaltigkeit – Ein persönlicher Erfahrungsbericht aus Sachsen,. In: Sächsische Carlowitz-Gesellschaft (Hrsg.) Die Erfindung der Nachhaltigkeit. S. 243–256, oekom, München

Fuller, R. Buckminster; Applewhite, E. J. (1975). Synergetics. New York: Macmillan.

Gerner, M. (2007) Fachwerk. Entwicklung, Instandsetzung, Neubau. DVA, München

Gleich, M. et al. (2000) Life counts. Berlin, Berlin

Goldberg D. E.; Deb, K.; Clark J.H. (1992) Genetic algorithms, noise, and the sizing of populations, Complex Systems, 6, S. 333–362.

Goldberg D. E.; Richardson, J. (1987). Genetic algorithms with sharing for multi-modal function optimization, in: (Grefenstette, 1987a), S. 41–49.

Goldberg, D. E.; Lingle, Jr. R. (1985) Alleles, loci, and the traveling salesman problem, in: (Grefenstette, 1985), S. 154–159.

Gomez, P.; Probst, G. (1999) Die Praxis des ganzheitlichen Problemlösens. Haupt, Bern, Stuttgart, Wien

Habenicht, (2005). Grundlagen, Technologie, Anwendungen. Springer, Heidelberg

Hammer, A. (2016) Climate Adaptive Building Shells for Plus-Energy-Buildings, Designed on bionic Principles. In: World Academy of Science, Engineering and Technology International Journal of Architectural and Environmental Engineering, Vol: 10, No: 2, 202–213, 2016

Harzheim, L. (2008) Strukturoptimierung. Grundlagen und Anwendungen. Harri Deutsch, Frankfurt a. M.

Heinrich-Böll-Stiftung et al. (2020) Insektenatlas 2020. 1. Aufl., Selbstverlag, ISBN 978-3-86928-215.2

Heinert, H. (1976) Grundlagen der Bionik. VEB deutscher Verlag der Wissenschaften, Berlin

Hertel, H. (1963) Struktur, Form, Bewegung. Krausskopf, Mainz

Herzog, T. (1995) Transluzente Wärmedämmung. Detail – Zeitschrift für Architektur + Baudetail, 1995, n. 1, v. 35, S. 32–39

Hickman, C. P. et al. (2008) Zoologie. Pearson, München

Holland J. H. (1992). Adaptation in Natural and Artifical Systems, MIT Press, Cambridge, MA

Holland J. H. (1973). Genetic algorithms and the optimal allocation of trials, SIAM Journal on Computing, 2(2), S. 88–105.

Holland J. H. (1969). A new kind of turnpike theorem, Bulletin of the American Mathematical Society, 75(6), S. 1311–1317.

Holmberg, K.; Erdemir, A. (2017) Influence of tribology on global energy consumption, costs and emissions. Friction, Sept. 2017, Volume 5, Issue 3, S. 263–284

Huhn, F. et al. (2015) Quantitative flow analysis of swimming dynamics with coherent Lagrangian Vortices. In: Chaos 25, 087405 (2015)

Industrieverband Klebstoffe e. V. (2016) Handbuch Klebtechnik. Springer Vieweg, Wiesbaden

Jodidio, P.; Gössel, P. (Hrsg.) (2016) Santiago Calatrava: Architekt, Ingenieur, Künstler, Taschen, Köln

Kegel, B. (2018) Epigenetik. Wie unsere Erfahrungen vererb werden. DuMont, Köln

Kramer, M. O. (1960) The Dolphins' Secret. New Scientist 7, S. 1118–1120, London

Kurmann, F. (2020) Digitale Bauweise mit großer Tragweite. VDI-Nachrichten, Nr. 4/5, 24.1.2020, 8

Küppers, J.-P.; Küppers, E. W. U. (2016) Hochachtsamkeit. Über unsere Grenzen des Ressortdenkens. Springer VS, Wiesbaden

Küppers, J.-P. und Küppers, E. W. U. (2015) Bedingt handlungsbereit. Die jüngste Migrationswelle und ihre Grenzen systemischer Krisenbewältigung in einer globalisierenden Welt. ZPB 3/2015, 110–121

Küppers, E. W. U. (2015) Systemische Bionik. Impulse für eine nachhaltige gesellschaftliche Weiterentwicklung. Reihe: Essentials, Springer Vieweg, Wiesbaden

Küppers, E. W. U. (2011a) Die systemische Kommune. AKP 1/2011, 52–54

Küppers, E. W. U. (2011b) Die Wirkungsnetz-Organisation – ein Modell für die öffentliche Verwaltung? apf 5/2011, 129–136

Küppers, E. W. U. (2011c) Systemische Denk- und Handlungsmuster einer neuen nachhaltigen Politik im 3. Jahrtausend. ZPB, July 2011, Volume 3, Issue 3–4, S. 377–398

Küppers, E. W. U. (2007a) Kleine Biegung, große Wirkung. Bionische Rohrbögen in der Lüftungstechnik. Chemie Technik, Sept. 2007, 118–120

Küppers, E. W. U. (2007b) Natureffiziente Lösungen erobern die Technik. Teil 1, HLH Bd. 58, Nr. 11, Nov.; Teil 2: HLH Ba. 58, Nr. 12, Dez.

Küppers, E. W. U. (2007c) Rohrbögen mit Mäander-Effekt[®]. Schiff & Hafen, Sept. 2007, 56–62

Küppers, E. W. U.; Heyser, W. (2004/2005) MANGO – Systemischer Baukasten umweltverträglicher Verpackungsmaterialien auf der Basis biologischer, funktional-optimierter Schalen und Hüllen. F&E-Projekt gefördert durch das BMBF, PTJ-Bio/Fkz-0311980

Küppers, E. W. U.; Tributsch, H. (2002) Verpacktes Leben – Verpackte Technik. Bionik der Verpackung. Wiley-VCH, Weinheim

Küppers, E. W. U. (2002a) Grenzflächen des Lebens – bionische Nutzen für die Verpackungstechnik? In: Baier, et al. (Hrsg.) Transparenz und Leichtigkeit. Symposium, 33–68, Universität Duisburg- Essen, Druck 2003

Küppers, E. W. U. (2002b) VR-Serie: Bionik. Teil I: Falten und Falttechniken der Natur. In: Verpackungs-Rundschau 2/2002, 75–76

Küppers, E. W. U. (2001) Bionik und Bauen. Lernen von den Baumeistern der Natur. In: Becker, P.-R.; Braun, H. (Hrsg.) (2001) nest Werk. Architektur und Lebewesen. Isensee, Oldenburg, S. 87–102

Küppers, U. (1995) Biokybernetische, systemverträgliche Siedlungsstruktur. Ansatz für eine ganzheitlich vernetzte Raumplanung. In: Schaur, E. (Hrsg.) (1995) Intelligent Bauen – Building with Intelligent. S. 130–143, IL41, ©Institut für leichte Flächentragwerke, Krämer, Stuttgart

Küppers, U. (1983) Randwirbelteilung durch aufgefächerte Flügelenden. Diss., Fortschritt-Berichte der VDI Zeitschriften, Reihe 7: Strömungstechnik, Nr. 81, 165 S.

Mancuso, S. (2018) Pflanzen Revolution. Wie die Pflanzen unsere Zukunft erfinden. Kunstmann, München

Mattheck, C. (2017) Enzyklopädie der Formfindung nach der Natur. KIT Karlsruhe, KS Druck GmbH, Kronau

Mattheck, C. (2010) Denkwerkzeuge nach der Natur. 1. Aufl. KIT Karlsruhe, KS Druck GmbH, Kronau

Meier, C. J. (2017) Vielzweckgreifer haftet wie ein Gecko. Max-Planck-Gesellschaft, Forschungsbeitrag v. 26.5.2017 (https://www.mpg.de/11315370/vielzweckgreifer-haftet-wie-ein-gecko?filter_order=L&research_topic)

Nachtigall, W.; Pohl, G. (2013) Bau-Bionik. 2. Aufl., (1. Aufl. 2003), Springer Vieweg, Berlin, Heidelberg

Nazareth, A. (2018) Bionic Architecture. Degree of Master of Architecture. Unitec Institute of Technology, 2018, Auckland

Negt, O. (2004) … es war für uns alle Neuland. Zu den Anfängen und Charakteristika der wissenschaftlichen Begleitung der Glockseeschule in Hannover – Ein öffentlicher Brief an die Herausgeber. In: Heiner Ullrich, Till-Sebastian Idel, Katharina Kunze (Hrsg.): Das Andere erforschen – Empirische Impulse aus Reform- und Alternativschulen (= Franz Hamburger, Marianne Horstkemper, Wolfgang Melzer, Klaus-Jürgen Tillmann [Hrsg.]: Schule und Gesellschaft. Band 32). 1. Auflage. Verlag für Sozialwissenschaften, Wiesbaden

Onusseit, H. (2008) Praxiswissen Klebtechnik. Band 1: Grundlagen, Hüthig, Heidelberg

Otto, F. (1988) Gestaltwerdung – Zur Formenentstehung in Natur, Technik und Baukunst. Müller, Köln

Papadopoulos, J. (2020) In der Falle von Moria. Das völlig überfüllte Lager auf der Insel Lesbos ist zum Symbol für die gescheiterte Flüchtlingspolitik der EU geworden. In: Le Monde diplomatique, Februar 2020, S. 10

Patek, S. N.; Caldwell, R. L. (2005) Extreme impact and cavitation forces of a biological hammer: strike forces of the peacock mantis shrimp Odontodactylus scyllarus. The Journal of Experimental Biology 208, 3655–3664

Patek, S. N. (2019) The Power of Mantis Shrimp Strikes: Interdisciplinary Impacts of an Extreme Cascade of Energy Release. Symposium: Integrative and Comparative Biology, January 3–7, Tampa, Florida, S. 1–13

Patek, S. N (2016) Die schnellsten Bewegungen von Lebewesen. Spektrum.de, 20.1.2016, https://www.spektrum.de/magazin/ultraschnelle-bewegungen-von-lebewesen/1382041 (Accessed on 15.2.2020)

Patzelt, O. (1972) Wachstum und Bauen. VEB Verlag für Bauwesen, Berlin

Rabe, D. et al. (2006) Microstructure and crystallographic texture of the chitin–protein network in the biological composite material of the exoskeleton of the lobster

Homarus americanus. Materials Science and Engineering: A Volume 421, Issues 1–2, 15 April 2006, Pages 143–153

Randers, J. (2012) 2052. oekom, München

Rasche, M. (2012) Handbuch Klebtechnik. Hanser, München

Rechenberg, I. (1994). Evolutionsstrategie '94, frommann-holzbog, Stuttgart.

Rechenberg, I. (1973) Evolutionsstrategie: Optimierung technischer Systeme nach Prinzipien der biologischen Evolution, frommann-holzbog, Stuttgart

Rechenberg, I. et al. (2009) Tribologie im Dünensand. Schlussbericht BMBF-Förderkennzeichen 0311967A, 1.2.2006–31.3.2009

Rojas-Chapana, J. A. et al. (2005) Colloidal Assembly and Functionalization of Pore channels in Polymer foils. J. o. Porous Materials 12: 215–224

Rosen, M. W. (1959) Water flow about a swimming fish. Thesis, M. of, Science, University of California, Copy No. 28, U. s. Naval Ordnance Test Station.

Schaur, E.(Hrsg.) (1995) Intelligent Bauen – Building with Intelligent. IL41, ©Institut für leichte Flächentragwerke, Krämer, Stuttgart

Schwefel H. P. (1995). Evolution and Optimum Seeking, Wiley & Sons, New York, NY

Schwefel H. P. (1975). Evolutionsstrategie und numerische Optimierung, Doktorarbeit, Technische Universität Berlin, Berlin.

Sinclair, U. (1980) Der Dschungel. März, Berlin und Schlechtenwegen, ebenso 1985, Rowohlt, Reinbek b. Hamburg

Sinclair, U. (1984) Öl. März, Herbstein, ebenso 1986, Rowohlt, Reinbek b. Hamburg

Smoliga, J. M. (2018) Reconsidering the woodpecker model of traumatic brain injury. www.thelancet.com/neurology Vol 17 June 2018, (Accessed on 14.2.2020) S. 500–501

Song, S.; Drotlef, D.-M.; Majidi, C.; Sitti, M. (2017) Controllable load sharing for soft adhesive interface on three-dimensional surfaces. PNAS, May 30, 114 (22) E4344–E4353

Spork, P. (2017) Der zweite Code. Epigenetik oder: wie wir unser Erbgut steuern können. Rowohlt, Reinbek b. Hamburg

Tegmark, M. (2017) Leben 3.0. Ullstein, Berlin

Tributsch, H. (1995) Bionik solarer Energiesysteme. In: Nachtigall, W.; Wisser, A. (Eds.) Biona Report 9, 147–170, Akademie der Wissenschaft, Mainz, Fischer, Stuttgart

Tsui, E. (1999) Evolutionary Architecture. Wiley & Sons, Canada

VDI-Nachrichten (2020) Es geht ums Überleben. Interview mit dem Architekten Werner Sobek,. In: VDI-N. 14. 2. 2020, S. 8–9

Vélez, S. (2000) Grown your own House. Vitra Design Museum, Weil am Rhein

Vester, F. (1991a) Ballungsgebiete in der Krise. vom Verstehen und Planen menschlicher Lebensräume. (Orig., 1983). dtv-Sachbuch Nr. 11332, München

Vester, F. (1991b) Ausfahrt Zukunft Supplement. Materialien zur Systemuntersuchung. Studiengruppe für Biologie und Umwelt GmbH, München

Wallace, I. J.; Hainline, C.; Lieberman, D. E. (2018) Sports and the human brain: an evolutionary perspective. Chapter 1. In: Handbook of Clinical Neurology, Vol. 158 (3rd series) Sports Neurology B. Hainline and R.A. Stern, Editors

Wang, L. et al. (2013) Biomechanism of impact resistance in the woodpecker's head and its application. In: SCIENCE CHINA Life Sciences, vol 56, No. 8, 715–719

Weicker, K. (2007) Evolutionäre Algorithmen, 2. Aufl., B. G. Teubner, Wiesbaden

Weisman, A. (2008) Die Welt ohne uns. Piper, München, Zürich

Wunderlich, K.; Gloede, W. (1977) Natur als Konstrukteur. Edition Leipzig, DDR

Yang, Y. et al. (2017) Biomimetic Anisotropic Reinforcement Architectures by Electrically Assisted Nanocomposite 3D Printing. Advanced Materials, 2017, 29, 1605750

Yi, L.; Hu, J.; Wu, Z. (2020) Fabrication of Coatings with Structural Color on a Wood Surface. Coatings, 10, 32, 1–12

Yuan, Y. et al. (2017) Bionic building energy efficiency and bionic green architecture: A review. In: Renewable and Sustainable Energy Reviews 74 (2017) 771–784

Zheng, Y. (2019) Bioinspired Design of Materials Surface. Elsevier, Amsterdam NL

Zuazua-Ros, A. et al. (2017) bio-inspired heat dissipation system integrated in buildings: development and applications. In: Energy Procedia 111 (2017) 51–60

Part III

Antagonist of Nature

We all should not be under the illusion, in view of the outstanding achievements that arise from evolution (Main Part I) and in view of the outstanding abilities and possibilities we possess to use these natural ingenious principles for our purposes of life and work (Main Part II), that all this will be accepted without resistance.

We see ourselves as a human species until today, until tomorrow and certainly far into the future less as learned students, as an integral part of an evolutionary development, but rather as *absolute beneficiaries of* everything that nature offers us. In many ways, we strive for short-term – and short-sighted – success at the expense and expense of nature in our technosphere. At the same time, we ignore the foresighted need to think about what it also means for us to do significant damage to nature in the medium to long term.

After all, we have managed to attribute to humanity its own geological era, the Anthropocene, but without being able to be proud of it.

To promote education in the minds of *all* people is a basic law for our survival. A related basic law is to recognize that mental and practiced vicious circle routines do not lead out of the self-inflicted catastrophe that is abundantly visible, among other things, in the earth-wide mountains of waste and garbage swirls in the oceans.

The destructiveness of the effective triangle of *politics, economics* and *finance* has a considerable share in the present state of the earth on which we have to live for a while yet.

Important *warners* and *admonishers* have long taken up the fight against the opponents or enemies of nature. Main Part III lets some of them have their say – and for once man is not a means to an end but himself at the centre of the action.

8

Fight or Perish!
A Critical Look at Our Planet, Society, People and Things

Abstract The title of this chapter is challenging! In "Fight or Perish!" it is about nothing less than the survivability of our societies on earth and the enemies, the numerically small but powerful "masters of mankind", who are still allowed to plunder the earth to their liking – without regard to human, animal and plant losses.

Three, in the author's view, thoroughly interconnected complexes are illuminated:

1. *Konrad Lorenz "Eight Deadly Sins of Civilized Mankind"*
2. *The global foresight on the environment* and
3. *Noam Chomsky's observation: why we must stand up to the masters of humanity.*

They all show the absolute necessity to act instead of react, to bring our earth back into the state of an increasingly dynamic equilibrium. This must be done without the pernicious effect of *"vicious spirals"* initiated by ourselves, interacting *"positive feedbacks"* with their baleful destructive forces on organisms.

Biocybernetics can help to contribute a lot to a sustainable progress, which is also accompanied by adapted growth, by increasing installation of "negative", system-stabilizing feedback loops. This system-stabilizing effect is not limited to nature itself, for it is a master at it and needs no instruction from us humans!

© The Author(s), under exclusive license to Springer Fachmedien Wiesbaden GmbH, part of Springer Nature 2022
E. W. U. Küppers, *Ingenious Principles Of Nature*,
https://doi.org/10.1007/978-3-658-38099-1_8

Conversely, the more promising path would be to guide societal or economic, social and ecological processes, in their existing – often unrecognized – interconnectedness, towards goals through forward-looking resilient strategies and sustainable progress.

The best school in which a young person can learn that the world has a meaning is the direct contact with nature itself (*Konrad Lorenz 1983*).[1]

8.1   Konrad Lorenz "Eight Deadly Sins of Civilized Mankind" from His and Today's Point of View: Excerpt

Communicated to listeners in advance in a series of lectures on Bavarian radio, from November to December 1970, the Austrian zoologist and winner of the Nobel Prize for Medicine, Konrad Lorenz, published the 112-page book on the *"Eight Deadly Sins of Civilized Mankind"* in the same year of his honor.

It was the time of the environmental movement, the anti-nuclear movement, which won over the whole of the Federal Republic of Germany because of its persistence and public presence. It was the time of the *"Atomkraft? – No Thanks"* logo and the energy industry's question-and-answer game, mocked by anti-nuclear activists: *"Where does the electricity come from? – With us, the electricity comes from the socket!"* The recommendable book by Joachim Radkau and Lothar Hahn (2013) on the rise and fall of the German nuclear industry also took an in-depth look at the 1970s.

In this socially and politically agitated time, Konrad Lorenz's small booklet appeared in 1973, in which he not only lamented the state of nature, but also addressed human nature, not least from the environment of ecology and genetics, but especially from the field of "comparative behavioural research", which he founded, whereby Lorenz's statements had, however, also received a lot of criticism. This happened, for example, with the expressed analogy between *"disease and society"* and *"diseases in the sense of medicine"* (Wuketis 1990, p. 202).

Table 8.1 summarizes Lorenz's "Eight Deadly Sins of Civilized *Mankind"* before going into more detail (italics, shaded) on some of them in the context of this book on *Ingenious Principles of Nature,* from that time (1970s) and today (2020s), although all eight "deadly sins" concern natural and social development together but in a differentiated way.

---

[1] From: Wuketits, F. M. (1990) Konrad Lorenz, Life and Work of a Great Naturalist, citation page.

Table 8.1 Konrad Lorenz' eight deadly sins of civilized mankind

1	*Structural properties and dysfunctions of living systems*
2	Overpopulation
3	*Devastation of the natural habitat*
4	*Humanity's race against itself*
5	Heat death of the feeling
6	*Genetic decay*
7	Tearing down tradition
8	Indoctrinability

8.1.1   Konrad Lorenz "First Deadly Sin of Civilized Mankind": Structural Properties and Dysfunctions of Living Systems

View from the 1970s (Lorenz 1993, pp. 11–18)

Ethology can be defined as that branch of knowledge which came into being by applying to the study of animal and human behaviour the questions and methods which have been natural and obligatory in all other biological disciplines since Charles Darwin. [...] Ethology, then, treats animal as well as human behaviour as the function of a system which owes its existence as well as its particular form to a historical development which has taken place in the history of the phylum, in the development of the individual and, in the case of man, in the history of culture. The genuine causal question as to *why* a particular system is so and not otherwise constituted can find its legitimate answer only in the natural explanation of this development.

Among the causes of all organic development, besides the processes of mutation and the recombination of genes, natural *selection* plays the most important role. It brings about what we call *adaptation*, a genuine cognitive process by which the organism assimilates information that is present in the environment and is of importance for its survival, m. a. W. by which it acquires knowledge about the environment. [...].

The analysis of the organic system underlying human social behaviour is the most difficult and ambitious task that natural science can set itself, for this system is by far the most complex on earth. [...].

A structural characteristic of all higher integrated organic systems is that of regulation by so-called control loops or homeostases (self-reinforcing control loops with so-called "positive feedbacks" lead at best to unstable equilibria, at worst to dysfunctions and the end of all activities, the author).

View from the Year 2020

Without going into detail about the human drives mentioned by Lorenz, such as " […] hatred, love, friendship, anger, loyalty, attachment, distrust, trust, etc. etc. […]" (ibid., p. 15), which are associated with certain behaviours and whose good or bad evaluation is of little value without an understanding of the system function of the whole system, we focus on the evolutionary factors, such as mutation, recombination, selection, etc. A process of adaptation, "[…] by which the organism assimilates information […]" (ibid., p. 11), as Lorenz put it, always occurs with the involvement of chance, which guides the adaptation of the organism to a new environmental situation.

Lorenz, however, could not yet know that an epigenetic influence, in the form of specific human behaviour, e.g. diet, vices such as smoking, etc., also affects the gene sequences of deoxyribonucleic acid (DNA) in the same way as genetic mutation and recombination. Here, environmental behavioral factors, which were still discarded in Lamarck's[2] time, act as acquired traits on the gene sequences of offspring.

The fact that a single counteracting "negative" feedback into an otherwise reinforcing "positive" control loop has a system-stabilizing effect has been known for some time by the mathematician and cyberneticist Norbert Wiener (1948). Technical – even more – organismic systems consist of a multitude of interconnected control loops whose dynamic sequence regulates highly complex organismic functional processes (see Küppers 2019, Chap. 7, 153 ff., Cybernetic Systems in Practice). Potential dysfunctions in an organismic system can be adaptively "regulated out" in this way and the organism can be restored to a state of homeostatic equilibrium.

Lorenz completely ignores the structural properties and dysfunctions as well as the behavioural patterns of plants as organisms. Today we know about the enormous survival-securing communication ability of stationary plants towards their conspecifics and alien organisms, which they master with a large sensory repertoire to ensure their survival (see Chap. 4).

---

[2] The Lamarckian theory of evolution is based on the fact that organisms can pass on acquired characteristics to their offspring. The founder of the theory was the French biologist Jean-Baptiste de Lamarck (1744–1829).

8.1.2 Konrad Lorenz "Second Deadly Sin of Civilised Mankind": Overpopulation

View from the 1970s (Lorenz 1993, pp. 19–22)

In the single organism one normally hardly ever finds a circle of positive feedback. Only life as a whole is allowed to indulge in this immoderation, hitherto apparently with impunity. Organic life has built itself into the stream of dissipating world-energy, it "eats" negative entropy,[3] it tears energy to itself and grows with it, and through its growth it is put in a position to tear more and more energy to itself and to do this the faster, the more it has already captured. That this has not yet led to overgrowth and catastrophe is due to the fact that the compassionless powers of the inorganic, the laws of probability, keep the multiplication of living things within bounds; but secondly, it is due to the fact that regulatory circuits have developed within the various kinds of living things. [...].

It is advisable to discuss the excessive multiplication of human beings first, if only because many of the phenomena to be discussed later ("deadly sins") are its consequence. All the gifts that come to man from his profound insights into the surrounding nature, the advances of his technology, his chemical and medical sciences, everything that seems likely to alleviate human suffering, works horribly and paradoxically to the ruin of mankind. It threatens to do precisely what almost never happens to otherwise living systems, namely, to suffocate within itself (namely, through overpopulation, the author). [...]

All of us who live in densely populated countries or even in large cities no longer know how much we lack general, warm and cordial love of humanity. One has to have been an uninvited guest in a house in a really sparsely populated country, where several kilometres of bad roads separate the neighbours from one another, in order to appreciate how hospitable and loving people are when their capacity for social contact is not constantly overtaxed. (the author knows this kind of hospitality from sparsely populated Namibia, Africa). [...].

Certainly, the cramming together of masses of people in modern cities is largely to blame for the fact that we are no longer able to see the face of our neighbor in the phantasmagoria of eternally changing, overlapping and blurring images of human beings. Our love of neighbor is so diluted by the masses of neighbors, of those who are all-too-close, that in the end it is no longer even traceable. [...].

---

[3] Negative entropy or negentropy, as opposed to entropy (heat, disorder) is absorbed by living things to create order through structures to keep entropy low, which in turn is exported to the environment. However, this uptake of order or import of negentropy is limited by the individual life span of the organism.

So we have to make a choice, that is, we have to "keep at bay" emotionally many other people who would certainly be just as worthy of our friendship. "Not to get emotionally involved" is one of the main worries of many city dwellers. […].

The further the massification of human beings goes, the more urgent becomes for the individual the necessity "not to get involved," and so today, especially in the largest cities, robbery, murder, and rape can go on in broad daylight and on densely crowded streets without a "passer-by" intervening. The cramming together of many people in a confined space leads not only indirectly to phenomena of dehumanization through exhaustion and the silting up of interpersonal relationships, it also triggers aggressive behavior quite directly. […].

The general unfriendliness that can be observed in all large cities is clearly proportional to the density of the crowds accumulated in certain places. In large railway stations or in the bus terminal in New York, for example, it reaches degrees that are appalling. Indirectly, overpopulation contributes to all ills and decay, […]. The belief that by appropriate conditioning a new breed of people can be created who are immune to the evil consequences of the closest confinement is, in my opinion, a dangerous delusion.

View from the Year 2020

In the face of increasing urbanisation over the last few years to decades, producing sprawling conurbations in some regions with tens of millions of inhabitants in narrow urban canyons; in the face of dormant state administrative bureaucracies and their managements which, in the absence of forward-looking perspectives, have exacerbated the struggle for acceptable, feel-good affordable housing by rigorously cutting social housing, as has happened in Berlin and elsewhere; in the face of the profitable investment of sprawling high-priced housing units for a relatively small elite in the population; given the increasing alienation of neighbours in so-called blocks and towers of flats, in miserable conditions and potential for danger, such as the Grenfell Tower in the London borough of Kensington, which burnt out in 2017; given the perceived increasing brutalisation of moral values with a lack of respect between people, which are currently attracting inglorious publicity in politics (threats to the murder of politicians) and sport (disparaging incitement against individuals by alleged football fans), Lorenz's statements have been amplified in their current extent and planned arbitrary aggressiveness.

At the very moment – and far beyond a humane state of affairs – catastrophes of great humane proportions are taking place in the refugee camps on the Greek islands, such as Chios, Samos and Lesbos (the author, together with a co-author, already described in 2015, at the time of the great wave of refugees that caught Germany in particular but not only unprepared, the conditions in the refugee camps that were already undignified at that time (Küppers and Küppers (2015)).

It was and is a wretched war at the expense of those in need of help, which power seekers and other directly and indirectly involved politicians and nations are waging in the Middle East. The situation intensified at the Syrian-Turkish and Greek-Turkish borders at the end of February and beginning of March 2020 due to the influx of refugees from the northern Syrian war zone around Idlib. The human suffering is clearly aggravated by the arbitrariness of despotic parties, after the opening of the Turkish border in the direction of the European mainland – Greek border as the external border of the European Union – to receive the refugees from the Syrian war at this border with barbed wire, water cannons and tear gas grenades. Very recently, arsonists who set fire to the completely overcrowded Moria camp on the Greek island of Lesbos, thus destroying even the most makeshift tents in which about 12,000 refugees are living, show the frustration of those seeking help on the one hand and the almost complete inability of the European Union to provide *sustainable*(!) aid on the other. (https://www.unhcr.org/dach/de/52533-unhcr-nothilfe-auf-moria-wird-weiter-verstaerkt.html,   https://www.proasyl.de/news/katastrophe-von-moria-soforthilfe-und-evakuierung-jetzt/, https://wirtschaft.com/aerzte-ohne-grenzen-gegen-migrationsplan-der-eu-kommission/, accessed 25.9.2020).

Our partial and global inability to deal with waves of refugees migrating from war, destruction and loss of homeland on the one hand, and global overpopulation on the other, shows abundantly clearly our mental weaknesses of foresighted and sustainable crisis management, more so than before.

But let's get back to the subject of overpopulation. In 1973, the total number of people on earth was 3925 billion. In 2020, it is already 7758 billion,[4] almost doubling within about 50 years!

With the increasing number of earth citizens, the period of time per year in which we can satisfy the earth through our hunger for global resources and the – despite all efforts still extremely insufficient nutrition of our constantly – in the regionally strongly varying – growing world population, without overloading the ecosystem of the earth, is also reduced.

"Earth Overload Day"[5] states that by then humanity can consume as much natural resources as the Earth can renew in a year.

---

[4] https://population.un.org/wpp/Download/Standard/Population (accessed 2020-03-02).
[5] https://germanwatch.org/sites/germanwatch.org/files/FAQ_Erdüberlastungstag_2019_1.pdf (accessed 2020-03-02).

In 1973, when Konrad Lorenz wrote his *"eight deadly sins of civilized man-kind"*, the nearly 4 billion human beings still had almost 11 months left for the annual consumption of resources that the earth renews every year. The *"Earth overload day" in* 1973 was at the end of November.

Last year in 2019, *"Earth Overload Day"* already moved forward to the end of July; in 2020, it moved back by about 3 weeks – due to the SARS-CoV-2 pademy – to August 22 (https://germanwatch.org/de/overshoot, accessed 9/25/2020).

Although the comparison is a statistical one, showing that globally we now need 1,75 Earths, according to American consumption even 4 Earths, and according to German "hunger for resources" still 3 Earths, in order to bring the regenerative capacity of the Earth back into balance, these figures must actually frighten everyone, especially the so-called "masters of humanity", as Chomsky calls them (see Sect. 8.3).

Whether it will eventually be possible to put the problems of humanity just described, which are becoming more intense and extensive rather than more attenuating and smaller, into the "metal hands" and AI algorithms full of artificial intelligence of robots or humanoids one day, as is already happening at some – friendly and hostile – border points in China and elsewhere (Küppers 2018, pp. 271–272), is more than doubtful – but not impossible – not least because of the high process dynamics and complexity.

8.1.3   Konrad Lorenz "Third Deadly Sin of Civilised Mankind": Devastation of the Natural Habitat

View from the 1970s (Lorenz 1993, pp. 23–31)
Konrad Lorenz begins by stating:

> It is a widespread misconception that nature is inexhaustible. Every species of animal, plant or fungus – for all three kinds of living beings belong to the great wheel – is adapted to its environment, and to this environment belong, of course, not only the inorganic constituents of a given locality, but quite equally all its other living inhabitants. All the living beings of a habitat are thus adapted to *one another*. […].
>
> In short, two forms of life can stand in a very similar relationship of dependence to each other as man does to his domestic animals and cultivated plants. The laws governing such interactions are often quite similar to those of the human economy, which is also expressed in the term that biological science has coined for the study of these interactions: it is called *ecology*. However, *one* eco-

nomic term, which will concern us here, does not appear in the ecology of animals and plants: it is that of *overexploitation.*

The interactions in the fabric of the many species of animals, plants and fungi that together inhabit a habitat and together make up the biotic community or biocenosis are tremendously diverse and complex. The adaptation of the various types of living organisms, which has taken place over periods of time whose scale corresponds to geology and not to human history, has led to a state of equilibrium that is as admirable as it is easily disturbed. Many regulatory processes safeguard this against the inevitable disturbances caused by weather and the like. All slowly occurring changes, such as those caused by the evolution of species or gradual changes in climate, cannot endanger the equilibrium of a habitat. Sudden impacts, however, even if apparently minor, can have unexpectedly large, even catastrophic effects. The introduction of an apparently harmless species can literally devastate vast areas of land, as has happened in Australia with the rabbit. This intervention in the equilibrium of a biotope has been caused by man. However, the same effects are in principle also conceivable without his intervention, albeit more rarely.

The ecology of man is changing many times faster than that of all other living beings. The pace is dictated by the progress of his technology, which is constantly accelerating in geometric proportion. Therefore, man cannot but cause profound changes and, all too often, the total collapse of the biocenoses in and of which he lives. [...].

As an exception to the human ecology just described, Lorenz names indigenous peoples of South America who live with nature and use only that which ensures their survival.

The haste of the present time (1970s, the author), [...] leaves no time for people to examine and consider before they act. Then the unsuspecting are still proud of being "doers", while they become perpetrators of nature and of themselves. Misdeeds are happening everywhere today in the use of chemical agents, e.g. in the destruction of insects in agriculture and fruit-growing, but almost as shortsightedly in pharmacopoeia.[6] Immunobiologists raise serious concerns about commonly used drugs. The psychology of the "must-have", [...], makes some branches of the chemical industry almost criminally reckless as far as the distribution of drugs is concerned, the effect of which is not at all foreseeable in the long run. As far as the ecological future of agriculture is concerned, but also with regard to medical concerns, there is an almost unbelievable lack of concern. Those who have warned against the ill-considered use of poisons have been discredited and silenced in the most disgraceful manner.

---

[6] Pharmacopoeia: official pharmacopoeia, list of official medicinal products with regulations on their preparation, composition, use or the like, https://www.duden.de/rechtschreibung/Pharmakopoee (accessed 03/03/2020).

By blindly and vandalically ravaging the living nature that surrounds and sustains it, civilized humanity threatens itself with ecological ruin. Once it begins to feel this economically, it may realize its mistakes, but very likely it will be too late. Least of all, however, she will realize how much damage she is doing to her soul in the course of this barbaric process. The general and rapidly spreading alienation from living nature bears much of the blame for the aesthetic and ethical brutalization of civilized man. Where is the *reverence* for anything to come from for the growing man, when everything he sees around him is man's work, and very cheap and ugly man's work at that? [...].

Among those who have to decide whether to build a road, a power station or a factory, thereby destroying forever the beauty of a whole, vast area of land, aesthetic considerations play no part at all. From the head of the local council of a small town to the Minister for Economic Affairs of a large state, there is complete unity of opinion that no economic – or even political – sacrifice should be made to the beauty of nature. The few conservationists and scientists who have open eyes to the incoming calamity are utterly powerless. Some parcels of land belonging to the municipality up on the edge of the forest receive increased sales value if a road leads to them, so the charming little stream that meanders through the village is captured in pipes, straightened and arched over, and already a beautiful village street has become a hideous suburban street.

View from the Year 2020

The age of the industrial revolution is bringing us the sixth great mass extinction (Kolbert 2015). According to estimates (Mora et al. 2011), there are around 8,seven million species on Earth, of which around 6,five million live on land and 2,three million in the water. The majority of these species are not yet known at all. Approximately 15,000 new species are discovered every year.

> Global biodiversity is the foundation of our lives for sustainable development and applications.

However, the latest IPBES report[7] from 2019 (Global Assessment on Biodiversity and Ecosystem Services) shows a trend of global species extinction (Fig. 8.1) that is frightening and in which we humans are significantly involved!

---

[7] https://ipbes.net/sites/default/files/2020-02/ipbes_global_assessment_report_summary_for_policy-makers_en.pdf (accessed 03/03/2020).

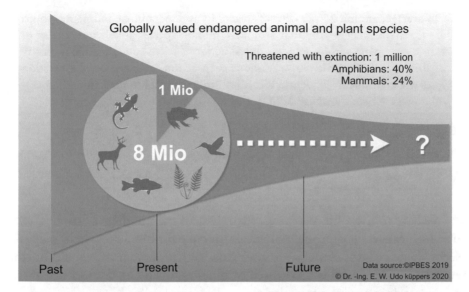

Globally valued endangered animal and plant species

Threatened with extinction: 1 million
Amphibians: 40%
Mammals: 24%

1 Mio

8 Mio

?

Past   Present   Future

Data source:©IPBES 2019
© Dr. -Ing. E. W. Udo küppers 2020

Fig. 8.1 Threatened species worldwide. (Source: IPBES Report 2019)

The key messages of the IPBES report of 2019 (here only excerpted on Complex A) are:

A: *Nature and its vital contributions to people, which together embody biodiversity and ecosystem functions and services, are deteriorating worldwide.*
    Nature and its vital contributions to people, which together embody biodiversity and ecosystem functions and services, are deteriorating globally.

A1: *Nature is essential for human existence and good quality of life. Most of nature's contributions to people are not fully replaceable, and some are irreplaceable.*
    Nature is indispensable for human existence and good quality of life. Most of nature's contributions to humans are not fully replaceable and some are irreplaceable.

A5: *Human actions threaten more species with global extinction now than ever before.*
    An average of around 25 percent of species in assessed animal and plant groups are threatened [...], suggesting that around one million species already face extinction, many within decades, unless action is taken to reduce the intensity of drivers of biodiversity loss. Without such action, there will be a further acceleration in the global rate of species extinction, which is already at least tens to hundreds of times higher than it has averaged over the past ten million years.

Today, more than ever before, more species are threatened with extinction worldwide due to human actions.

On average, around 25 per cent of species in assessed animal and plant groups are threatened, suggesting that around 1 million species are already at risk of extinction, many within decades, unless action is taken to reduce the intensity of drivers of biodiversity loss. Without such action, global species extinction rates will continue to accelerate, already at least ten to a hundred times higher – on average – than they have been in the last 10 million years.

> The worldwide destruction of species is mainly caused by humans. It thus also reduces the essential prerequisite of using known and unknown ingenious principles of nature for our survival.

We have to live with this perspective. Even if the numbers of species, each of which carries at least one biological principle for bionic research, may still seem huge. With a view to the new geochronological epoch of the Anthropocene,[8] which is also caused by humans, the dying of species diversity can – but does not have to – also proceed exponentially.

Lorenz' keyword of "overexploitation" is currently more topical than ever. Alarming is the numbers of forest clearings[9] for corn, soy or palm oil fields. According to the report of *the Food and Agriculture Organization of the United Nations* in 2010, a global reduction in deforestation was noted for the decade 2000–2010, but still alarming remains the deforestation process in individual countries.

The organization "Save the Rainforest" e. V.[10] came up with new alarming figures in 2019:

The global destruction of forests has reached a dramatic peak of almost 30 million hectares per year – despite all the promises of sustainability and billions of euros spent on forest protection. The world's forests are being destroyed primarily for our consumption of meat, soy, palm oil, timber and mineral resources such as iron, aluminium, gold and coltan. 29,4 million hectares of forest, an area almost as large as the UK and Ireland combined, were cut down in 2017, reports Global Forest Watch (GFW).[11] That's almost as much as in 2016, when

---

[8] The Anthropocene is an Earth historical epoch whose atmospheric, biological, and geological processes and life-threatening consequences have been shown to be caused by humans.

[9] https://www.fao.org/news/story/en/item/40893/icode/ (accessed 03/03/2020).

[10] https://www.regenwald.org/news/8950/30-millionen-hektar-pro-jahr-weltweite-abholzung-auf-rekordniveau (accessed 03/03/2020).

[11] https://blog.globalforestwatch.org/data-and-research/deforestation-is-accelerating-despite-mounting--efforts-to-protect-tropical-forests-what-are-we-doing-wrong (accessed 03/03/2020).

29,7 million hectares[12] were the most trees to fall victim to clearing since the online service began keeping records.

A good half of this affects the forests on the equator. Almost 15,8 million hectares of tropical trees disappeared in 2017 – an area half the size of Poland. The sad frontrunners are Brazil with 4,52 million hectares, the Democratic Republic of Congo with 1,47 million hectares and Indonesia with 1,3 million hectares.

> With the overexploitation of our oxygen-giving and carbon dioxide-binding forests, we are at the same time destroying an immeasurable wealth of biodiversity.

Forest clearances, whether in the South American rainforest, in Africa or Asia or Indonesia, have the goal of creating monocultures to feed animals and humans, in order to harvest masses of food through high automation and profits in this way, with the active support of the genetic engineering and pharmaceutical industry – keyword glyphosate – and other insecticide and pesticide producing companies, without the disturbing "weeds".

The vicious spiral in which seed/insecticide/pesticide manufacturers and farmers find themselves is obvious:

→ Seeds → Cultivation in monocultures → Pest and weed destruction →Soil leaching → Biodiversity eradication → Temporary yield reduction in harvests → More seeds → More … etc.

> It is obvious and a cybernetic law that cyclically increasing vicious circles can move in two directions: First, toward exponential growth and collapse; second, toward exponential decay.

Anyone who has ever had the opportunity to study the biodiversity-rich "underworld in the soil", permeated by organismic, useful networks, a little more closely, is immediately aware of all that is being lost – often irretrievably – with the rigorous destruction of this natural living world by the increasing *greed of* the *"masters of humanity"* for growth and profit. Robert McFarlane (2019) has intensively studied this "Underland" and ascribed three valuable tasks to it for all epochs and cultures:

---

[12] https://www.theguardian.com/environment/ng-interactive/2018/jun/27/one-football-pitch-of-forest--lost-every-second-in-2017-data-reveals (accessed 03/03/2020).

1. *Protect precious things:* Memories, valuable materials, news, endangered lives.
2. *Produce valuable things:* Information, wealth, metaphors, minerals, visions.
3. *Dispose of harmful things:* waste, trauma, poison, secrets.

It is unmistakable that there is more to the *"Unterland"* than pure ecology.

The devastation and destruction of the natural habitat, which Konrad Lorenz calls the third deadly sin of civilized mankind, seems to have not yet reached its peak – better: its low point. The current rapid progress of the orgies of destruction by only one species – man – against an estimated 8,seven million species[13] on our earth, most of which are still unknown, is the "Judas reward" for the *"masters of mankind"* to often ruthlessly attack their own livelihood and that of others in order to *"enrich"* themselves.

8.1.4 Konrad Lorenz "Fourth Deadly Sin of Civilised Mankind": Mankind's Race Against Itself

View from the 1970s (Lorenz 1993, pp. 32–38)
Lorenz repeats the statement from his "first deadly sin" of the necessity of *"negative feedbacks"* in living systems, in order to then go into a special kind of *"positive feedbacks"*.

A special case of positive feedback occurs when individuals of the *same* species enter into competition with each other, which exerts an influence on their development through *selection*. *In* contrast to that caused by extra-species environmental factors, *intra-specific* selection causes changes in the genome of the species concerned which not only do not increase their chances of survival, but in most cases are clearly detrimental to them. […].

My teacher Oskar Heinroth used to say in his drastic way, "Next to the wings of the argus cock, the work rate of modern mankind is the stupidest product of intraspecific selection." This statement was distinctly prophetic at the time it was made, but today it is a gross understatement, a classic "understatement." In the Argus, as in many animals with analogous formations, environmental influences prevent the species from escalating through intraspecific selection into monstrous developmental paths that ultimately lead to catastrophe. No such salutary regulating forces are at work in the cultural development of mankind: it has – to its misfortune – learned to master all the powers of its extra-species

---

[13] https://www.scinexx.de/news/geowissen/87-millionen-arten-leben-auf-der-erde/   (accessed 04/03/2020).

environment, but knows so little about itself that it is helplessly exposed to the satanic effects of intraspecific selection. "Homo homini lupus" – "man is the predator for man" – is just as much an "understatement" as Heinroth's famous saying.

What is good and useful for humanity as a whole, even what is good and useful for the individual, has already been completely forgotten under the pressure of interpersonal competition.

As value is felt by the overwhelming majority of people living today (1970 years, the author) only that which is successful in the compassionless competition and suitable to outdo the fellow human being. Every means that serves this purpose appears deceptively as a value in itself. One may define the crushing error of *utilitarianism*[14] as the confusion of means with ends. Money is originally a means; the vernacular still knows this; one says, for instance, "He has, after all, the means." But how many people are there today who understand you at all when you try to explain to them that money is not in itself a value? Exactly the same applies to time: "Time is money" says to anyone who considers money to have an absolute value that for every second of time saved the same applies. [...].

The gain of half an hour is in the eyes of all a value in itself, which no sacrifice can be too great to gain. Every automobile factory must see to it that the new type is a little faster than the preceding one, every road must be widened, every curve extended, ostensibly for the sake of greater safety, but in reality only so that one can drive a little faster – and thus more dangerously.

One must ask what is doing greater damage to the soul of mankind today: the blinding greed for money or the grueling haste. (Italics by the author)

In this as in other statements or determinations of Lorenz, the term "today" can be mapped one-to-one to Lorenz's 1970s as well as to the present time of the 2020s. The recognizable handling of time and money has even drastically intensified up to the present!

Whichever of the two (money and time, i.e.) it is, it is in the sense of those in power (the "masters of mankind", i.e.) of all political directions to promote both and to increase those motivations up to hypertrophy[15] (supernormal growth, i.e.) which drive man to competition. [...].

The hurrying man is certainly not only lured by greed; the strongest lures would not be able to induce him to such energetic self-harm; he is *driven,* and

---

[14] Utilitarianism describes a type of purpose-driven, ethical utilitarianism, with the goal of enhancing or maximizing the well-being of all involved.

[15] Hypertrophy is used in medicine to describe abnormal growth of organismic systems or substances; here the term is related to the economic quantities of money and time.

what drives him can only be fear. Anxious haste and hurrying fear contribute to deprive man of his most essential qualities. One of them is *reflection*. [...].

A being who *ceases to* reflect is in danger of losing all these specifically human qualities and accomplishments. One of the most malignant effects of haste, or perhaps directly of haste-producing fear, is the manifest inability of modern people to be alone with themselves for even a short time. They avoid any possibility of self-reflection and contemplation with a fearful assiduity, as if they feared that reflection might hold up to them an almost ghastly self-portrait [...].

Once, while walking in the woods, my wife and I unexpectedly heard the rapidly approaching whine of a portable radio being carried on the luggage rack by a lone cyclist about 16 years old. My wife remarked, "He's afraid he might hear the birds singing!". I think he was just afraid of being in danger of running into himself for a moment. [...].

In addition to commercial intraspecific selection for an ever accelerating pace of work, there is a second dangerous circular process at work, which Vance Packard has called attention to in several of his books, and which is followed by a progressive increase in human *needs*.

Vance Packard described Lorenz called "second dangerous circle process", in his 1957 book *"The Hidden Persuaders"*, German (1958 in Econ-Verlag, Düsseldorf): *"Die geheimen Verführer"*, with the subtitle: "Der Griff nach dem Unterbewussten in jedermann". The power of advertising is revealed as a deceptive game with people's supposed needs.

What in Lorenz's "Race of Mankind with Itself" also takes place in socially critical influences in modern industrial societies, partly up to the present, Packard already worked out in detail and with foresight in 1964 in *"The Defenceless Society"*.

View from the Year 2020

The perception and practice of a certain feeling of power of one person over others is multifaceted in the political as well as economic, monetary or political and general social environment – and not only since yesterday or the day before yesterday. That practiced power of a few always provokes counter-power is also not new. In view of the increasingly destructive and disappearing basis of life, nature, on the other hand, the seemingly carefree demonstration of power by a few, the clinging to personal power structures, takes on a new significance.

The lack of time for reflection or self-reflection lamented by Lorenz can also be felt and experienced in many ways today. The construction of so-called status symbols, such as passenger cars, continues to obey short-sighted economic goals, such as maximizing sales and profits, despite sometimes massive efforts to convince buyers otherwise. The evidence for this assertion is literally on the roads, which can be seen every day as perilous motorways inviting speeding, traffic jams for miles on end, congested inner-city traffic routes full of exhaust fumes, and so on. The claim to power of the owners, to stay with the automobile industry, is furthermore underpinned by underhand manipulations and deceptions of buyers, as the so-called *"diesel scandal" has* recently shown until today.

Each practiced fast *protagonist* is juxtaposed with a thoughtful slow *antagonist*, such as:

- *Speed (risk-taking, error-proneness)* before *slowness (sustainability, attentiveness, error prevention)*
- *economic work compression* before *physiologically healthy work activity per time unit*
- *digital fast consumption rush* before *analog thoughtful consumption consideration*
- *Suggested weekend time off to ensure work performance* before holistically organised life-work time
- *Acceleration* before *deceleration*

Value-based compromises between both actors, employers and employees, commodity providers and commodity users, etc., would already be a first sustainable step forward in Lorenz's *"race of humanity for itself."* But as always, the *wish* is *the father of the thought.*

Exactly 30 years ago, the psychologist and business economist Oswald Neuberger (1990) wrote a remarkable article on the position of people in the work process (see Sect. 6.2). The statement that entrepreneurs – then as now – like to make, that people in "their" company are naturally the centre of events, could only be understood then and can only be understood now with a great deal of economic liberalism. Quite a few facts speak for the fact that the working human being is rather understood as a MEANING than as a CENTER in the working process. The well-known, often postulated and practiced *"work-life-balance"process*, the creation of a work-life balance, strengthens the human being as a means for companies much more than the human being would be as the centre. Countless examples of a pseudo-successful weekend *"work-life balance" promotion* could be listed. Business-minded managers who want to relax from their everyday stress for a short time over the weekend in *"work-life-balance"* resorts, or who let themselves relax through specific physical exercises prescribed by resort employees, provide the *"work-life-bal-ance"* resort with monetary advantages rather than themselves through sustainably effective holistic physical and mental training. Last but not least, the diverse, tact-giving sporty Bespaßungsaktionen in week-long holidays also increase the cash registers of the providers than anything else.

The fast-moving pace of time today hardly leaves enough room for a longer, regular *recovery* or *regeneration of* the human being as a holistic organism. Body and mind, work and leisure time cannot be separated from each other – as is suggested in a continuous sprinkling process of advertising experts! This insight also requires a bit of reflection.

The more the compression of work per time increases and the more these specifications are demanded by companies in times of digitalization, robotization, cobotization (cobot = collaborative robots that perform work together with humans) and artificial intelligence algorithms, the longer humans will remain a means to an end. On the other hand, it will almost certainly also be the case that humans will be replaced by robots and no adequate work will be available as compensation, except for their own work in free time (see Küppers 2018). Whether new work communities with cobots, inferior replacement work because of robots, or work in prescribed free time, all variants lead to dilemmas for affected people under current economic capitalist and globalizing labor relations! The consequences are likely to have enormous socially relevant effects, if further burdens of unaffordable housing, social dependence of the living situation, migration problems and many more are added.

Can we, as a sentient, foresighted species, still steer the path of accelerated progress, which is fraught with great risks, into a sustainable future, while accepting all the disadvantages that have already taken place and those that are additionally discernible? We must!

Many scholarly authors and commentators have addressed the all-encompassing problem of *time and acceleration*, or *time and deceleration,* such as sociologist, political scientist and time researcher Hartmut Rosa (2018) in "Unavailability (Preserving Unrest)", "Resonance", 2016; "Acceleration", 2005, "Acceleration and Alienation" 2003.

A quarter of a century ago, the French media critic Paul Virilio (1932–2018) predicted an unbridled acceleration of information processing and coined the term *"racing standstill"*. Internet, high-performance trading on the stock exchanges by computer algorithms, etc., ultimately lead to an inevitable collapse, like that of the world economy and financial economy in 2008/2009. Extraordinary and incalculable events, like the current infectious disease COVID-19 in 2020, put entire cities, regions, countries, even continents in quarantine.

In this surreal, frightening situation of the present, who remembers the 1998 Nobel Prize winner in literature José Saramago and his novel "The City of the Blind" (1997, original 1995).[16] Quote on the back of his book:

A traffic light in a nameless city jumps to green. A car nevertheless comes to a halt. The driver has suddenly gone blind. The friendly helper who brings the blind man home and then takes possession of his car suffers the same fate. Blindness spreads like an epidemic. The state reacts brutally. The blind are interned in an empty asylum where they are left to fend for themselves. But there is a sighted woman among them who has only faked the disease to stay with her husband. With her help, the escape might succeed …

Striking communicative and organisational similarities, or more precisely: short-sighted mistakes in political and social action, with the current COVID 19 pandemic cannot be denied, at least in part, even if today's epidemic, which has become a pandemic, involves completely different, more drastic boundary conditions and risks.

In the essays "The Negative Horizon" (1996, first edition 1989), as well as "Frenetic Standstill" from 1992, the French philosopher and critic of the media society Paul Virilio dealt with the laws of human history as a process of acceleration, in the present time of gigantic information networks racing towards its end point. The essay describes the impending end state of this fear-inducing acceleration in the circumferentially dominating telecommunications. It describes the delusional compulsion to be present everywhere and at all times thanks to electronic telecommunications, the seduction of simultaneous participation in everything, the experience of history-less instantaneousness in observation. Virilio diagnosed this state of media ghettoization as electronic apartheid, as coma (ibid., cover text).

Media editor Christian Meier (2017) comments:

A quarter of a century ago, Paul Virilio predicted the age of unbridled acceleration. Also triggered by the direct transmission of images in real time. An apocalyptic vision. […]

The positive consequences of progress, which Virilio by no means rejects, have been replaced by propaganda, the propaganda of never-ending progress. A clear hint in the direction of Silicon Valley, the breeding ground of the acceleration of our thoroughly digitalized society. From there we are steadfastly promised world improvement. Whoever wants to analyse progress, however, Virilio said, must also examine the accidents of progress. His insight is much quoted:

---

[16] I thank my son Jan-Philipp for helpfully pointing out this analogy.

"The invention of the ship was at the same time the invention of the shipwreck." (italics added by the author).

The *frenzied standstill of a society* that controls time and space in a highly technological way is thus working to erase itself.

What could be more obvious than to deal with the *discovery of slowness*, as Nadolny (1983). Reheis (1998), Geißler (1999), Baier (2000) and many others have done.

Nadolny's voyage of discovery novelistically rewrites the sea voyage of the English polar explorer John Franklin (1786–1847) in such a way that "[...] the course of life becomes a subtle study of time, and slowness becomes the art of making sense of the rhythm of life," as it says in the preface to the book. Some particular discoveries about Nadolny's slowness, which will undoubtedly remain valid for a long time to come and are exemplary of socially sustainable processes, follow chronologically:

There is nothing to add to these wise and progressive statements, even more than 37 years after publication.

Overview
- p. 157
  We ourselves are the chance!
- p. 174
  Make man better? Only three things can do that:
  1. The study of the past
  2. The healthy way of life in nature
  3. In sickness the medicine.
- p. 199
  I take my time before I make a mistake.
- p. 208 ff.
  Fast and Slow. The fast can be put into survey professions that are subject to the acceleration of the age: Members of Parliament, stockbrokers, flawless economists. Slow people, on the other hand, are left to solitary professions: craftsmen, doctors, painters, researchers. The slower work is the more important and sustainable!
- p. 218.
  Between exaggeration and understatement is 100%.
- p. 220.
  Strength can also be something other than speed.
- p. 221
  I've learned to keep looking stupid until I'm smart. Or until the others look even stupider than I do.

- p. 222
  Attrition tactics through extreme politeness, constant repetition of always the same arguments and the complete ignoring of any sense of time.
- p. 226.
  Peace where people approach each other slowly rather than quickly.
- p. 283
  Discovery and slow observation go together.
- p. 296
  Only those who are well prepared recognize the alarm signs. Politics can hardly be very different from navigation!
- p. 301
  In navigation, one must determine the starting position as accurately as the destination.
- p. 308
  Two people must be at the head – not one and not three. One of them must manage the business and keep up with the impatience of the questions, requests and threats of the governed. The other has calmness and distance, he can say no at the crucial point!
- p. 315
  Not a pose of overview, but overview from observation of details. Navigation.
- p. 329
  New schools should teach permanence without being boring. Students must learn to discover. That is precisely what today's schools cannot do!
  Bad schools prevent anyone from seeing more than the teacher. Teachers need to be explorers.
- p. 338
  Learning and seeing is more important than education. More of the spirit of our navigators must enter the school.

From the only four of eight described Lorenz's "deadly sin of civilized mankind", which in my opinion come closer to the book title than the four others, it is easy to see how over several decades of social development, the core statements of the four described "deadly sin of civilized mankind" have hardly changed to the advantageous, rather to the disadvantage for the majority of people on our planet. Do we have to endure this "price of inequality", as the Nobel Prize winner for economics Joseph Stiglitz (2012) put it? The fact that a different society with different structures is possible was also described by Stiglitz in Chap. 10 of his book "The Price of Inequality" from his economic perspective. To go into this in more detail would lead us away from the actual topic of the book. Therefore, we move on to a second set of topics on "*Struggle or Perish!*"

8.2   UNEP-GEO-6: Global Environment Outlook

The *UNEP Geo-6* Global Environment Outlook *Report* 2019 (original: Global Environment Outlook Geo-6 Healthy Planet, Healthy People, UN-Environment 2019) highlights the fundamental importance of a "healthy planet" for our health and well-being. This is not least due to the fact that we humans have in part long since exceeded the ecological limits of our earth, but we still largely engage in *"business as usual"*, a very normal *"business as usual"*, as if we had another earth up our sleeve as a resource.

Figure 8.2 shows on the left, in a composite, human influences which, through intervening dynamics, promote the *"disease"* or *"destruction" of* our planet and the associated health burdens on organisms (humans, animals and plants).

In contrast, the transformation of our way of life is outlined on the right in Fig. 8.2, which can, and given the realities of our state of nature and the environment, must, lead to a possible recovery of planet Earth with healthy people!

According to the *UNEP-GEO-6 report,* the decisive decisions for a nature worth preserving and sustainable developments in politics and economy are influenced by five "Earth systems" and five "drivers", which can be seen

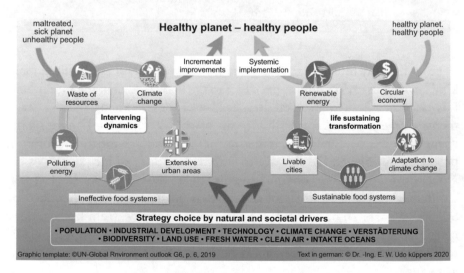

Fig. 8.2   UN-GEO 6, Global Environmental Outlook (2019), Development Pathways

summarized in Fig. 8.2 below. The decisive fork in the road in Fig. 8.2 actually leaves only one way to go, consistently the right(!) path outlined on the right in Fig. 8.2.

Despite all this, however, uncertainties remain, which have developed in particular from the decades of entrenched processes of a globalizing economy that is consistently focused on economic advantages and the inequality of human life paths that is linked to this. Such unequal living conditions of a majority of work-dependent people versus a minority of work-giving people virtually challenge a struggle for sustainable converging living conditions. *"Fight or perish!"* is therefore the inevitable consequence, which is described in Sect. 8.3 from the perspective of the polymath Noam Chomsky, who clearly perceives the problems on our earth and also names them publicly.

The struggle against the exploitation of our planet is not a recent phenomenon. As far back as 72 years ago (Osborn 1948, *"Our Plundered Planet"* and Vogt 1948, *"Road to Survival"*) and probably several hundred years ago and earlier, as civilizations formed and exploited the Earth's raw materials for their own, the ability of humans to live and survive on an increasingly plundered planet was threatening their very existence. Today, at the beginning of the third millennium of our era, in the year 2020, the battle against human activities destroying our livelihood is no less being waged. Only it has been given a new – earth-historical – name by the Dutch meteorologist Paul Crutzen (2002): *Anthropocene* (Zalasiewicz et al. 2019; Renn and Scherer 2015).

Whether we humans are actually capable of learning, in the face of continuing environmental destruction in the sea, on land and in the air, remains the great future uncertainty.

> We know what we need to do for the health of our earth now, but future uncertainty presents us with a dilemma.

8.3 Noam Chomsky, "Why We Must Stand Up to the Masters of Mankind"

Preliminary Remark to the Esteemed Readers
Noam Chomsky is one of the few polymaths of our time whose statements, whatever they refer to, carry weight. Even if you might get the impression that some of Chomsky's subsequent statements have little in common with the

"Ingenious Principles of Nature", from a systemic perspective connecting strands are discernible here and there. As little as man can detach himself from nature in his mechanized environment, so little is nature – measured over the relatively short life periods of humans compared to the billions of years of nature's development – completely independent of human influence on it. The fact that even on the subject of freedom of the press or the "freedom of speech", as a quotation further below puts it, a bond can be forged with nature and its genius becomes apparent from the perspective of interconnected contexts.

Therefore, go boldly into this chapter and recognize and reflect on the often invisible and, at first glance, seemingly incongruous connections between the "Ingenious Principles of Nature" and the manifold human activities. You are assured of added value.

The back cover of Chomsky's "Fight or Perish!" reads:

> Why does so much inequality continue to prevail in our world? Why don't the 99 percent revolt against the elites, the "masters of humanity"? Is humanity on the brink of self-extinction?

With these and other questions, Noam Chomsky (2018) goes into an interview with journalist Emran Feroz, from whose content excerpts are presented that directly and indirectly concern nature and its destruction by humans. By way of introduction, Feroz introduces readers to Chomsky's personality (ibid., pp. 7–8):

> In the Renaissance era, there was the "uomo universale" – polymaths who had mastered various fields and were considered experts in art, science, and numerous other areas. Leonardo da Vinci was one of them. For some people, even one of the last. Such thinkers were not only present in the Occident, but also in the Orient. Philosophers, mathematicians, physicians, poets and theologians were among them. Such all-rounders, whose personality is held in high esteem and weight in wide circles, are hardly to be found in our present post-modern age. One has the impression that people are dumbing down and moving ever closer to the abyss as they pursue their self-destructive activities in the haze of globalization and capitalism. One person who has opposed all this for decades is Noam Chomsky. For many people, there is no question that Chomsky is one of the most important intellectuals in the world, and possibly even in modern human history. If there are any people in our world who can be associated with the "uomo universale" of the Renaissance, then Chomsky is without a doubt

one of them. Already at a young age, he revolutionized linguistics with his theories, such as the so-called Chomsky Hierarchy, which classifies and ranks formal grammars and languages. Chomsky's scientific work continues to play an important role in both mathematics and computer science to this day. The mathematical formalization of languages was also one of the foundations of computational linguistics and machine language translation.

Asked about Chomsky's often used perspective of an uninvolved observer from outside the earth, an alien, also with regard to our political behaviour and the consequences, he answers (ibid., pp. 16–20):

The world is a complex place, full of complex interactions and considerable diversity. Let us focus on what is called "the West" and some other generalizations that are striking.

What the Alien would see are societies that have gone through a generation of neoliberal policies that have elevated the market to the center of society ("business first").

As expected, these have led to a weakening of social ties and public institutions – especially trade unions – which are linked to stagnation as well as the decline of the majority and a very strong concentration of wealth and the consequent deterioration of functioning democracies. One result of this is that societies tend towards plutonomies and become a haven of increasing insecurity. [...].

Such policies create a *vicious circle* (italics mine) of familiar mechanisms: the concentration of wealth leads to an even greater concentration of political power, which in turn leads to policies that deepen the rifts between the ruling elite and the precariat and further undermine democracy. Results in people's political behavior are anger, fear, a search for a scapegoat, and contempt for institutions. We call this populism and are currently seeing it in elections around the globe [...].

When asked about the "masters of humanity" and whether extraterrestrials would also clearly recognize them, Chomsky responds (ibid., pp. 19–20):

Adam Smith had no problem identifying those he called the "masters of mankind." In his day, these were English merchants and manufacturers who were the "chief architects of policy" and ensured that their interests received the most attention, no matter how severe the impact on others, including the people of England. The greatest victims of the English's blatant injustice, however, were elsewhere – for Smith, particularly in British India. Much has changed in two and a half centuries. but the basic principles have remained the same.

Today's masters of humanity come from the international investor class, the highly concentrated and interlocked corporate sector with a rapidly growing component of finance capital, especially after they were unleashed by President Nixon's dismantling of the Bretton Woods system[17] and then by the massive deregulation of the crash bailout system[18] to dominate the neoliberal era. They dictate policy, though not without internal dissent, and they strive to ensure that their own interests are served, regardless of the impact on others. They do this with much success, especially in recent years.

Truly critical intellectuals are usually marginalized, denigrated, or denied employment – such as a job – in their own society. Such tendencies have been evident throughout history. (ibid., p. 26)

When asked by Feroz whether the so-called refugee crisis is a natural disaster or the result of fatal political decisions, Chomsky responds (ibid., p. 55):

It includes an element of *natural disasters* (italics mine). The terrible drought that has visited Syria and massively affected large sections of society is a consequence of global warming. This in turn is not really natural. The crisis in Darfur was partly a result of entire regions becoming desert, driving nomadic populations into populated areas. The horrific famines in Central Asia are also linked to the *attacks on our environment* (italics mine) that took place during the Anthropocene era. (Italics added by the author):

Industrialization destroys basic living conditions, and it will continue to do so as long as no one curbs it.

Freedom of the press is a great good in democracies, that is well known. That public denunciation of secret government actions, such as deeply humanly reprehensible military state actions of the USA and England in the Iraq war in 2003, by the Australian Julian Assange, founder of the internet platform Wikileaks, or the deliberate approval of the destruction of nature, as happened in the Brazilian rainforest in 2019, by the president Jair Bolsonaro and other such government actions more, do not please the authors, is also known.

That we must also "rise up against the masters of humanity" out of such power actionism was made clear by Chomsky in the subtitle of his book "Struggle or Perish".

On freedom of the press in the free world, Chomsky quotes not only Mark Twain but also George Orwell a 2017 text whose statements are also indicative of the state of our nature and environment today (ibid., pp. 111–120):

Mark Twain is reported to have once said (italics mine):

---

[17] The Bretton Woods system was created after World War II as a new international monetary system with exchange rate bands and the dollar as an anchor currency to link fixed and flexible exchange rates (see Bordo 1993).

[18] Crash bailout system = collapse bailout system.

"By God's grace, we in this country possess three incredibly precious things: freedom of conscience, freedom of speech, and the wisdom to never claim either one."

In his unpublished introduction to *Animal Farm* – dedicated to "Literary Censorship" in free England – George Orwell gave a reason for that cleverness Twain was referring to. There is, he wrote, a general, tacit agreement that it is not necessary to state certain facts. This tacit agreement brings about a "veiled censorship" based on "a strict doctrine of faith, a collection of ideas which all reasonable men accept without any questioning," and "anyone who questions these prevailing ideas is silenced with surprising efficiency"-even without "any official prohibition."

We can permanently observe the exercise of this wisdom in free societies. Exemplary for this is the American-British invasion of Iraq (as mentioned above, by the author).

Let us return to our biggest problem facing us, nature and the environment: climate change, because it has a direct influence on everything we do. Feroz's question to Chomsky, how we can realistically do something about the flight of many people from climate change, he answers as follows (ibid., p. 149):

The use of fossil fuels must be ended quickly, while the use of renewable energies must be massively increased. The same applies to research into renewable energy. The preservation of nature must be considered the most important task. There must also be a far-reaching critique of the capitalist model of exploitation of people and resources. It is, after all, this model that could deal the death blow to our species.

A final comment by Chomsky refers to the "fundamental right to education" mentioned earlier by the author. Feroz attributes this to Chomsky as a life task and he responds (ibid., pp. 159–162):

The current state of the education system has both positive and negative elements. An educated public is necessary for a functioning democracy. The path towards it can sometimes be advanced, but sometimes it is hindered. It is an important task to move the levers of education in the right direction. […].

As soon as it comes to climate change, the most important issue in the history of mankind, forty percent of the population (of the U.S., n.d.) think that the whole thing is not a problem. Since Christ will return in the next few decades anyway. This is just one symptom of many that highlights the pre-modern characteristics of culture and society in this country. […] (Italics d. d. A.):

Elites do not want functioning democracies. They want a society where people are scared and intimidated. Their main concern should be paying the next rent, political passivity is desired.

Chomsky's views are clear: Those who do not fight against the impending doom of humanity, which is not least promoted by the massive all-encompassing destruction of networked biodiversity, despite individual positive advances, have already lost. How intelligent will we be – with and without "Artificial Intelligence" – to reduce the massive survival problems facing us and steer progress towards a nature-friendly future?

To Konrad Lorenz be reserved the concluding sentence, which sums up nature and human engineering:[19]

No smart economy without smart ecology.

References

Baier, L. (2000) Keine Zeit. 18 Versuche über die Beschleunigung. Kunstmann, München

Bordo, M. D. (1993) The Bretton Woods International Monetary System: An Overview. In: Bordo, M. D; Eichengreen, B. (1993) A Retrospective on the Bretton Woods System. The University of Chicago Press, S. 5.

Chomsky, N. (2018) Kampf oder Untergang! Warum wir gegen die Herren der Menschheit aufstehen müssen. Westend, Frankfurt a. M.

Crutzen, P. J. (2002) Geology of mankind. Nature 415, 23

Geißler, K. A. (1999) vom Tempo der Welt – und wie man es überlebt. Herder, Freiburg, Basel, Wien

Kolbert, E. (2015) Das sechste Sterben. Wie der Mensch Naturgeschichte schreibt. Suhrcamp, Berlin

Küppers, E. W. U. (2018) Die humanoide Herausforderung. Leben und Existenz in einer anthropozänen Zukunft. Springer Vieweg, Wiesbaden

Küppers, E. W. U. (2019) Eine transdisziplinäre Einführung in die Welt der Kybernetik. Lehrbuch, Springer Vieweg, Wiesbaden

Küppers, J.-P.; Küppers, E. W. U. (2015) Bedingt handlungsbereit Die jüngste Migrationswelle und ihre Grenzen systemischer Krisenbewältigung in einer globalisierten Welt. Zeitschrift für Politikberatung 7 (3/2015):110–121

Lorenz, K. (1993) Die acht Todsünden der zivilisierten Menschheit. R. Piper, München

Meier, M. (2017) Er sagte den rasenden Stillstand voraus. Welt, 4.1.2017

McFarlane, R. (2019) Im Unterland. Eine Entdeckungsreise in die Welt unter der Erde. Penguin, München

Mora, C. et al. (2011) How Many Species Are There on Earth and in the Ocean? PloS Biology, 9 (8), e1001127, August 2011

---

[19] S. Wuketits, F. M. (1990, p. 215).

Nadolny, S. (1983) die Entdeckung der Langsamkeit. Piper, München, Zürich

Neuberger, O. (1990): Der Mensch ist Mittelpunkt. Der Mensch ist Mittel. Punkt. Personalführung, 1, 3–10.

1948 Osborn, F. (1948) Our Plundered Planet. Little, Brown and Comp. New York, USA. Deutsche Ausgabe: 1950, Unsere ausgeplünderte Erde, Pan-Verlag, Zürich, Switzerland.

1972 Packard, V. (1972) Die wehrlose Gesellschaft. Droemer Knaur, München, Zürich, Original (1964) The Naked Society, David McKay, USA

Packard, V. (1958) Die geheimen Verführer. im Econ, Düsseldorf, Original (1957) The Hidden Persuaders, ig-Pub. New York

Radkau, J.; Hahn, L. (2013) Aufstieg und Fall der deutschen Atomwirtschaft. Gesellschaft für ökologische Kommunikation mbh, München

Reheis, F. (1998) Die Kreativität der Langsamkeit. Neuer Wohlstand durch Entschleunigung. Primus, Darmstadt

Saramago, J. (1997) Die Stadt der Blinden. Rowohlt, Reinbek b. Hamburg

Stiglitz, J. (2012) Der Preis der Ungleichheit. Wie die Spaltung der Gesellschaft unsere Zukunft bedroht. Siedler, München

Renn, J; Scherer, B. (Hrsg.) (2015) Das Anthropozän. Matthes & Seitz, Berlin

Rosa, H. (2003) Beschleunigung und Entfremdung: Entwurf einer kritischen Theorie spätmoderner Zeitlichkeit. Suhrkamp, Berlin

Rosa, H. (2016) Resonanz. Eine Soziologie der Weltbewegung. Suhrkamp, Frankfurt a. M.

Rosa, H. (2018) Unverfügbarkeit (Unruhe bewahren). Residenz, Salzburg

Rosa, H. (2005) Beschleunigung. Die Veränderung der Zeitstruktur in der Moderne. Suhrkamp, Frankfurt a. M.

UN Environment (2019) Global Environment Outlook Geo-6 Healthy Planet, Healthy People Cambridge University Press, Cambridge, UK

UN-IPBEST (2019) The global Assessment Report on Biodiversity and Ecosystems Services. IPBEST Secretariat, Bonn, Germany

Verilio, Paul (1996) Der negative Horizont. Bewegung, Geschwindigkeit, Beschleunigung. Fischer, Frankfurt a. M.

Verilio, Paul (1992) Rasender Stillstand. Essay. Edition Akzente, Hanser, München

Vogt, W. (1948) Road to Survival. William Sloane Ass. Wincheater, MA, USA

Wiener, N. (1948) *Kybernetik. Regelung und Nachrichtenübertragung im Lebewesen und in der Maschine.* Zweite, revidierte und ergänzte Auflage. Econ-Verlag, Düsseldorf 1963 (287 S.), Original: Wiener, N. (1948) *Cybernetics or Control and Communication in the Animal and the Machine.* Übersetzt von E. H. Serr, E. Henze, Erstausgabe: MIT-Press, USA

Wuketits, F. M. (1990) Konrad Lorenz. Leben und Werk eines großen Naturforschers. Piper, München, Zürich.

Zalasiewicz, J. et al. (Eds.) (2019) The Anthropocene as a Geological Time Unit. A Guide to the Scientific Evidence and Current Debate. Cambridge University Press, Cambridge, UK

9

Summary and Outlook

Abstract The résumé summarizes six main concerns of this work and leads into an outlook that is characterized by a new hierarchy of needs of the current Earth age Anthropocene.

9.1 Summary

What can be learned by you, dear reader, as the essential insight from the texts, quotations and illustrations contained in the three subdivided parts:

(i) *The inexhaustible wealth of evolutionary adaptive solutions*
(ii) *Beyond the exhaustive wealth of technospheric maximum solutions*
(iii) *antagonist of nature*

have been presented? From my point of view, there are six main concerns:

1. Those who – like evolution – for billions of years, under the most stringent quality criteria, have been producing individual and collective, networked peak performances for sustainable progress on our planet Earth, cannot be entirely wrong with their development strategy and effective principles.
2. Humans, with their self-constructed and practiced strategies of progress, are moving in a technospheric space that is largely at odds with the principles of nature; – indeed, even more so, is increasingly destroying nature – our fundamental basis of life.

3. The definition of a new earth-historical age, the Anthropocene, and its consequential problems of the earth, which are globally burdened by us humans, only shows the short-sighted misguided approaches to survival, which we have initiated through careless technical progress – without taking the time to sufficiently reflect on its results and consequences.

4. On the way of our "progress" we even play – at least a few among us – with the ancient and again renewed idea of making ourselves superfluous someday as a replacement by robots, humanoids, or cyborgs! Behind this is also the vision of a digital progress that, according to computer scientist Ray Kurzweil (2014), spreads to a so-called singular superpower over everything and postulates a new *humanity 2.0,* whatever that entails.

   I believe that the evolutionary power of nature is serene enough over billions of years of experience and practice to counter and ultimately dominate even this – still fictitious – singular attack on humanity that humans paradoxically initiated themselves.

5. Misguided power and organisational structures lead to a vicious circle (positive feedback), causing further waste of resources and thus the destruction of life, habitats and livable spaces in many different ways. It is said that every crisis is also an opportunity for something new and sustainable. Some of us also fight incessantly, sometimes vehemently against the minority of the "masters of mankind" and for these new chances of a better living environment. However, the branch on which we are *all* sitting – without any exception – is becoming increasingly rotten.

6. Humans and the Earth need a new symbiotic partnership with new strategic goals. These must not be oriented towards what is given as the target maximum and at the same time be in bondage to the catastrophic aftercare principle, but which follow an adaptive minimum of resources, with an optimal – systemic or networked or biocybernetic – precautionary principle that is fault-tolerant, forward-looking and thus sustainable.

9.2   Outlook: The New Hierarchy of Needs in the Digitalizing Anthropocene

The main concerns listed in Sect. 9.1 can be linked in a forward-looking way to a strategic pyramid of needs, which the author sees as a continuation, in a new Anthropocene age, of the hierarchy of needs of *Abraham Maslow* (1908–1970) (Maslow 1943, 1981; see also Koltko-Rivera 2006) and the theory of value change from the 1970s by *Ronald Ingelhart* (Ingelhart 1977), which is based on Maslow's hierarchy of needs. Other sources include Ingelhart (1982, 1998).

The theory of the *"hierarchy of needs* in the digitalizing *Anthropocene"* is presented to the new *Anthropocene* era, which is also characterized by an increasing *digitalizing* influence *with* significant changes in people's working and living conditions (Fig. 9.1).

The hierarchy of needs of a new earth-historical age shown in Fig. 9.1 shows at its base the essential prerequisite for *securing existence,* which is connected with the human need to work, coupled with a remuneration that is assigned less to the capitalist cost-benefit calculation and more to the value of the work itself. For this purpose, new evaluation schemes have to be created, which transform the type of work, person, age, performance and other criteria more into a reasonable cost-time framework, which leaves the human being in the centre of events and *does not* abuse him – as so often – as a means to an end.

On the next level of needs towards the top of the pyramid is the freedom and security of people, which is increasingly threatened by the dynamics of globalization, the world-spanning communicative networks and the intransigence of individual people in power. The *freedom and security of* a society is thus highly vulnerable in all its facets. Both basic needs can only be secured in

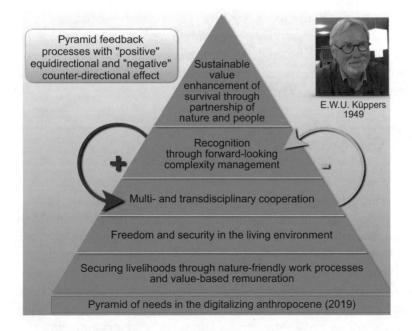

Fig. 9.1 Hierarchy of needs as a pyramid according to E. W. U. Küppers (2019)

the long term, i.e. sustainably, through persistent cooperation instead of dangerous confrontation.

This leads directly to the subsequent level of need for *multi- and transdisciplinary cooperation* at all levels within a community and across societies. The latter requires a universally or majority recognized earth-wide steering body, such as a new *"United Nations"* with new structures and competencies in action. These assertive necessary competencies for cooperative rather than confrontational action are oriented to mental and practiced patterns of thought and action that follow the indisputable dynamics of life. It is about nothing less than sustainable solutions through *anticipatory complexity management*. Recognition through motivation and reward, consistently understanding complexity as a solution and not as a problem and acting accordingly, is the core of the penultimate level towards the highest level of the hierarchy of needs as a pyramid.

This highest level leads to an interpenetration of biospheric survival strategies of the highest quality with the creative, sustainable value-preserving strategies of humans in their technosphere, in a composite of ecological, economic, social or societal environment. The survivability of all organisms, through systemic use of their individual and collective strengths would be the ultimate goal. So much for a brief description of the new anthropocene organization of needs.

All five levels are interconnected by positive *and* especially negative feedback circulation processes (see circle symbols next to the pyramid in Fig. 9.1), which ensure stabilizing progressive system dynamics of practiced tasks and solutions. This theory is characterized by the will of the people involved and the idea that the path to survival is the goal.

9.3 Oh: Mankind!

We have reached the end of what we hope will be for you an exciting reading journey through the ingenious world of nature, which can only make us marvel at the dexterity and sophistication with which plant and animal organisms manage their lives and further development. Cross-species cooperation – rather than confrontation – is the means of success here. Both, however, are cleverly intertwined.

With creativity and wise foresight, we humans can learn much from the principles of nature for our own further development. Where our path into the future will lead us in the present turbulent, very dynamic times is uncertain. The fact that we have to make enormous efforts with every further step in order to overcome, or at least limit, the self-inflicted crises of the greatest magnitude is not only evident from the dangerous climate change.

Aggregating *"devilish control loops"*, as Jay Wright Forrester once called them, have the power to break social boundaries in our lives. They are our constant companions; they are the irrational result of our often short-sighted and misguided thinking and acting. Social divisions into rich and poor, working and unemployed, with and without health care, etc., stretch diabolically across all continents. The astonishing naivety and stupidity with which some "pseudo-lords of humanity" not only pursue monocausal economics and power politics, but also in very recent times confront the dangerous deadly Sars-Cov-2 virus that has taken its place worldwide, is therefore not surprising.

It is not the worst idea to turn to the past from time to time and to look with a historical eye at crisis situations similar to the present one, not least in order to orientate oneself on strong personalities who knew what they wanted and did, in the sense of cooperative behaviour, for the good of societies.

One of these personalities was the former French Foreign Minister Robert Schumann.[1] He laid the foundation for today's European Union (EU) in his speech on 9 May 1950, five years after the end of the war, when he proposed the creation of a European Coal and Steel Community (ECSC).

One of his much quoted sentences, with which he begins his speech, is of such power that it shines far into the future and certainly makes not a few of today's responsible people in politics and society aware of their own short-sighted and misguided vicious circle activities:

> *The peace of the world cannot be preserved without creative efforts commensurate with the magnitude of the threat.*

References

Ingelhart, R. (1998) Modernisierung und Postmodernisierung. Kultureller, wirtschaftlicher und politischer Wandel in 43 Gesellschaften, Campus, Frankfurt a. M.

Ingelhart, R. (1982) Die stille Revolution. Vom Wandel der Werte, Athenaeum, Berlin

Ingelhart, R. (1977) The Silent Revolution: Changing Values and Political Styles among Western Publics. Princeton: Princeton University Press.

---

[1] https://europa.eu/european-union/about-eu/symbols/europe-day/schuman-declaration_de   (accessed 5/10/2020).

Koltko-Rivera, M. E. (2006) Rediscovering the Later Version of Maslow's Hierarchy of Needs: Self-Transcendence and Opportunities for Theory, Research, and Unification. Review of General Psychology, vol. 10, No. 4, 302–317

Küppers, E. W. U. (2019) Private records, first published in Küppers, E. W. U. (2020) Geniale Prinzipien der Natur, Springer, Wiesbaden

Kurzweil, R. (2014) Menschheit 2.0: die Singularität naht. Lolabooks, Berlin

Maslow, A. (1943) A Theory of Human Motivation. In Psychological Review, (Vol. 50 #4, Seite 370–396)

Maslow, A. H. (1981) Motivation und Persönlichkeit. (Originaltitel: Motivation and Personality, Erstausgabe 1954, übersetzt von Paul Kruntorad) 12. Auflage, Rowohlt, Reinbek bei Hamburg

Glossary

Algorithm Algorithm is a unique, well-defined sequence of computational rules or operations that lead to the solution of a problem. Algorithms can be implemented in computer programs. Thus, given a defined input, a defined output or solution is obtained. The value of a given algorithm determines its performance, the accuracy of its results, its scope, its compactness, and the speed at which it operates. (Source: Penrose, R. (2002) Computational thinking. Spektrum Akademischer Verlag, Heidelberg, Berlin, p. 16).

Anergy Anergy is that part of energy which cannot be converted into work, i.e. is not technically usable (→ **exergy**).

Angel Circle (→ Vicious Circle) Angel Circle (→ Vicious Circle) are terms from system dynamics. They are used in connection with feedback connections or dynamic flow variables between system elements. Example: The inflow of a water flow into a tank increases the water level in the tank. If the inflow continues, this causes the water in the tank to overflow. This results in a vicious circle: the more water flows in, the more likely an overflow will occur. This results in consequential problems such as flooding, cleaning work, etc. If the inflow of water in the tank is regulated by an outflow, a dynamic water level is established – depending on the water outflow quantity – which stabilises the entire system and prevents overflow. Consequential problems do not take place due to this *"angel circle"*, because every change of the inflow can be controlled by a change of the outflow.

Anthropocene The term goes back to the Dutch climatologist J. P. Crutzen and the US biologist E. F. Stoermer and refers to the current geological epoch shaped by humans, following the previous epoch of the *Holocene, which* lasted approximately 11,700 years. The probable beginning of the Anthropocene has not yet been definitively determined and fluctuates between the middle of the eighteenth

E. W. U. Küppers, *Ingenious Principles Of Nature*, https://doi.org/10.1007/978-3-658-38099-1

century (invention of Watt's steam engine) and 1945 and subsequent years (effects of the atomic bombs).

Biocybernetic principles They were established in the 1970s by the cyberneticist Frederic Vester (1925–2003). He recognized that natural systems obey certain laws, which he called the basic rules of survivable systems. In total, there are eight basic rules:

1. Negative feedback must dominate over positive feedback
2. The system function must be independent of growth
3. The system must be function-oriented and not product-oriented
4. Use of existing forces according to the Jiu-Jitsu principle *(redirection of force)* instead of fighting according to the boxer method *(force against force)*
5. Multiple use of products, functions and organizational structures
6. Recycling – use of closed-loop processes for waste and waste heat recovery
7. Symbiosis – Mutual use of diversity through coupling and exchange
8. Biological design of products, processes and organizational forms through feedback planning with the environment.

(Source: Ballungsgebiete in der Krise: Vom Verstehen und Planen menschlicher Lebensräume, study commissioned by the Federal Ministry of the Interior as part of the environmental research programme, dtv, Munich 1976/1991).

Biocybernetics, biocybernetic Biological cybernetics. It deals with the control and regulation processes in living organisms and the ecosystem.

Biodiversity, biological diversity According to the UN Convention on Biological Diversity (CBD), biodiversity refers to "the variability among living organisms from all sources, including, inter alia, terrestrial, marine and other aquatic ecosystems and the ecological complexes of which they are part". Thus, it encompasses diversity within as well as between species, in addition to the diversity of ecosystems themselves. According to this definition, biodiversity also consists of genetic diversity. (Source: https://de.wikipedia.org/wiki/Biodiversität/; (accessed 15.03.2020), https://www.mpg.de/biodiversitaet (accessed 15.03.2020), see also Wilson, E. O. (1995): 'The value of diversity: the threat to species richness and human survival'. Piper, Munich).

Biogeonics Biogeonics is a compound word made up of the syllables **biology, geology** and bionics and indicates that both biological and geological influences underlie a bionic or system bionic development. For example, geological bank structures have an influence on the meandering course of rivers, which in turn are integrated into a biological environment including biological organisms in the flow of the rivers.

Bionics (→ Systemic Bionics) Bionics (→ Systemic Bionics) is an interdisciplinary scientific discipline that makes use of proven biological product, process and organisational principles in order to transform them into technical applications.

Climate Change Climate Change means a clearly perceptible change in mean global warming over years and decades, with an increase in extreme weather, such as overheated summers, storm surges, floods, tornadoes, etc. The inconvenient truth of global warming due to climate change, which is clearly evident in regions of permafrost such as Siberia, the Arctic and Antarctica, can no longer be denied. Although we know not only about the self-caused nature and environmental problems, but also how valuable dynamic balances of nature, for example that of the self-purifying power of our waters, are for our lives, how extremely useful huge forest areas are as CO_2 sinks and at the same time ensure the survival of an inscrutable network of plants and animals – despite all this we still actively participate in the cost-driven economic destruction of our nature and environment. We act against our better knowledge. (Source: Böckmann, C. (2017) An inconvenient truth. VDI-Nachrichten, 3 March 2017, Issue 09 2017; Brasseur, G. P.; Jacob, D.; Schuck-Zöller, S. (eds.) (2017) Climate change in Germany. Development, consequences, risks and perspectives. Springer Spektrum, Heidelberg).

Communication It is the mutual exchange of data, information and knowledge. The psychologist Schulz von Thun has dealt intensively with communication or "talking to each other" (Vol. 1–3, 1998, first edition 1981, Rowohlt, Reinbek b. Hamburg) and has worked out from his "communication diagnosis" a so-called *communication square* – a psychological model of interpersonal communication (Vol. 1, p. 30). It states that every type of message transmission (utterance) contains four aspects:

1. the self-revelation of the message transmitter (sender)
2. the appeal to the message receiver
3. the factual content of the message itself
4. the relationship between message transmitter (sender) and message receiver

The complexity of an interpersonal communication is thus unmistakable, which leads to a statement by the media scientist Berhard Pörksen: "The quality of communication determines the quality of our lives" (Pörksen and Schulz von Thun (2014) Kommunikation als Lebenskunst. P. 18).

Complexity It is a term that cannot be defined uniformly. More than 50 definitions from different disciplines show this. In general, complexity can be described in terms of complexity, dynamics, changeability, mutual dependencies – interdependencies – of influencing variables. A system is complex if it has many influencing variables, these interact with each other, usually show a non-linear behavior and change over time. Complex systems with a large number of variables are difficult to understand and control as a whole, which is why a reduction of complexity for the purpose of better manageability is often a means of making the system manageable for people. The danger, however, is that hidden parts of the complex system are not taken into account with sufficient precision when finding specific solutions, with the result that unexpected – often problem-accumulating – consequences are brought about.

Control loop Control loop is characterized as a closed-loop action which, influenced by external disturbance variables, dynamically moves towards a certain target given by the reference variable. If the difference between the specified reference variable as the target and the actual measurable controlled variable tends towards zero, a certain stability of the control loop has been achieved. The → **negative feedback** is the decisive component of the control loop to keep it in a dynamic equilibrium or steady state in biological/technical control processes by controlling or adjusting it.

Cybernetics, cybernetic According to the founder of cybernetics Norbert Wiener (1894–1964), cybernetics is the science of controlling and regulating living organisms and machines. Examples: The human organism regulates its body temperature – over all seasons – almost constantly at 37 °C. The room temperature is regulated by the system radiator, thermostat, outside temperature to a pleasant 20 °C room temperature.

Deadly Sins of Civilized Mankind, Lorenz's Eight In chronological order: (Source: Lorenz, K. (1973) Die acht Todsünden der zivilisierten Menschheit. Piper, Munich)

1	*Structural properties and dysfunctions of living systems*
2	Overpopulation
3	*Devastation of the natural habitat*
4	*Humanity's race against itself*
5	Heat death of the feeling
6	*Genetic decay*
7	Tearing down tradition
8	Indoctrinability

Deglobalization Deglobalization describes the economically controlled withdrawal of states and state formations from international market links or agreements. One aim of this strategy is to strengthen the country's own economy, which goes hand in hand with trade barriers or tariffs against imports. The opposite of deglobalization is globalization.

Destructivity Destructivity describes a state of mind of people that is directed towards the destruction of people and things. The psychoanalyst Erich Fromm (1900–1980) described destructiveness as "malicious aggression" (destructiveness, cruelty, murderousness, etc.). He analysed it as a human passion or character structure; but at the same time as a trait that is intensified in capitalist societies. The opposite of destructiveness is constructiveness. (Source: Fromm, E. (1977) Anatomie der menschlichen Destruktivität, Rowohlt, Reinbek bei Hamburg).Highly destructive are companies that seek short-term and short-sighted economic gains, at the expense of rainforest destruction and at the expense of endemic human, animal and plant life.

Ecosystem (→ Technosystem) Ecological communities describes the totality of populations in a given habitat. A community and its inanimate (abiotic) environment act together as an ecological system or ecosystem. The term *biogeocenosis* has the same meaning as the term ecosystem. (Source: Odum, E. P. (1991) Principles of Ecology, p. 39, Spektrum der Wiss., Heidelberg).

Efficiency, ecological efficiency Also *ecological efficiency*, term for the size ratio of various balances in the use of food (nutrition, food) in individuals, populations and ecosystems (food chain, food web, food pyramid). For example, the ratio of available energy bound in food (bioenergy) to that bound in endogenous substance can be expressed. The main computational variables are, within a trophic level (food relationships, trophic), the ratio of production to respiration (P/R; respiration) or to newly formed biomass (P/B), and between different trophic levels, the ratio of production values (P_n/P_{n-1}), also called *food chain efficiency*. Ecological efficiency is the ratio of available energy (e.g., energy bound in food) to energy bound in endogenous matter of an individual, population, or ecosystem. The size ratio of certain components yields the respective ecological efficiency. (Source: https://www.spektrum.de/lexikon/biologie/oekologische-effizienz/47456, accessed on 16.03.2020).

Efficiency, economic efficiency Efficiency describes the ratio of benefits to costs that is required to achieve a certain result. Efficiency can thus be understood as a criterion that describes whether a goal can be achieved in a certain way. Comparable German terms are effectiveness or also economy. The concept of efficiency is closely related to the concept of *"ecological"* *"sufficiency"* in sustainability research, a consumption of energy and raw materials that is as low as possible. Both terms are of great importance for sustainable consumption. Economic efficiency is a "decision criterion that selects from several economically equally effective measures (economic accuracy) those that are associated with the lowest economic costs (also called cost efficiency)". Cost efficiency requires that the *"marginal abatement costs" of* all companies are identical, which is achieved by different abatement quantities. Climate example on *marginal abatement costs:* "The costs incurred in abating an additional ton of greenhouse gases relative to current levels." (Source: https://www.enzyklo.de/Begriff/Grenzvermeidungskosten/, accessed 03/16/2020).

Emergence, emergent This is the emergence of new properties and structures in systems that result from the interaction of individual elements of the system but cannot be predicted by analyzing the properties of individual system elements. Historically handed down and popular is the sentence of the philosopher Aristotle (384–322 BC): The whole is more than the sum of its parts.

Energy = anergy plus exergy This is the ability to perform work. Energy is divided into a portion called "*exergy*", which is converted into work without limitation, and a portion called "*anergy*", which is not converted into work (→ *First Law of Thermodynamics*). The unit of measurement for energy is the joule [J].

Entropy (→ Negentropy) Entropy (→ Negentropy) or "*disorder*" is a fundamental "*state variable of thermodynamics*" with the unit of measurement [Joule/ Kelvin]. Simply formulated, the body whose temperature is higher contains more entropy. In the case of touching bodies of unequal temperature, entropy always flows from the warmer to the colder body until a temperature equalization takes place.

Epigenetics, epigenetic "[...] is the study of *mitotically* and/or *meiotically* heritable changes in the deoxyribonucleic acid (DNA) sequence." [...]. *Mitosis* is the normal cell division that results in asexual reproduction in unicellular organisms and growth and renewal of their tissues in multicellular organisms. [...]. The second form of cell division is called *meiosis* or *reduction division*. It takes place only in the formation of germ cells, that is, of ova and spermatozoa, and is thus exclusively in the service of sexual reproduction. Its tasks is to reduce the double set of chromosomes of the body's cells to half, so that when the egg and sperm fuse, a double set can be created again. (Source: Kegel, B. (2018) Epigenetics. pp. 81–82, DuMont, Cologne).

The transmission of *acquired characteristics* from generation to generation or grandparent generation to grandchild generation, in connection with epigenetics, which had not yet been researched at Lamarck's time (1744–1829), was postulated by him in his theory of evolution.

Equilibrium, here: thermodynamic equilibrium (→ non-equilibrium) Biotic communities and ecosystems are not "superorganisms", but systems that are not in thermodynamic equilibrium and are capable of *self-organization* [...]. (Source: Odum, E. P. (1991) Principles of ecology, p. 199, Spektrum der. Wiss., Heidelberg).

Evolution, evolutionary Evolutionary consideration of the change in heritable characteristics of living organisms in a population over generations. New findings also lead to evidence for "*epigenetic*" – individual changes in living organisms that are not causally based on genetic mutation.

Evolutionary Developmental Biology – Evo-Devo Evolutionary developmental biology (evo-devo) explores the mechanistic relationships between the processes of individual development and phenotypic change during evolution. Although evo-devo is widely accepted and has revolutionized our understanding of how organisms evolved, major implications for the theoretical basis of evolution are often overlooked. (Source (original English): Müller, G. B. (2007) Evo-devo: extending the evolutionary synthesis. In: Nature Reviews, Genetics, vol. 8, Dec. 2007, p. 943–949).

Exergy (→ Anergy) This is the fraction of energy that is converted into work.

Feedback, negative – Balanced Feedback In circulation processes or control sequences, the components that dynamically stabilize the overall control system and do not lead to overload (destruction) or standstill (inactivity) are called feedbacks. There are many feedbacks in biological as well as technical circulation processes. Circuits with an odd number of negative feedbacks (an even number of negative feedbacks would cancel out the balancing stabilizing effect) are also called "angel circles".

Feedback, positive – Reenforced Feedback In closed-loop processes or control sequences, those components are called "vicious loops" which dynamically strengthen the overall control system in their effect up to overload (destruction) or dynamically weaken it up to standstill (inactivity). Circuits with any number of positive feedbacks are also called "vicious circles".

Genotype (→ Phenotype) Dualism in modern biology is thoroughly physicochemical and arises from the fact that organisms have a genotype and a phenotype. The genotype is made up of nucleic acids. The phenotype is formed based on the information conveyed by the genotype and consists of proteins, lipids, and other macromolecules. (Source: Mayr, E. (1998) This is Biology, p. 126, Spektrum Akademischer Verlag, Heidelberg, Berlin). Random genotypic changes (variabilities) affect phenotypic flexibility (adaptability in the environment).

Globalization Stands first of all for international division of labour, which has existed for two millennia. Thus globalisation is nothing new, only its capitalist form, forced by productivity increases and growing profitability pressure, worries more and more people. Those whom it frightens see globalisation as accelerating job losses at home. Those who assert the necessity of globalization point to the necessary growth that brings changing employment patterns. Both assertions are true in this generalization. Analytically, both are unhelpful. The following reflections on one of the most striking fighting terms should shed some light on this.

- Globalisation means an increasingly interconnected world economy. This is causing ongoing structural change in national economies. It puts the question of national competitiveness and the competitiveness of national location factors on the agenda in propaganda terms. Globalization describes an economic process that derives its dynamics from changing investment, production and distribution decisions of companies. Economic globalization is a historical process of economic expansion.

- Progressive globalization is measured by increasing world market integration. The growing internationalization of companies, especially through mergers and the existence of a global financial market, is one indicator. As a consequence of globalisation, there is growing competition between nation states for the location of companies, or for the localisation of their business activities. Today, not a week goes by without the publication of a location ranking of countries, cities and regions. Hard location factors are presented, which primarily means the respective national tax system, the so-called labour costs and the expected subsidies for investments. A frequently mentioned soft location factor is the so-called quality of life, under which housing and culture are subsumed. [...].

- So globalisation is just another name for the urge of capital to grow worldwide because it has become tight at home in terms of growth: "The German market no longer allows for organic growth". So globalization is nothing new. It is just that more and more people are being caught up in the capitalisation of their living conditions. [...].

Conclusion: Globalization generally stands today as a term for the capitalist form in which the living conditions of the people on the globe have aligned, are aligning and will align. As a fighting term, globalization is perfectly suited to create false hopes ... Expectations and fears. (Source: Vögele (2007) Das Elend der Ökonomie, pp. 113–124, Rotpunktverlag, Zurich).

Gross Domestic Product – GDP –, economic It includes the total value (more correctly: the total costs) of all goods and services produced domestically within a year. *GDP is a purely productive economic measure* and does not take into account the costs incurred for social, environmental or external areas of a country. GDP therefore only very incompletely reflects the actual performance of a country.

Gross Primary Production – BPP –, ecological It denotes the amount of light that is converted into chemical energy per unit of time by photosynthesis. Not all of the molecules produced are stored as organic matter, however, because the primary producers use some of it as fuel for their own cellular respiration. Thus, **net primary production (NPP)** is equal to gross primary production minus the amount of energy used by producers in respiration (**R** from respiration): **NPP = BPP → R**.

Hierarchy of needs in the digitalizing Anthropocene, after Küppers See detailed description in Sect. 9.2.

Hierarchy of needs, according to Maslow As a social-psychological model, the US-American psychologist Abraham Maslow developed his ideas of human needs according to a certain structure, starting with the basic needs or needs for securing one's existence, such as nutrition, sleep, reproduction, etc., via basic needs for health, work, housing, etc.; via needs in the social context, such as social contacts and participation, via personal needs of independence, physical and mental strength, etc., up to the highest level of self-actualisation with the development and expansion of one's personality and creativity. (Source: see Sect. 9.2, bibliography Maslow, A. (1943, 1981) and Koltk-Rivera, M. E. (1943)).

Hot Spots, here Biodiversity Hot Spots Hot Spots are those habitats or biotopes on earth in which the greatest density of species diversity prevails and which are therefore particularly endangered by the destructive, economically controlled interventions of humans.

Humanoid Humanoid are robots with human-like appearance and functions.

Impact network Impact network is a dynamic functional connection between different elements (subjects, objects, tools, etc.), which together form a unit. **Impact network analyses** therefore attempt to improve problems that arise in an often complex impact network context by means of suitable modelling as a whole towards a specific goal. The goal should meet the requirements of sustainability in the forestry or biocybernetic sense, taking into account not only economic production and service but also possibilities of symbiosis, multiple use, biologically inspired work design, and so on. Typical of this is the construction of a new housing estate, a commercial yard, an urban extension, etc. The sensitivity model according to Prof. Vester° (Source: Vester, F.; Hesler, A. v. (1980) Sensitivity model. Umlandverband Frankfurt) is an early example of practised holistic impact network analysis. (See also Küppers, E. W. U. (2013) Denken in Wirkungsnetzen. Tectum Marburg). Impact network thinking means: recognising interrelationships, evaluating holistically, solving sustainably.

Lords of Mankind, Chomsky's Chomsky uses this term as a subtitle in his book on "Struggle or Perish!" while pointing out that the term was coined by the economist Adam Smith (1723–1790) (ibid., pp. 19–20):

"Adam Smith had no problem identifying those he called "masters of mankind." In his day, these were English merchants and manufacturers who were the "chief architects of policy" and ensured that their interests received the most attention, no matter how severe the impact on others, including the people of England. The greatest victims of the English's blatant injustice, however, were elsewhere – for Smith, particularly in British India. Much has changed in two and a half centuries, but the basic principles remain the same.

Today's masters of humanity come from the international investor class, the highly concentrated and interlocked corporate sector with a rapidly growing component of finance capital, especially after they were unleashed by President Nixon's dismantling of the Bretton Woods system[1] and then the massive deregulation of the crash bailout system[2] to dominate the neoliberal era. They dictate policy, though not without internal dissension, and they strive to ensure that their own interests are served, regardless of the impact on others. This is done with much success, especially in recent years."

"Truly critical intellectuals are usually marginalized, denigrated, or denied employment – such as a job – in their own society. Such tendencies have been evident throughout history." (ibid., p. 26).

---

[1] The Bretton Woods system was created after World War II as a new international monetary system with exchange rate bands and the dollar as an anchor currency to link fixed and flexible exchange rates (see Bordo 1993).

[2] Crash bailout system = collapse bailout system.

Meander Effect® Here: **Flowing meanders** are phenomena of nature within a water cycle, which shape the course of free-flowing rivers in such a way that they generate a minimum of entropy. Characteristic of meandering watercourses are the typical overshooting curves, whose bionic counterparts are used in special shaped parts of technical pipe or channel systems. The enormous flow deflection losses in pipe systems of air-conditioning technology, the pipe transport of solids, liquids or multi-component conveyed goods are predestined for the use of energy-efficient moulded parts with the meander effect®.

Metabolism, biological Another term for this is **metabolism**, is a process that produces intermediate and end products in the organism through chemical-physical conversion processes of chemical substances such as food, water and oxygen.

Mutation (→ Selection) Here: **Gene mutations** are random changes in genetic material. They occur during cell division due to copying errors. The duplication (replication) of DNA molecules during cell division and the emergence of germ cells is a remarkably precise process, but errors do occasionally occur.

Negentropy (→ Entropy) It is entropy with negative sign. We ourselves are *"islands of life"*, like all organisms by the way, because an inner order (negentropy) is maintained by our metabolic processes, which we feed with workable energy in the form of solid and liquid food and export their "waste products" as non-usable energy for work into the environment. The Austrian physicist Erwin Schrödinger (1887–1961) coined the term in his 1944 book: "What is Life?".

Non-equilibrium, here: thermodynamic non-equilibrium (→ equilibrium) Organisms and ecosystems maintain their highly organized low-entropy [...] state by converting high-value energy into lower-value energy (for example, in the respiration of carbohydrates, (the majority of which consist of sugar molecules, d. A.)). Living systems and the entire biosphere are what Ilya Prigogine has called "far-from-equilibrium systems" (far-from-equilibrium systems) with powerful "dissipative structures" that pump out the disorder. (Prigogine 1972). (Source: Prigogine, I.; Nicoles, G.; Babloyantz, A. (1972) Thermodynamics and Evolution. In: Physics Today, 25/11 (1972) pp. 23–38 and 25/12 (1972) pp. 138–141).

Obsolescence, planned Describes the economically controlled short-lived nature of products, the opposite of resource and product efficiency.

Optimization, evolutionary Optimization, evolutionary is a group of technical algorithmic methods modelled on natural development strategies of organisms that evolve by mutation, selection, crossing-over and other mechanisms. Evolutionary algorithms (EA) with the subgroups Genetic Algorithms (GA) and Evolutionary Strategies (ES) characterize the two main groups of technical-evolutionary optimization methods, which go back to the US-American computer scientist J. H. Holland (1929–2015) (GA) and I. Rechenberg as well as H. P. Schwefel (ES), among others.

Phenotype (→ Genotype) See Genotype.

Power of habit (→ routine cycle) It is generally shows our dependence on preferences, such as the consumption of sugary sweets, which is also manipulated by targeted advertising effects that reveal themselves as mental models in the form of long-term harmful routine loops, among other things, and which we find difficult to escape. For example, a minimal impulse (smells of sweets from a chocolate shop) is often enough to drive us single-mindedly to buy a bar of chocolate, even though we don't actually want to, but reward ourselves with the taste of eating the sweet.

The routine cycle develops after repetitions into a vicious cycle with all foreseeable consequences.

Impulse and *incentive* leads to *routine* or *habit*, leads to expected *reward*, leads to *new impulse* and *incentive*, and so on. The enormous difficulties in breaking the entrenched and entrenched routines of a → **vicious circle** are evident in many variations in societies. Only a few seem to succeed in replacing harmful routines with sustainably beneficial routines. One of these was Hanns Carl von Carlowitz (1645–1714), who brought the concept of forestry sustainability into being after decades of cutting down trees, selling them as mere commodities, and thereby destroying entire forests, with clear social disadvantages.

Routine cycles as → **vicious circles** can be recognized in many ways in today's societies, whether as addiction routine circles, politically misguided action routine circles, economic cost routine circles, communication routine circles, and many more.

Rebound effect Here: **Effect of economics** is, for example, the consequence of an ecologically improved, nature-friendly product or process whose ecological advantages are subsequently cancelled out by the quantity of goods produced. In other words, the rebound effect occurs when the savings effect of increasing efficiency improvements of, for example, fuel- or exhaust-reducing passenger cars is beneficial for the individual vehicle and the environment, but due to the higher number of energy-efficient passenger car purchases or due to more kilometres driven per time, their overall consumption becomes more inefficient again and, not least, the environment is burdened more than before.

Robot Robot is an electromechanically controlled machine that, through algorithmic programming, imitates human movement patterns or performs tasks that were previously performed by humans. In this context, a further development is taking place through so-called **cobots – collaborating robots –** in which work processes are carried out jointly through human-cobot cooperation.

Routine Circuit See → **Force of habit**.

Second Law of Thermodynamics (2nd HS) This says something about the direction and irreversibility (irreversibility) of a process. The state variable → **entropy** can also be derived from the 2nd HS. Since organisms are → **open systems** and thus exchange energy (as well as matter and information) with the environment, → **entropy** is also transported across the system boundary and produced inside the system. Therefore, the entropy change of a biological system consists of two parts.

Accordingly, the entropy balance is: "The → **entropy of** a system changes by inflow or outflow of entropy across the system boundary and by entropy production inside it." (Lucas 2007, p. 200).

Selection (→ Mutation) Here: **natural selection. In** his time, Darwin did not yet know the influence of genetics on selection. He concentrated exclusively on the selection of the phenotype, on the "survival of the fittest" brought into play by Herbert Spencer. Natural selection is – as we know today – a process that fundamentally takes place in two successive steps (cf. Mayr 2003, p. 152):

Step 1: Random variations emerge

"Mutation of the zygote (diploid cell after fertilization, d. A.) from its formation to death; meiosis (maturation division, d. A.) with recombination by crossing-over (mutual exchange of corresponding chromosome segments, d. A.) in the first division and random migration of chromosomes in the second (the reduction division); any random elements in mate choice and fertilization."

Step 2: Deterministic aspects affect survival and reproduction

"Greater success of certain phenotypes during their life cycle (selection by survival); non-random mate choice and all other aspects that cause the reproductive success of certain phenotypes to increase (sexual selection). In the second step, large-scale random elimination occurs simultaneously."

Self-organisation "[…] (is) the result of the → **Second Law of Thermodynamics** – with the consequence that living systems show an increasing → **complexity** because of, and not in spite of, the expenditure of → **entropy.** The "overall strategy" involves a decrease in → **entropy** (disorder), an increase in information (order), an increasing ability of the ecosystem to withstand disturbances unchanged (resistance stability), and an increasingly efficient utilization of energy and nutrients." (emphasis mine) (Source: Brooks, D. R.; Wiley, E. O. (1986) Evolution as Entropy. University of Chicago Press, Chicago, from: Odum, E. P. (1991, p. 208)).

Self-regulation Here: **organismic.** Self-regulating systems are at the same time always self-organizing, in that the functions of a system – for example a forest – maintain themselves or the system adapts to new changes. Closely related to the phenomenon of self-regulation is the cybernetic process of → **feedback**, in which disturbance to the system – for example, local tree destruction due to bark beetle infestation – can be compensated for over time. Incidentally, our body's own metabolism is also subject to clear controlled regulatory processes of self-regulation.

Self-replication Here: **genetic.** On DNA (DNA) replication, Karp (2005, p. 679) remarks: "A fundamental feature of the living is reproduction. Reproductive processes can be observed at all levels: Living things reproduce by sexual or asexual reproduction, genetic material duplicates itself by DNA replication. But the apparatus that duplicates DNA also goes into action in another process: repairing genetic material that has suffered damage. [...] The ability to self-duplicate may have been one of the first fundamental features to develop in the evolution of the simplest life forms."

Standstill, raging A quarter of a century ago, the French media critic Paul Virilio predicted an unbridled acceleration of information processing and coined the term "*racing standstill*". Internet, high-performance trading on the stock exchanges by computer algorithms, etc. ultimately lead to an inevitable collapse, such as that of the global economy and financial sector in 2008/2009. Extraordinary and incalculable events such as the infectious disease COVID-19 in 2020, put entire cities, countries and continents in quarantine.

Strategy of forward-looking, prudent thinking and action – long term farseeing It is particularly evident in the holistic perception of economic, ecological and social interlinked target variables, such as sustainable growth, appropriate economic profits and sustainable profitability, in which consequences and consequential costs (*external costs*) determine one's own development process as part of the overall strategy. This mental and operational approach is adapted to be growth-optimised, socially compatible and, last but not least, environmentally friendly in the sense of forestry sustainability.

Strategy of monocausal short-sighted thinking and acting – short term missent This is particularly evident in the one-sided focus on maximising economic targets such as growth, profit and profitability, without taking into account the harmful consequences and consequential costs that are generally socialised (*external costs*) in their own development process. This mental and operational approach is often highly environmentally destructive and far beyond sustainability in the forestry sense.

Survival of the Fittest This was mentioned by Charles Darwin in his theory of evolution and means nothing other than that the organism that has the best chance of survival is the one that adapts to the dynamic environment. Herbert Spencer, an inglorious contemporary of Darwin, very misleadingly reinterpreted Darwin's "survival of the fittest" as "survival of the best adapted individuals", which gained dubious fame in connection with Social Darwinism, a theory in which aspects of Darwinism are mapped onto social developments.

Sustainability It is a principle from nature. According to this principle, resources are only used to the extent that they serve further development. An incomprehensible network of material, energetic and communicative processes has led over millions of years to the continued existence of countless species in nature up to the present day. Mankind is on the way (→ **Anthropocene**) to irretrievably destroy this unique network of sustainability. The term sustainability originated in forestry and was described by Johann Hanns Carl von Carlowitz (1645–1714) in 1713 in his work "Silvicultura oeconomica". According to this, only so many trees should be felled and young trees replanted that a continuous use of the raw material wood is guaranteed. The RIO Convention *Sustainability* from 1992, a UN agreement on biological diversity, states that the protection of diversity also includes the diversity of species, genetic resources and the diversity of habitats and ecosystems, and that these are strongly linked to social and economic characteristics. (Source: https://de.wikipedia.org/wiki/Biodiversitäts-Konvention;   (accessed   on 18.03.2020)).

Sustainability Cube It is a quality space from → **Systemic Bionics** that links progress on the three quality space axes (1) *natural model*, (2) *technical implementation* and (3) *networked sustainability*.

Sustainability dilemma We are prisoners of our own mental models. The sustainability dilemma is that today's sustainability is associated with completely different mental models and practical goals! The power of habit or the entrenched course of our thought routines still largely controls our actions even when new routines are required due to new insights, new perspectives or new goals. Carlowitz mastered this change from solidified routine (clearing the forest as a supplier of wood without preventing its environmental impact) to new innovative routine (clearing the forest as a supplier of wood in a forestry sustainable way). The new mental model led to a strong sustainability in the forest, both to ensure electoral profits and to reduce the shortage problem of wood and more. Those who link sustainability, for example, to the mass production of goods of various kinds have not understood the value of ancestral forest sustainability.

Sustainability Paradox The *sustainability paradox* is an expression of a product- or process-oriented quality improvement for environmental relief, which, however, is cancelled out by the quantity of the products – and even causes the environmental impact to increase again! This is the well-known and notorious growth driver → **rebound effect.** Currently, it can be seen in many products, such as passenger cars, whose energy consumption decreases from generation to generation, but is cancelled out again by increasing car weight, digital "upgrading", and so on. This is indeed productive "sustainability" in car manufacturing, but with extremely weak sustainability, far beyond the Carlowitzian idea.

Systemic bionics (→ Bionics) Systemic bionics bases the biological-technical transformation process on the systems approach. In this context, bionic solutions are not viewed in isolation and with a focus on details of the networked nature, but rather these are included in the bionic solution process. The aim is to do justice to natural sustainability in the technosphere as well through holistic solution strategies. (Source: Küppers, E.W. U. (2015) Systemische Bionik. Springer, Wiesbaden).

Systems, open Systems, open are capable of processing and transforming input variables of energy, substances and information and making them available again to the system environment as output variables.

Technosystem (→ Ecosystem) Technical/technological working communities describes the totality of participating human and material products, procedures and organizations in a given working space. Technosystems are open systems and thus dependent on incoming and outgoing personnel and material resources.

Theory of evolution according to Darwin Charles Darwin's "On the origin of species by means of natural selection, or the preservation of favoured races in the struggle for life" John Murray, London 1859, or "The origin of species by means of natural selection, or the preservation of favoured races in the struggle for existence" Klett-Cotta, Stuttgart (2018), describes a total of five independent theories:

Theory 1:	Variability of species (the basic theory of evolution)
Theory 2:	Descent of all living beings from common ancestors (evolution by branching)
Theory 3:	Gradual course of evolution (gradualism, d. A.) (no jumps, no discontinuities)
Theory 4:	Reproduction of species (emergence of biological diversity)
Theory 5:	Natural selection

A few years after the publication of Darwin's *Origin of Species*, Theory 1 and 2 had become generally accepted. Theories 3–5 could not come together until much later, when the so-called "synthesis of evolutionary research" took place. It "[...] led to general agreement, and the molecular biology revolution of the following years meant a further strengthening for the Darwinian doctrine [...]" (Darwin 2018, p. 11).

Theory of evolution according to Lamarck This theory of evolution of the French botanist, zoologist and developmental biologist was based on the diversity of species through acquired characteristics of organisms. Lamarck's notion that organisms strive to change on their own to achieve something (attractive) was clearly rejected by Darwin. His natural selection had to proceed imperceptibly; as slowly as the earth-historical processes of weathering or the geological reshaping of landscapes and continents. For some years now, Larmarck's approach to a theory of evolution has again been brought to the forefront of evolutionary interest by → **epigenetics**.

Theorem of Thermodynamics, First (1st HS) This is also known as the *law of conservation of energy*. It states that the energy of a thermodynamic system changes only through the supply or removal of energy across the system boundaries. In other words, energy is neither created nor destroyed. It follows that energy = exergy + anergy = constant. (Source: Lukas, K. (2007) Thermodynamics, p. 147, 6th ed., Springer, Berlin, Heidelberg).

Tragedy of the commons The "tragedy of the commons" refers to a social science and evolutionary theory model according to which freely available but limited resources are not used efficiently and are threatened by overexploitation, which also threatens the users – in this case the people – themselves. This pattern of behaviour is also examined by game theory. Among other things, it examines the question of why individuals in many cases stabilize social norms through altruistic sanctions despite high individual costs.

Transformation process, analogue-digital is currently happening through the strong influence of digital tools in analogue process flows that have been functioning for decades in industries, work and leisure sectors. The short-term, often short-sighted, little systemically thought-out and practiced pressure of digitalization proponents on the participating persons and processes is enormous and leads to equally enormous dislocations, uncertainties, new dangers and risks in socio-technical systems.

Transformations, biosphere-technosphere (→ Bionics → Systemic Bionics) See explanations on bionics and systemic bionics.

Value change, theory of, according to Ingelhart Ingelhart's theory of the change in values from a materialistic to a post-materialistic, highly technical society is based on Maslow's hierarchy of needs and also names five developmental levels, from the basic level: "remuneration according to performance", through flexibility, teamwork, promotion and recognition, to the top level, which comprises the meaning of work. There are two hypotheses that Ingelhart has elaborated in his theory of value change: (1) the scarcity *hypothesis* (desire for scarce goods) and (2) the *socialisation hypothesis* (prosperity in youth strives for new cultural rather than material values). (For sources, see Sect. 9.2, bibliography Ingelhart, R. (1998, 1982, 1977)).

Value of an organism or community (cost of an organism or community) Monetization under capitalism, in which every product and service is given a cost to eventually be sold in a highly dubious sum figure to end up as → **gross domestic product – GDP** – has long since overwhelmed the biosphere, nature with its interconnected organisms, and the geosphere, with its wealth of raw materials. For example, the so-called ***value***, but actually only monetized by a banal cost factor of a tree or forest: by the marketable cost of the number of solid cubic meters of wood! What is the cost of a bee or a hive, including the pollination services of the organisms?

What costs can be calculated for fungal networks in the soil to maintain organismic networks? This and many other questions of economic interest are doomed to failure from the outset. Why? Because no human being can calculate the actual value, let alone the cost, of organismic services. The past and still ongoing bee mortality gives only a small insight into the inadequate ability of humans in dealing with interconnected nature (see among many other publications on bee mortality Schuh, H. (2007) Die Biene, das Geld und der Tod. Die Zeit, 24.05.2007, No. 22).

Vicious Circle (→ Angel Circle) See explanation of angel circle.

Index

Printed in the United States
by Baker & Taylor Publisher Services